Innovation in Energy Security and Long-Term Energy Efficiency II

Innovation in Energy Security and Long-Term Energy Efficiency II

Editor

Manuela Tvaronavičienė

MDPI • Basel • Beijing • Wuhan • Barcelona • Belgrade • Manchester • Tokyo • Cluj • Tianjin

Editor
Manuela Tvaronavičienė
Vilnius Gediminas Technical University
Lithuania

Editorial Office
MDPI
St. Alban-Anlage 66
4052 Basel, Switzerland

This is a reprint of articles from the Special Issue published online in the open access journal *Energies* (ISSN 1996-1073) (available at: https://www.mdpi.com/journal/energies/special_issues/Energy_Security_Energy_Efficiency_2020).

For citation purposes, cite each article independently as indicated on the article page online and as indicated below:

LastName, A.A.; LastName, B.B.; LastName, C.C. Article Title. *Journal Name* **Year**, *Volume Number*, Page Range.

ISBN 978-3-0365-4661-2 (Hbk)
ISBN 978-3-0365-4662-9 (PDF)

Contents

energies

MDPI

Article

The Causal Connection of Natural Resources and Globalization with Energy Consumption in Top Asian Countries: Evidence from a Nonparametric Causality-in-Quantile Approach

Hafezali Iqbal Hussain [1,2], Muhammad Haseeb [1], Manuela Tvaronavičienė [3,4,*], Leonardus W. W. Mihardjo [5] and Kittisak Jermsittiparsert [6,7]

1 Taylor's Business School, Taylor's University, 1 Jalan Taylors, Subang Jaya 47500, Selangor, Malaysia; hafezali.iqbalhussain@taylors.edu.my (H.I.H.); muhammad.haseeb@taylors.edu.my (M.H.)
2 University of Economics and Human Sciences in Warsaw, Okopowa 59, 01-043 Warsaw, Poland
3 Vilnius Gediminas Technical University, Sauletekio 11, 10223 Vilnius, Lithuania
4 General Jonas Zemaitis Military Academy of Lithuania, Silo 5a, 10322 Vilnius, Lithuania
5 Bina Nusantara University, 11530 Jakarta, Indonesia; mihardjo@gmail.com
6 Department for Management of Science and Technology Development, Ton Duc Thang University, Ho Chi Minh City 758307, Vietnam; kittisak.jermsittiparsert@tdtu.edu.vn
7 Social Research Institute, Chulalongkorn University, Bangkok 10330, Thailand
* Correspondence: manuela.tvaronaviciene@vgtu.lt

Received: 29 March 2020; Accepted: 19 April 2020; Published: 5 May 2020

Abstract: Given the significance of energy conservation as a prime objective of environmental sustainability, countries all around the world are keen to identify significant factors that lead to the augmentation of energy utilization. Considering the rising emphasis of economies in utilizing natural resources to attain higher levels of globalization, the current research was aimed at investigating how the returns of natural resources and globalization affect energy consumption in top Asian economies. In doing so, the study emphasized the nonlinear relationship among the variables and applied the novel nonparametric method of causality in quantile to identify the quantile-based causal connection of natural resources and globalization on the returns and volatility of energy utilization in selected Asian countries. Moreover, the presence of nonlinearity in the variables was tested by the Brock-Dechert-Scheinkman test (BDS test), which confirmed that all variables showed nonlinear behavior. Furthermore, the findings of quantile cointegration confirmed a nonlinear long-run relationship of natural resources and globalization with energy utilization. The prime findings of causality in quantile revealed that the returns of natural resources and globalization had a significant causal effect on the returns of energy consumption in all countries. On the other hand, the volatility in energy consumption concluded no causal association with the returns of natural resources and globalization in any of the studied Asian countries. The findings are beneficial for the policymakers to formulate policies that will help to reduce the level of energy consumption.

Keywords: natural resources; globalization; energy; nonparametric causality in quantiles

1. Introduction

Natural resources are the resources that are bestowed on a country by nature, without any particular investment by the country, and they are considered one of the most influential determinants of the economic development of a country [1–5].

Therefore, most of the investments of resource-abundant countries are normally focused on extraction, which itself is a huge and critical area for natural scientists, environmental economists, and practitioners [6]. Resource-abundant countries fall into two categories: (1) Richly endowed, where countries transform their operations for the advancement of the capacity and capability of various industries and direct the extraction of the natural resources towards the betterment of the economic development and financial health of the country [7]. (2) Narrowly focused, where countries make themselves and their respective economies more dependent on the natural resources by focusing most of the industries in the same field thus increasing the level of dependency on the presence of the natural resources [8–10].

Natural resources enable direct foreign investment to a country, and that has a tendency to improve the standard of living of the locals [11,12], however, it also adversely affects the potential growth of the country in the long term [13], thus making the host country intensely dependent on the consumption and extraction of the natural resources; this is referred to as a resource curse [14–18].

Moreover, most researchers are in agreement with respect to the adverse association between economic development and natural resources of a country [19–21]. Researchers also suggest that by improving institutional quality, having a good incentive program to the personnel engaged in the respective activities [22], and strengthening the political stability [23], the country can counter the phenomenon of being trapping in a resource curse.

Globalization refers to the phenomenon where countries are consolidated by means of intercountry and direct investments, which make the countries grow together by nurturing their potential and capabilities and fostering economic development [24].

Conventionally, the term was used for the development of communication and transportation infrastructure that creates connectivity among the countries, however, the scope of the term has broadened [2]. By strengthening foreign trade among countries, globalization can enhance the capabilities, capacity, and efficiency of the manufacturing countries and contribute significantly to achieving economies of scales [24–26]. Moreover, availability of abundant natural resources also leads to an increase in globalization, because the countries that have resources can transform them into a usable form that can be exported to the counties lacking those resources [27]. Therefore, natural resources also increase foreign trade, foreign investments, and foreign exchange, which increase globalization and improve the efficiency of energy consumption [2,28].

Despite of the generation of value-added services and goods by the consumption of energy, energy consumption damages the environment and adversely contributes to global climate change [29]. For instance, oil- and gas-oriented companies while transforming the natural resources into usable forms consume most of the energy, and hence pollute the environment [30,31]. Moreover, natural resources extraction leading to an increased energy consumption can ultimately harm the global environment. Therefore, protecting and safe guarding the environment is of significant importance [32–35].

Asian countries, on one hand, are progressive and fast growing regions in terms of industrialization, urbanization, globalization, and population. On the other hand, they significantly contribute adversely to global pollution, climate change, and environment degradation [36,37].

Despite Asian countries being a major contributor to pollution and climate change, the current literature mainly focuses on the regions of the Organisation for Economic Co-operation and Development (OECD) countries and Asian countries have received less focus [38,39]. Therefore, this led to the motivation of the present study to examine the dynamic effects of natural resources and globalization on energy consumption in top Asian countries.

The remainder of this paper is arranged as follows: The next section presents evidence from the literature, followed by a discussion of the methodology of the present study. Next, estimations and findings of the study are discussed. Finally, conclusions are drawn and recommendations are given to the practitioners and policy makers for future guidance in devising strategies based on the findings of the present study.

2. Literature Review

The theoretical connection of globalization asserts that as the economy is globalized, its level of energy utilization changes. Being the prime source of energy, nearly 65% of worldwide carbon emanations come from the burning of fossil-fuels [40]. Following the apparent link, many studies believe that globalization is a crucial aspect of motivating energy usage. However, the rise or decline in the levels of energy depend on the net effect of multiple factors within the globalization. Generally, the upsurge in global output and income levels are linked with enhanced consumption and manufacturing along with the elimination of trade barricades due to enhanced globalization. This in turn amplifies energy usage and leads to a positive association [41]. On the other hand, the negative link between globalization and energy usage can also be attributed to new ventures leading to innovation spillover, which can assist in reducing energy usage by supporting energy efficient research [42]. Hence, the energy efficiency derived from importing skills as a result of globalization can reduce power usage, however, a rise in efficiency can only offset a portion and not all of the energy needs of an economy.

Similarly, the theoretical link between globalization and energy usage is also supported by the popular scale effect, technique effect, and composition effect [42]. The assumption of the scale effect asserts the positive connection between globalization and energy by stating that the expansion of fresh industries and economic activities augments energy usage [43,44]. Moreover, the alternative presumption of technique effect proclaims that a rise in globalization often empowers the economies to lower their energy utilization through importing progressive technologies and inflows of capital and supports the negative connection among the variables without obstructing the economic structure [45]. Finally, the composition effect in globalization confers the changes in energy intensity to the alteration in a country's industrial structure [46]. For instance, the emergence of globalization empowers a production shift from agriculture to industry and ultimately to the service industry, and thus changes the economic composition towards the sectors that demand lower energy use [40].

By linking higher levels of consumption and production to the environment, it is believed that the demand shift results in more eco-intensive processing that enlarges the environmental burden by putting more pressure on energy-insensitive processes [47]; this hinders the country's prospects of sustainable development. Conventionally, the dependence on natural resource consumption has been argued to persist the resource curse as it hinders sustainability of the economy by hindering growth [48]. In this regard, it is believed that countries having sufficient natural resources experience diminutive growth relative to those of resource-scarce economies [49]. Alternatively, many recent studies contradicted the concepts of a resource curse and found natural resources to be a blessing for countries [50]. Moreover, the growing dependence of economies on the utilization of natural resources demands reliable and sustainable access to numerous natural resources, such as forestry, water, minerals, productive land, and essential metals [51]. However, given the rate at which resource depletion is occurring, the physical accessibility of such natural resources is considered challenging. The dependence of many economies on minerals and metal extraction by consuming excessive amount of energy leads to environmental destruction [52]. In response, green sources of energy have provided a fine solution to fulfil energy needs with minimal pressure on the environment, but the installation and completion of green projects are often fossil-based and energy-dependent [53], while certain green sources such as biomass, wood, waste, and plants also enhance the pressure on natural resources and their sustainable access [54]. Hence, based on the above reasons, the link of globalization and natural resources with energy consumption is considered rather complex and needs to be re-evaluated in different time-series settings with the utilization of advanced methods to ensure the reliability of the derived findings [51].

Empirical Studies

Numerous research studies in the prevailing literature evaluated factors that contribute to environmental destruction [55]. Among them, energy consumption is regarded as a vital stimulator of environmental pollution [56]. The role of energy is inevitable for growth, however, by acknowledging the damaging consequences of energy use in climate change, the current economies are motivated to recognize the in-depth aspects of energy dependence so as to combat the environmental impact of energy use in the process of atmospheric downfall [57]. Following the adverse impact of energy dependence on environmental quality [58], particularly in damaging the ozone layer and emitting greenhouse gases [59], many studies have sought to identify the antecedents of energy usage in both developed [60,61] and developing economies [62,63].

For instance, Shahbaz et al. [40] examined the causal effect between globalization on energy utilization in twenty-five advanced countries. For this, the authors utilized the data from the timespan of 1970–2014. The empirical results of the study found significant causal associations among the variables in 14 out of 25 economies. Specifically, the results demonstrated the positive causal link between globalization and energy utilization in the majority of the studied economies except for in the UK and USA, where the rise in globalization was found to decrease energy usage. Lastly, the study reported that in the economies of Italy, New Zealand, Finland, Greece, Ireland, Portugal, Austria, Spain, Iceland, and Denmark, globalization did not exert significant influence on energy utilization. Also, Shahbaz et al. [41] examined the link between globalization and energy utilization in the economies of Ireland and the Netherlands. For this, the authors utilized the time-series quarterly data from the timespan of 1970–2015. The empirical results of the study found a significant impact of globalization on energy usage in the long run. Interestingly, the results stressed that a rise in globalization carried an upsurge in energy utilization in both of the highly globalized economies that were studied. However, the results failed to validate the existence of a short-term link among the considered variables.

Furthermore, in India, Shahbaz, Mallick, Mahalik, and Sadorsky [42] investigated the connection of globalization with energy utilization. For this, the authors utilized the time-series data from the timespan of 1971–2012 by applying the Autoregressive-Distributed Lag (ARDL) approach to perform empirical examination. The empirical outcomes of the study confirmed the presence of a significant link between globalization and energy utilization. Moreover, the results stressed that an increase in globalization curtails the adversities to the environment by lowering the level of energy usage in the Indian economy. For the economies of Indonesia, Malaysia, and Thailand, Azam et al. [63] analyzed the factors affecting energy utilization. For this, the authors utilized the time-series data from the timespan of 1980–2012. The empirical results of the study found a significant impact of globalization in the form of trade liberalization and foreign direct investment (FDI) in enhancing energy usage in the considered economies by documenting the presence of positive relationships among the variables.

In Bangladesh, Murshed, Tul-Jannat, and Amin [64] analyzed the impact of globalization on energy utilization in Pakistan. For this, the authors utilized the time-series data from the timespan of 1980–2015. The empirical results of the study stated that globalization had no causal relationship with energy usage in Bangladesh. Utilizing the indicator of trade openness to measure globalization, Shahbaz, Loganathan, Sbia, and Afza [65] also examined the link between globalization and energy utilization in Malaysia between 1970 and 2011. Similar to the findings of Azam et al. [63], the empirical results of the study also confirmed the positive influence of trade openness on energy usage in Malaysia. Bringing the impact of renewable energy to the globalization–energy nexus, Koengkan, Poveda, and Fuinhas [66] recently analyzed the impact of globalization on renewable energy utilization in Latin America. For this, the authors utilized the panel data for ten Latin American economies from 1980 to 2014. The empirical results of the study found a significant impact of globalization on renewable energy usage, highlighting that the rise in globalization led to an augmentation of the utilization of renewable energy in the studied economies.

Shahbaz et al. [67] investigated the relationship between globalization and energy utilization by validating the possibility of an environmental Kuznets link. For this, the authors utilized the mix panel data of eighty-six economies from the timespan of 1970–2015. The empirical results of the study validated the environmental Kuznets curve (EKC) hypothesis in a majority of the studied economies. Specifically, the results found that in sixty-four economies, the rise in globalization enhanced energy usage initially but ultimately resulted in increased energy efficiency and decreased energy consumption.

In another panel estimation, Rahman and Miah [27] examined the influence of numerous sources of energy on the level of globalization in a panel of twenty-six economies between 1990 and 2010. The findings documented that utilization of nonrenewable sources for power generation resulted in decreasing globalization in the studied economies. Alternatively, the adoption of green sources of energy were found to exert a positive influence on globalization in the considered sample.

Identifying the causal connection between green energy, fossil-based energy, and natural resources, Bekun, Alola, and Sarkodie [68] analyzed the panel of sixteen European economies. For this, the authors utilized the panel data from the timespan of 1996–2014. The empirical results of the study found that natural resources exert a unidirectional causal effect on both renewable and nonrenewable energy consumption. In Ghana, Kwakwa, Alhassan, and Adu [33] examined the link between the extraction of natural resources and energy utilization and carbon emanation. For this, the authors utilized the time-series data from the timespan of 1971–2013. The empirical results of the study found a significant impact of natural resources on energy as well as carbon discharge in Ghana. Specifically, the results reported that a rise in the extraction of natural resources enhanced the energy utilization and environmental degradation in the country.

3. Methodology

In the present study, an innovative hybrid approach for the identification of nonlinear based causality was employed, which was discussed and proposed by Balcilar et al. [1] and was founded on the frameworks proposed by Jeong et al. [69] and Nishiyama et al. [70]. The energy consumption is denoted by $\mathrm{y_t}$ and natural resources and globalization are denoted by x_{1t} and x_{2t}, respectively. As discussed by Jeong et al. [69] in the θ-th quantile, y_t, will not be effected by x_{1t} and x_{2t}, respectively, in terms of lag vector $\{y_{t-1}, \ldots, y_{t-p}, x_{1t-1}, \ldots, x_{1t-p}, x_{2t-1}, \ldots, x_{2t-p}\}$, whereby:

$$Q_\theta\big(y_t y_{t-1}, \ldots, y_{t-p}, x_{1t-1}, \ldots, x_{1t-p}, x_{2t-1}, \ldots, x_{2t-p}\big) = Q_\theta\big(y_t y_{t-1}, \ldots, y_{t-p}\big) \quad (1)$$

Moreover, the prima facie effect by x_{1t} and x_{2t}, respectively, on $\mathrm{y_t}$ in the θ-th quantile in terms of $\{y_{t-1}, \ldots, y_{t-p}, x_{1t-1}, \ldots, x_{1t-p}, x_{2t-1}, \ldots, x_{2t-p}\}$ is:

$$Q_\theta\big(y_t y_{t-1}, \ldots, y_{t-p}, x_{1t-1}, \ldots, x_{1t-p}, x_{2t-1}, \ldots, x_{2t-p}\big) \neq Q_\theta\big(y_t y_{t-1}, \ldots, y_{t-p}\big) \quad (2)$$

where $Q_\theta(y_t\bullet)$ represents the θ-th quantile of the y_t subject to t and $0 < \theta < 1$. Moreover, the conditional distribution function of $\mathrm{y_t}$ is denoted by $Y_{t-1} \equiv (y_{t-1}, \ldots, y_{t-p})$, $X_{1t-1} \equiv (x_{1t-1}, \ldots, x_{1t-p})$, $X_{2t-1} \equiv (x_{2t-1}, \ldots, x_{2t-p})$, $Z_t = (X_{1t}, X_{2t}, Y_t)$, $F_{y_t|Z_{t-1}}(y_t|Z_{t-1})$, and $F_{y_t|Y_{t-1}}(y_t|Y_{t-1})$, where Z_{t-1} and Y_{t-1} are given, respectively. It is supposed that the conditional distribution $F_{y_t|Z_{t-1}}(y_t|Z_{t-1})$ is absolutely continuous in terms of $\mathrm{y_t}$ for all Z_{t-1}. Furthermore if it is represented that $Q_\theta(Z_{t-1}) \equiv Q_\theta(y_t|Z_{t-1})$ and $Q_\theta(Y_{t-1}) \equiv Q_\theta(y_t|Y_{t-1})$ will have $F_{y_t|Z_{t-1}}\{Q_\theta(Z_{t-1})|Z_{t-1}\} = \theta$ with a probability of 1. Therefore, as per the Equations (1) and (2), the hypotheses that need to be tested are shown below:

$$H_0 : P\big\{F_{y_t|Z_{t-1}}\{Q_\theta(Y_{t-1})|Z_{t-1}\} = \theta\big\} = 1 \quad (3)$$

$$H_1 : P\big\{F_{y_t|Z_{t-1}}\{Q_\theta(Y_{t-1})|Z_{t-1}\} = \theta\big\} < 1 \quad (4)$$

As per the framework proposed by Jeong et al. [69], the measure is used to compute the distance, which is $J = \{\varepsilon_t E(\varepsilon_t|Z_{t-1}) f_Z(Z_{t-1})\}$, whereas the regression error is represented by ε_t and the function of marginal density of Z_{t-1} is represented by $f_Z(Z_{t-1})$. The regression error is calculated when the null hypothesis is found true as presented in Equation (3), which is possible only in the scenario where

$E[1\{y_t \le Q_\theta(Y_{t-1})Z_{t-1}\}] = \theta$ or equivalently $1\{y_t \le Q_\theta(Y_{t-1})\} = \theta + \varepsilon_t$, whereas the indicator function is represented by $1\{\bullet\}$. The distance function as specified by Jeong et al. [69] is shown below:

$$J = E\left[\left\{F_{y_t|Z_{t-1}}\{Q_\theta(Y_{t-1})|Z_{t-1}\} - \theta\right\}^2 f_Z(Z_{t-1})\right] \quad (5) \tag{5}$$

As shown in Equation (3), it is imperative to consider that $J \ge 0$; only when H_0, as shown in Equation (5), becomes true will $J = 0$, whereas as shown in Equation (4), the values $J > 0$ are underneath the alternative hypothesis that is H_1. Moreover, the feasible test statistic based on the kernel function as discussed by Jeong et al. [69] for J can be shown as follows:

$$\hat{J}_T = \frac{1}{T(T-1)h^{2p}} \sum_{t=p+1}^{T} \sum_{s=p+1,\ s\neq t}^{T} K\left(\frac{Z_{t-1} - Z_{s-1}}{h}\right)\hat{\varepsilon}_t\hat{\varepsilon}_s \tag{6}$$

In Equation (6), the kernel function is represented by $K(\bullet)$, having bandwidth h, sample size is represented by T, lag-order is shown by p, the estimated unknown regression error is calculated by $\hat{\varepsilon}_t$, which is computed as:

$$\hat{\varepsilon}_t = 1\left\{y_t \le \hat{Q}_\theta(Y_{t-1})\right\} - \theta \tag{7}$$

In Equation (7), the θ-th quantile of y_t is calculated by $\hat{Q}_\theta(Y_{t-1})$, whereby (Y_{t-1}) is given. Moreover, $\hat{Q}_\theta(Y_{t-1})$ can be calculated through the kernel method, which is based on a nonparametric method and is shown below:

$$\hat{Q}_\theta(Y_{t-1}) = \hat{F}^{-1}_{yt|Y_{t-1}}(\theta Y_{t-1}) \tag{8}$$

where the estimator of the Nadarya–Watson kernel is represented by $\hat{F}_{yt|Y_{t-1}}(y_t Y_{t-1})$ and is given as:

$$\hat{F}_{yt|Y_{t-1}}(y_t Y_{t-1}) = \frac{\sum_{s=p+1,\ s\neq t}^{T} L\left(\frac{Y_{t-1}-Y_{s-1}}{h}\right)1(y_s \le y_t)}{\sum_{s=p+1,\ s\neq t}^{T} L\left(\frac{Y_{t-1}-Y_{s-1}}{h}\right)} \tag{9}$$

In Equation (9), the kernel function is represented by $L(\bullet)$ and the bandwidth is represented by h.

While extending the framework proposed by Jeong et al. [69], a test was developed for the 2nd moment. For this purpose, the nonparametric approach, which is quantile causality based on Granger as suggested and proposed by Nishiyama et al. [70], was used. While computing the higher-order moment, the illustration of causality is assumed as:

$$y_t = g(Y_{t-1}) + \sigma(X_{1t-1})\varepsilon_t + \sigma(X_{2t-1})\varepsilon_t \tag{10}$$

where the noise process is represented by ε_t, and stationery conditions are satisfied through unknown functions, which are $g(\bullet)$ and $\sigma(\bullet)$. Moreover, it should be noted that the aforementioned description is not in accordance with the testing of Granger type causality from x_{1t} and x_{2t} to y_t, respectively, however, it has the possibility for detection of nonlinear predictive power, which is computed from x_{1t} and x_{2t} to ${y_t}^2$ whereas $\sigma(\bullet)$ is the function of general nonlinearity. Therefore, for explaining the variation through Granger causality, the explicit description of squares of X_{1t-1} and X_{2t-1} are not required. Moreover, the hypotheses forms of Equation (10) for the purpose of explanation of variation are shown below:

$$H_0 : P\left\{F_{y_t^2|Z_{t-1}}\{Q_\theta(Y_{t-1})|Z_{t-1}\} = \theta\right\} = 1 \tag{11}$$

$$H_1 : P\left\{F_{y_t^2|Z_{t-1}}\{Q_\theta(Y_{t-1})|Z_{t-1}\} = \theta\right\} < 1 \tag{12}$$

In order to test the null hypothesis as shown in Equation (11), a feasible test statistic was obtained and y_t was replaced in Equations (6)–(9), with ${y_t}^2$. Moreover, the issue related to the causality was resolved by incorporating the methodology proposed by Jeong et al. [69], in which conditional causality in the 1st moment (mean) denotes causality in the 2nd moment (variance). Therefore, in order to resolve this concern, the causality in the scenario of the higher order moments was examined and evaluated by utilizing the model as shown below:

$$y_t = g(X_{1t-1}, X_{2t-1}, Y_{t-1}) + \varepsilon_t \tag{13}$$

Therefore, the quantile causality based on higher order is stated as:

$$H_0 : P\left\{F_{y_t^k|Z_{t-1}}\{Q_\theta(Y_{t-1})\big|Z_{t-1}\} = \theta\right\} = 1 \quad for\, k = 1,\, 2, \ldots, K \tag{14}$$

$$H_1 : P\left\{F_{y_t^k|Z_{t-1}}\{Q_\theta(Y_{t-1})\big|Z_{t-1}\} = \theta\right\} < 1 \quad for\, k = 1,\, 2, \ldots, K \tag{15}$$

By integration of the whole framework, y_t is caused by x_{1t} and x_{2t} Granger in the quantile θ up to the Kth moment by employing Equation (14), whereby for each k, the test static using Equation (6) is constructed. Moreover, since the statistics are correlated mutually, for the combined null hypothesis shown in Equation (14), it is extremely difficult to join the diverse statistics for each $k = 1, 2, 3 \ldots K$ [70]. Therefore, as per the recommendations proposed by Nishiyama et al. [70], this issue was efficiently addressed by including a modified method based on sequential testing. At the first stage, the nonparametric Granger causality in the 1st moment is $k = 1$. When rejecting of null hypothesis fails at $k = 1$, it will not directly jump to evaluate the noncausality at the 2nd moment and, therefore, the test for $k = 2$ can be constructed. In this manner, the existence of causality in variance, and/or causality in mean can be tested.

The practical application of testing of causality by means of quantiles specifies three essential choices: (1) bandwidth, which is represented by h, (2) kernel type for $K(\bullet)$ and $L(\bullet)$ as shown in Equations (6) and (9), and (3) the lag order, which is represented by p. The determination of lag order was done by employing Schwarz Information Criterion (SIC) underneath vector autoregression (VAR), including natural resources, globalization, and energy consumption. The least square cross-validation method was used for the selection of bandwidth values. Lastly, the Gaussian kernels were employed for computing the $K(\bullet)$ and $L(\bullet)$.

4. Data Analysis and Interpretation

The aim of this study was to investigate the dynamic causal effect of natural resources and globalization on utilization of energy in top Asian countries. In so doing, the recent study used natural resources' rent (% of gross domestic product (GDP)) as a proxy of natural resources (NAR); overall globalization index, as represented by GLO (which consists of economic, social, and political globalization); and utilization of energy (EGY), which is a measure of per capita of Kg of oil equivalent for top Asian nations including China, India, Indonesia, Malaysia, and Thailand. The selection of these nations was made as they have the most natural resources among all the Asian nations. The data for NAR and EGY was gathered from the World Bank, however, the information for GLO was collected from the KOF Swiss Economic Institute. Yearly information was gathered for 1970–2018. As the aforementioned methodology needs long time-series information [71], hence, the recent study transformed the yearly information into quarterly information by selecting a quadratic match-sum technique. This technique is beneficial once lower frequency information is transformed into higher frequency information, as it corrects the seasonality problem and connects the end-to-end deviation in the sample period. This technique was also suggested in previous research [72,73]. Beginning with the fundamental test, the descriptive statistics are presented in Table 1.

The findings of the descriptive statistics confirmed that the mean (average) coefficients of NAR were positive for all top Asian countries. The biggest coefficient of mean for NAR was for Malaysia at 18.884 (ranged from 7.163 to 37.570) followed by Indonesia at 10.614 (ranged from 3.718 to 33.658). The smallest coefficient of mean of NAR was for Thailand at 1.725 (ranged from 0.562 to 3.785) followed by India at 2.985 (ranged from 0.834 to 7.351). China had the middle coefficient of mean for NAR at 6.202 (ranged from 0.811 to 19.232). On the other hand, the highest coefficient of mean for GLO was for Malaysia at 65.411 (varied from 47.769 to 79.615) followed by Thailand at 51.130 (varied from 32.576 to 69.129). The lowest coefficient of mean for GLO was for China at 46.262 (varied from 23.034 to 61.994) followed by India at 42.420 (varied from 31.081 to 58.274). In addition, the middle coefficient of mean for GLO was for Indonesia at 49.384 (varied from 32.022 to 63.315). Finally, the mean coefficients of EGY were also positive for all countries. The leading coefficient of mean was for Malaysia at 1637.991 (fluctuated from 523.576 to 3003.456) followed by China at 1013.589 (fluctuated from 464.933 to 2236.730). The lowest mean coefficient of EGY was for India at 390.043 (fluctuated from 267.309 to 636.570) followed by Indonesia at 584.885 (fluctuated from 297.306 to 883.918). The middle coefficient of mean of EGY was for Thailand at 979.875 (fluctuated from 360.594 to 1991.594).

Table 1. Results of descriptive statistics for top Asian economies.

Countries	Mean	Min	Max	Std.Dev.	Skew	Kurtosis	JB test	*p* Value
Panel A: Natural Resources (% of GDP)								
China	6.202	0.811	19.232	4.714	1.289	4.136	14.874	0.001
India	2.985	0.834	7.351	1.199	0.933	5.478	18.039	0.000
Indonesia	10.614	3.718	33.658	5.955	1.913	7.271	61.649	0.000
Malaysia	18.844	7.163	37.570	8.737	0.436	3.907	13.664	0.001
Thailand	1.725	0.562	3.785	0.833	0.454	2.235	22.644	0.000
Panel B: Globalization Index (including Political, Social, and Economic Globalization)								
China	42.262	23.034	61.994	14.348	0.097	3.366	15.074	0.000
India	42.420	31.081	58.274	10.221	0.441	3.546	15.425	0.000
Indonesia	49.384	32.022	63.315	9.994	0.033	3.447	14.532	0.001
Malaysia	65.411	47.769	79.615	10.494	−0.103	4.578	13.873	0.001
Thailand	51.130	32.576	69.129	12.855	0.037	5.410	14.750	0.001
Panel C: Primary Energy Consumption (Per capita of Kg of Oil Equivalent)								
China	1013.589	464.933	2236.730	551.683	1.142	2.911	9.803	0.007
India	390.043	267.309	636.570	108.122	0.812	2.606	15.236	0.000
Indonesia	584.885	297.306	883.918	203.876	−0.006	3.446	24.527	0.000
Malaysia	1637.991	523.576	3003.456	810.742	0.175	4.610	23.851	0.000
Thailand	979.875	360.594	1991.594	537.228	0.436	3.800	24.126	0.000

Source: Author Estimations

Furthermore, the present study reported the skewness and kurtosis values, which were positive in almost all cases, however, the value of kurtosis was more than 3, which indicates a presence of nonlinearity in the dataset. In addition, the present research further applied the Jarque-Bera (JB) test to check the normality of the variables in all countries. The JB test statistics were statistically significant, which means NAR, GLO, and EGY were not normally distributed in all countries. The findings of the JB test also indicated a presence of nonlinearity among the variables for all countries. Hence, there is a need to apply a proper test to affirm the nonlinearity among the selected variables in the dataset [74]. In the current study, this was done by applying the BDS test for nonlinearity [75]. The findings of this test are reported in Table 2. The outcomes provide enough evidence to reject the null hypothesis of residual at different inserted dimensions (m), for the entire set of cases measured. The outcomes provide enough evidence of nonlinear association among NAR, GLO, and EGY in all countries. Therefore, the methods that focus on linear assumptions cannot be considered reliable and robust. Those outcomes provide adequate evidence of nonlinear relationships in the dataset.

Table 2. Results of BDS test for nonlinearity.

Country	$m = 2$	p-Value	$m = 3$	p-Value	$m = 4$	p-Value	$m = 5$	p-Value	$m = 6$	p-Value
Natural Resources Rent Equation Residual										
China	37.315	0.000	41.507	0.000	45.242	0.000	50.024	0.000	55.590	0.000
India	46.211	0.000	49.595	0.000	53.241	0.000	57.809	0.000	64.220	0.000
Indonesia	36.192	0.000	41.791	0.000	45.884	0.000	51.871	0.000	58.125	0.000
Malaysia	56.875	0.000	62.212	0.000	67.707	0.000	74.446	0.000	83.163	0.000
Thailand	79.621	0.000	85.632	0.000	92.794	0.000	102.972	0.000	116.824	0.000
Globalization Equation Residual										
China	44.535	0.000	47.897	0.000	51.903	0.000	57.596	0.000	65.344	0.000
India	31.812	0.000	34.797	0.000	37.871	0.000	41.640	0.000	46.516	0.000
Indonesia	20.243	0.000	23.375	0.000	25.665	0.000	29.013	0.000	32.511	0.000
Malaysia	25.847	0.000	27.740	0.000	29.780	0.000	32.335	0.000	35.921	0.000
Thailand	20.872	0.000	23.216	0.000	25.305	0.000	27.980	0.000	31.093	0.000
Energy Consumption Equation Residual										
China	16.954	0.000	18.234	0.000	19.759	0.000	21.927	0.000	24.876	0.000
India	12.111	0.000	13.247	0.000	14.417	0.000	15.852	0.000	17.709	0.000
Indonesia	7.707	0.000	8.899	0.000	9.770	0.000	11.045	0.000	12.377	0.000
Malaysia	9.840	0.000	10.561	0.000	11.337	0.000	12.310	0.000	13.675	0.000
Thailand	7.946	0.000	8.838	0.000	9.634	0.000	10.652	0.000	11.837	0.000

Source: Authors Estimation. Note: m denotes the embedding dimension of the BDS test. p-value is the probability of obtaining results as extreme as the observed results of a statistical hypothesis test, assuming that the null hypothesis is correct.

In the next phase, the present study applied two novel unit root tests: augmented Dickey–Fuller (ADF) and Zivot and Andrew structural break unit root test. These tests were used to affirm the stationary features for NAR, GLO, and EGY in all countries. The findings of both unit root tests are displayed in Table 3. The results of unit root tests confirmed that NAR, GLO, and EGY all showed nonstationary behavior at level series and became stationary at the first different series. Put simply, the outcome confirmed that all variables had a unique order of integration at the first difference series. After the unique order of integration, the present research applied a nonlinear cointegration method called quantile cointegration [76]. This method was used to investigate the long-run nonlinear connection between NAR, GLO, and EGY. The findings are displayed in Table 4. The findings reported α and δ coefficient values for both NAR and GLO models with EGY in all selected top Asian countries. Moreover, the findings of quantile cointegration also reported the critical value at 1%, 5%, and 10%. The results confirmed that all the calculated values were greater than the critical values, suggesting a rejection of the null hypothesis. In general, the outcomes confirmed a long-run nonlinear relationship between NAR with EGY and GLO with EGY in all top Asian countries.

In the final phase, the present research applied nonparametric causality in quantiles [1]. The aim of this method was to examine the causal connection of NAR and GLO with the return and volatility of EGY across different quantile distributions. The findings are reported in Figure 1. Every graph represents return values (mean) and volatility (variance) along with the critical value of 5% and 10%, respectively. Moreover, there are two axes in Figure 1, the horizontal axis (x-axis) explains the quantiles while the vertical axis (y-axis) describes the test results (t-stats value). The blue horizontal line symbolizes the 95% critical value of t-stats, whereas the orange dashed line exemplifies the 90% critical value of the test. The green dashed line explains the results for energy consumption returns and the yellow dark line describe the results for volatility.

Table 3. Results of unit root test.

Variables	ADF (Level)	ADF (Δ)	ZA (Level)	Break Year	ZA (Δ)	Break Year
Panel A: Natural Resources Rents						
China	0.184	−3.853 ***	−1.075	2008 Q2	−11.594 ***	1996 Q2
India	−1.483	−4.549 ***	−2.049	2010 Q1	−9.593 ***	2015 Q3
Indonesia	−2.044	−3.069 ***	−1.531	1984 Q4	−7.591 ***	2008 Q1
Malaysia	−0.338	−5.124 ***	−0.916	1999 Q2	−10.583 ***	2004 Q4
Thailand	−1.684	−4.616 ***	−2.021	2012 Q3	−6.005 ***	2016 Q4
Panel B: Globalization						
China	−0.472	−4.146 ***	−2.473	2014 Q1	−6.483 ***	2001 Q2
India	1.271	−3.892 ***	0.583	2009 Q1	−4.584 ***	2006 Q2
Indonesia	0.325	−4.093 ***	−0.482	2001 Q2	−6.482 ***	1995 Q3
Malaysia	−0.937	−5.483 ***	−1.486	1997 Q4	−7.482 ***	1984 Q1
Thailand	−2.081	−3.918 ***	−2.483	1988 Q4	−5.002 ***	2015 Q4
Panel C: Energy Consumption						
China	−0.931	−5.382 ***	−1.583	1976 Q4	−6.413 ***	2007 Q1
India	−2.081	−10.413 ***	−2.321	1984 Q4	−11.275 ***	2007 Q2
Indonesia	−1.226	−6.147 ***	−1.894	2000 Q3	−6.326 ***	2014 Q2
Malaysia	−0.269	−5.091 ***	−0.943	2016 Q2	−5.961 ***	1992 Q1
Thailand	−0.852	−4.381 ***	−1.035	2017 Q1	−5.039 ***	2011 Q3

Note: The values in the table specify statistical values of the ADF and ZA tests. The asterisks ***, **, and * represent level of significance at 1%, 5%, and 10%, respectively.

Table 4. Results of the quantile cointegration test.

Model	Coefficient	Supremum Norm Value	Critical Value at 1%	Critical Value at 5%	Critical Value at 10%
China					
Modelling between Energy and Natural Resources					
Model	Coefficient	Supremum Norm Value	Critical Value at 1%	Critical Value at 5%	Critical Value at 10%
EGY_t vs. NAR_t	α	3503.962	1688.804	1260.819	442.479
	δ	702.358	368.555	206.518	180.155
Modelling between Energy and Globalization					
Model	Coefficient	Supremum Norm Value	Critical Value at 1%	Critical Value at 5%	Critical Value at 10%
EGY_t vs. GLO_t	α	1209.021	582.712	435.038	152.675
	δ	242.344	127.168	71.258	62.161
India					
Modelling between Energy and Natural Resources					
Model	Coefficient	Supremum Norm Value	Critical Value at 1%	Critical Value at 5%	Critical Value at 10%
EGY_t vs. NAR_t	α	4736.817	1448.406	792.264	388.926
	δ	1289.183	482.926	287.88	207.423
Modelling between Energy and Globalization					
Model	Coefficient	Supremum Norm Value	Critical Value at 1%	Critical Value at 5%	Critical Value at 10%
EGY_t vs. GLO_t	α	6444.329	1970.523	1077.857	529.124
	δ	1753.904	657.009	391.654	282.195

Table 4. *Cont.*

Model	Coefficient	Supremum Norm Value	Critical Value at 1%	Critical Value at 5%	Critical Value at 10%
Indonesia					
Modelling between Energy and Natural Resources					
Model	Coefficient	Supremum Norm Value	Critical Value at 1%	Critical Value at 5%	Critical Value at 10%
EGY_t vs. NAR_t	α	2878.828	1892.718	1156.16	776.155
	δ	1688.849	975.646	572.689	249.982
Modelling between Energy and Globalization					
Model	Coefficient	Supremum Norm Value	Critical Value at 1%	Critical Value at 5%	Critical Value at 10%
EGY_t vs. GLO_t	α	4641.523	3051.622	1864.071	1251.391
	δ	2722.924	1573.03	923.344	403.046
Malaysia					
Modelling between Energy and Natural Resources					
Model	Coefficient	Supremum Norm Value	Critical Value at 1%	Critical Value at 5%	Critical Value at 10%
EGY_t vs. NAR_t	α	6497.219	1892.942	1275.087	704.464
	δ	3677.977	1464.707	850.557	442.59
Modelling between Energy and Globalization					
Model	Coefficient	Supremum Norm Value	Critical Value at 1%	Critical Value at 5%	Critical Value at 10%
EGY_t vs. GLO_t	α	2315.195	674.524	454.36	251.026
	δ	1310.597	521.928	303.084	157.711
Thailand					
Modelling between Energy and Natural Resources					
Model	Coefficient	Supremum Norm Value	Critical Value at 1%	Critical Value at 5%	Critical Value at 10%
EGY_t vs. NAR_t	α	4434.05	2267.235	1617.547	1241.176
	δ	1871.99	1301.375	636.36	389.006
Modelling between Energy and Globalization					
Model	Coefficient	Supremum Norm Value	Critical Value at 1%	Critical Value at 5%	Critical Value at 10%
EGY_t vs. GLO_t	α	4968.139	2540.327	1812.383	1390.678
	δ	2097.475	1458.128	713.011	435.863

EGY From NAR **EGY From GLO**

(i) China

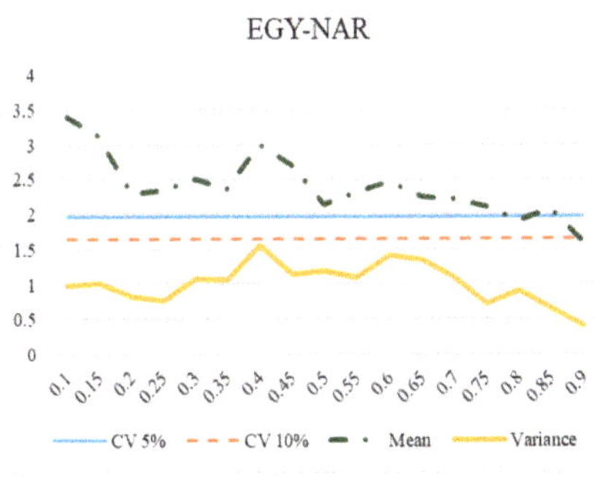

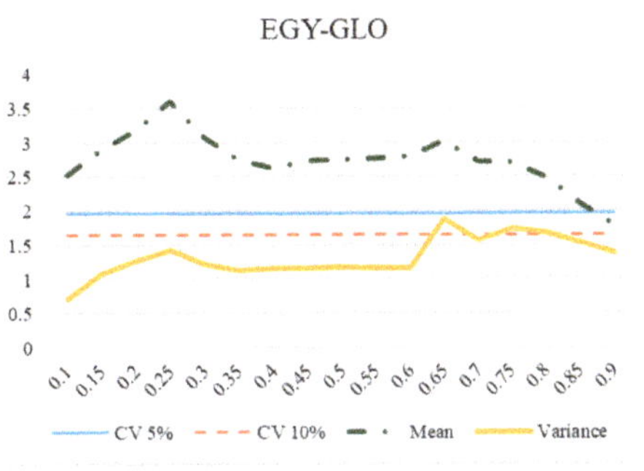

Figure 1. *Cont.*

(ii) India

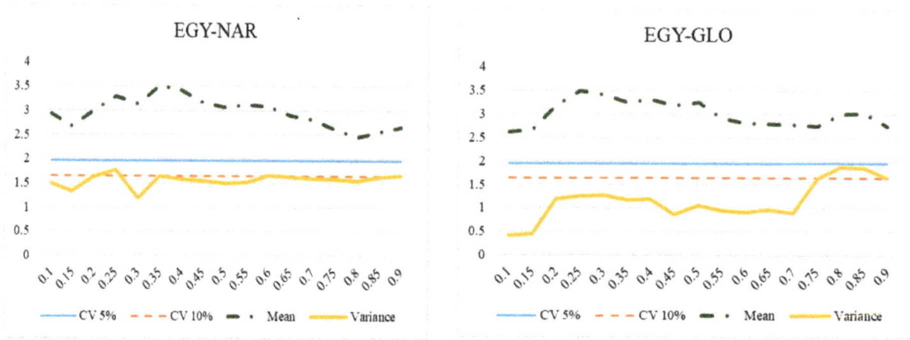

(iii) Indonesia

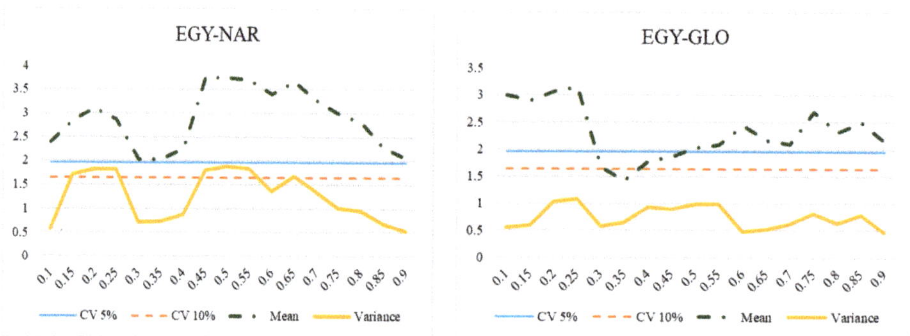

(iv) Malaysia

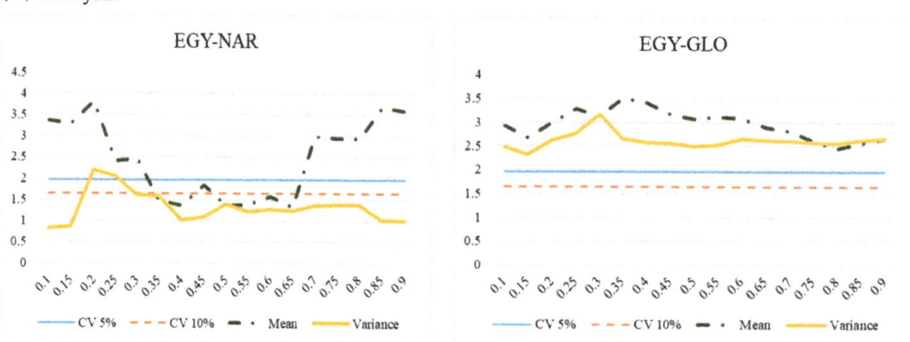

Figure 1. *Cont.*

(v) Thailand

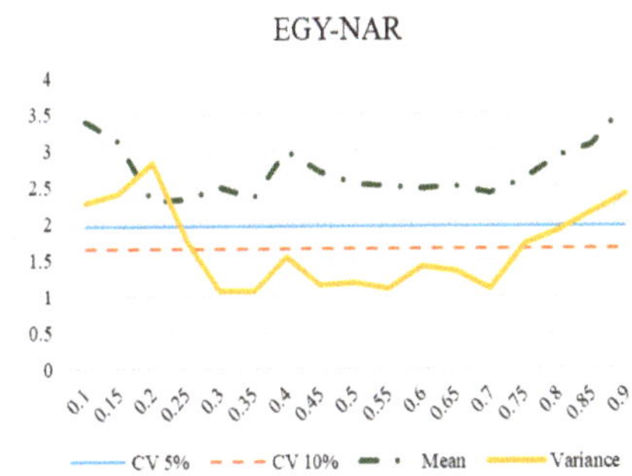

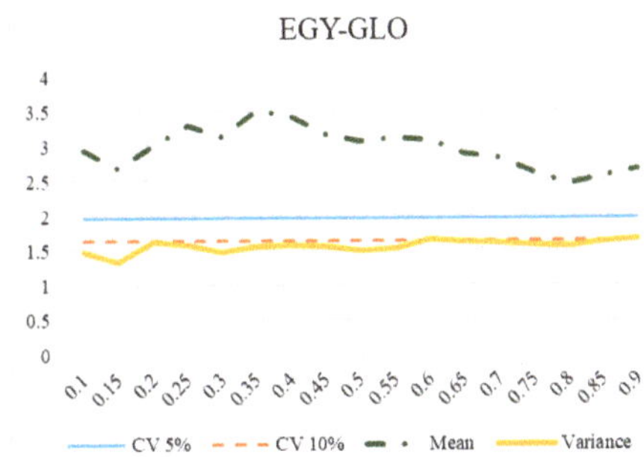

Figure 1. Results from nonparametric causality in quantiles.

In the case of China, the effect of NAR returns was significant for all lower, middle, and upper quantiles of the provisional distribution of EGY. The effect became strong and more significant in the lower quantiles with the t-stat value of approximately 3.4. The current study observed no causal effect of returns of NAR on the volatility of EGY in the case of the China. Moreover, the causal effect on volatility was not asymmetric, because they were insignificant for the upper, middle, and lower quantiles of the conditional distribution. On the other hand, the effect of GLO returns was also significant for all quantiles of distribution on the returns of EGY. The effect became more noteworthy in the low quantiles with the t-stats coefficient of approximately 3.56. However, the current research found no causal connection between returns of GLO and volatility of EGY. Put simply, the findings confirmed that returns of NAR and GLO had a significant causal impact on the returns of EGY at all quantiles.

In the situation of India, the impact of NAR returns was noteworthy for all quantiles of the returns of EGY. Moreover, the impact was strong and more significant at lower and middle quantiles of the returns of EGY. However, the present research did not find any causal effect of NAR returns on the volatility of EGY. In addition, the effect of returns of GLO was significant on the returns of EGY, however, the effect of the returns of GLO was significant on the volatility of EGY only at the upper quantiles at the 10% level of significance. In general, the returns of NAR and GLO had significant causal impact on the returns of EGY across all quantiles. The findings of causality from NAR and GLO to EGY are very interesting in the case of Indonesia. The findings confirmed that the returns of NAR had causal impact on the return and volatility of EGY at lower to middle quantiles. On the other hand, the influence of returns of GLO had a causal impact on the returns of EGY but there was no evidence of a causality from returns of GLO to volatility of EGY in Indonesia. In the case of Malaysia, the effect of NAR returns was substantial on the returns of EGY only at the extreme quantiles (i.e., lower and upper). However, the returns of NAR had a significant causal impact on the volatility of EGY on the lower middle quantiles. On the other hand, the effect of GLO returns had a significant causal impact on the returns and volatility of EGY. Put simply, the returns of NAR had a causal impact on EGY returns only, whereas the returns of GLO had a causal connection with EGY returns and volatility. In the case of Thailand, the returns of NAR were significant for all quantiles of returns of EGY. Moreover, the effect was strong and more prominent at both extreme quantiles (i.e., lower and higher). However, the study found no causal connection between the returns of NAR and volatility of EGY. On the other hand, the effect of GLO on EGY returns was significant across all quantiles of distribution. The effect was more powerful at the lower middle quantiles. In addition, the study found no causal impact of GLO returns on the volatility of EGY in Thailand.

5. Conclusions

Natural resources enable direct foreign investment to a country, and that has a tendency to improve the standard of living of the locals [11,12], however, it also adversely affects the potential growth of the country in the long term [13], thus making the host country intensely dependent on the consumption and extraction of the natural resources; this is referred to as a resource curse [14–18].

Moreover, most researchers are in agreement with respect to the adverse association between economic development and natural resources of a country [19–21]. Researchers also suggest that by improving institutional quality, having a good incentive program to the personnel engaged in the respective activities [22], and strengthening the political stability [23], the country can counter the phenomenon of being trapping in a resource curse.

Keeping this in mind, the current research aimed to investigate the causal effect of natural resources and globalization on the returns and volatility of the utilization of energy in top Asian economies. The findings of nonparametric methods of causality in quantile confirmed that the returns of NAR and GLO had a significant causal effect on the returns of EGY in the selected sample. These results are consistent with the earlier studies of Shahbaz et al. [41] and Shahbaz et al. [42] that investigated the globalization–energy link and Bekun, et al. [68] and Kwakwa, et al. [33] that examined the natural resource–energy association. On the other hand, the present study found no causal connection from the returns of NAR and GLO to volatility of EGY in any of the studied Asian countries.

The results shed greater insights on the level of energy dependence in the process of globalization and the utilization of natural resources in Asian economies. Knowing the positive role of globalization and natural resources in the growth of prospering countries [77–79], the findings of the current study implied higher challenges for government practitioners and policy makers, given the worldwide emphasis on energy conservation. The study suggested the need of implementing adequate environmental reforms and eco-friendly business practices to support energy efficiency in the course of integrating industries and allocating investments in local and foreign businesses. Moreover, the change of energy mix from coal-based to renewables can also satisfy the inevitable need of energy in industries without disrupting environmental sustainability. The current research is limited to only top Asian countries. Moreover, the present research is also limited to bivariate analysis, therefore, future research could apply multivariate advanced econometrics such as nonlinear ARDL, quantile ARDL, and a multiple wavelet coherence approach.

Author Contributions: All authors contributed equally. All authors have read and agreed to the published version of the manuscript.

Funding: This research received no external funding.

Conflicts of Interest: The authors declare no conflict of interest.

References

1. Balcilar, M.; Bekiros, S.; Gupta, R. The role of news-based uncertainty indices in predicting oil markets: A hybrid nonparametric quantile causality method. *Empir. Econ.* **2017**, *53*, 879–889. [CrossRef]
2. Hassan, S.T.; Xia, E.; Khan, N.H.; Shah, S.M.A. Economic growth, natural resources, and ecological footprints: Evidence from Pakistan. *Environ. Sci. Pollut. Res.* **2019**, *26*, 2929–2938. [CrossRef] [PubMed]
3. Baltgailis, J. The issues of increasing the effectiveness of teaching comparative economics. *Insights Reg. Dev.* **2019**, *1*, 190–199. [CrossRef]
4. Vigliarolo, F. Economic phenomenology: Fundamentals, principles and definition. *Insights Reg. Dev.* **2020**, *2*, 418–429. [CrossRef]
5. Zeibote, Z.; Volkova, T.; Todorov, K. The impact of globalization on regional development and competitiveness: Cases of selected regions. *Insights Reg. Dev.* **2019**, *1*, 33–47. [CrossRef]
6. Bhuiyan, M.A.; Jabeen, M.; Zaman, K.; Khan, A.; Ahmad, J.; Hishan, S.S. The impact of climate change and energy resources on biodiversity loss: Evidence from a panel of selected Asian countries. *Renew. Energy* **2018**, *117*, 324–340. [CrossRef]

7. Wu, S.; Li, L.; Li, S. Natural resource abundance, natural resource-oriented industry dependence, and economic growth: Evidence from the provincial level in China. *Resour. Conserv. Recycl.* **2018**, *139*, 163–171. [CrossRef]
8. Yongping, S.; Chusheng, Y. Abundant natural resources and industry structure distortion: Influence mechanism and multi-dimensional measure. *Soc. Sci. Nanjing* **2012**, *6*, 1–8.
9. Garayev, A.I.O.; Tanriverdiev, S.M.O.; Hajiyev, N.Q.-O. Analysis of the oil, price and currency factor of economic growth in Azerbaijan. *Entrep. Sustain. Issues* **2019**, *6*, 1335–1353.
10. Ustiuzhanin, A.A.; Liman, I.A.; Kiselitsa, E.P.; Shilova, N.N.; Leyman, T.I. The ruble exchange rate and the price of oil: Assessment of the degree of dependence, its causes and ways of overcoming. *Entrep. Sustain. Issues* **2019**, *7*, 121–132. [CrossRef]
11. Shahabadi, A.; Feyzi, S. The relationship between natural resources abundance, foreign direct investment and environmental performance in selected oil and developed countries during 1996–2013. *Int. J. Resist. Econ.* **2016**, *4*, 101–116.
12. Mikita, M.; Kolcun, M.; Spes, M.; Vojtek, M.; Ivancak, M. Impact of electrical power load time management at sizing and cost of hybrid renewable power system. *Pol. J. Manag. Stud.* **2017**, *15*, 154–162. [CrossRef]
13. Wang, K.; Wu, M.; Sun, Y.; Shi, X.; Sun, A.; Zhang, P. Resource abundance, industrial structure, and regional carbon emissions efficiency in China. *Resour. Policy* **2019**, *60*, 203–214. [CrossRef]
14. Song, M.; Wang, J.; Zhao, J. Coal endowment, resource curse, and high coal-consuming industries location: Analysis based on large-scale data. *Resour. Conserv. Recycl.* **2018**, *129*, 333–344. [CrossRef]
15. Gerelmaa, L.; Kotani, K. Further investigation of natural resources and economic growth: Do natural resources depress economic growth? *Resour. Policy* **2016**, *50*, 312–321. [CrossRef]
16. Badeeb, R.A.; Lean, H.H.; Clark, J. The evolution of the natural resource curse thesis: A critical literature survey. *Resour. Policy* **2017**, *51*, 123–134. [CrossRef]
17. Ahmed, K.; Mahalik, M.K.; Shahbaz, M. Dynamics between economic growth, labor, capital and natural resource abundance in Iran: An application of the combined cointegration approach. *Resour. Policy* **2016**, *49*, 213–221. [CrossRef]
18. Tyo, A.; Jazykbayeva, B.; Ten, T.; Kogay, G.; Spanova, B. Development tendencies of heat and energy resources: Evidence of Kazakhstan. *Entrep. Sustain* **2019**, *7*, 1514–1524. [CrossRef]
19. Sala-i-Martin, X.; Doppelhofer, G.; Miller, R.I. Determinants of long-term growth: A Bayesian averaging of classical estimates (BACE) approach. *Am. Econ. Rev.* **2004**, *94*, 813–835. [CrossRef]
20. Sala-i-Martin, X.X. *I Just Ran Four Million Regressions*; National Bureau of Economic Research: Cambridge, MA, USA, 1997.
21. Sachs, J.D.; Warner, A.M. *Natural Resource Abundance and Economic Growth*; National Bureau of Economic Research: Cambridge, MA, USA, 1995.
22. Robinson, J.A.; Torvik, R.; Verdier, T. Political foundations of the resource curse. *J. Dev. Econ.* **2006**, *79*, 447–468. [CrossRef]
23. Van Alstine, J.; Neumayer, E. The environmental Kuznets curve. *Handb. Trade Environ.* **2010**, *2*, 49–59.
24. Latif, Z.; Latif, S.; Ximei, L.; Pathan, Z.H.; Salam, S.; Jianqiu, Z. The dynamics of ICT, foreign direct investment, globalization and economic growth: Panel estimation robust to heterogeneity and cross-sectional dependence. *Telemat. Inform.* **2018**, *35*, 318–328. [CrossRef]
25. Luo, Y.; Bu, J. How valuable is information and communication technology? A study of emerging economy enterprises. *J. World Bus.* **2016**, *51*, 200–211. [CrossRef]
26. Widiyanti, M.; Sadalia, I.; Irawati, N.; Hendrawaty, E. Determining firm's performance: Moderating role of CSR in renewable energy sector of Indonesia. *Pol. J. Manag. Stud.* **2019**, *19*, 432–441.
27. Rahman, S.M.; Miah, M.D. The impact of sources of energy production on globalization: Evidence from panel data analysis. *Renew. Sustain. Energy Rev.* **2017**, *74*, 110–115. [CrossRef]
28. Miao, C.; Sadalia, I.; Irawati, N.; Hendrawaty, E. Natural resources utilization efficiency under the influence of green technological innovation. *Resour. Conserv. Recycl.* **2017**, *126*, 153–161. [CrossRef]
29. Asumadu-Sarkodie, S.; Owusu, P.A. Carbon dioxide emissions, GDP, energy use, and population growth: A multivariate and causality analysis for Ghana, 1971–2013. *Environ. Sci. Pollut. Res.* **2016**, *23*, 13508–13520. [CrossRef]
30. Hilson, G. The environmental impact of small-scale gold mining in Ghana: Identifying problems and possible solutions. *Geogr. J.* **2002**, *168*, 57–72. [CrossRef]

31. Nasirudeen, A.; Allan, A. Managing the impacts of mining on Ghana's water resources from a legal perspective. *J. Energy Nat. Resour. Manag. (JENRM)* **2014**, *1*, 156–165.
32. Kwakwa, P.A. On the determinants of electricity power losses: Empirics from Ghana. *OPEC Energy Rev.* **2018**, *42*, 3–21. [CrossRef]
33. Kwakwa, P.A.; Alhassan, H.; Adu, G. Effect of natural resources extraction on energy consumption and carbon dioxide emission in Ghana. *Int. J. Energy Sect. Manag.* **2019**, *14*, 20–39. [CrossRef]
34. Shakhovskaya, L.; Petrenko, E.; Dzhindzholia, A.; Timonina, V. Market peculiarities of natural gass: Case of the Pacific Region. *Entrep. Sustain. Issues* **2018**, *5*, 555–564. [CrossRef]
35. Tvaronavičienė, M.; Baublys, J.; Raudeliūnienė, J.; Jatautaitė, D. Global energy consumption peculiarities and energy sources: Role of renewables. In *Energy Transformation Towards Sustainability*; Elsevier: Amsterdam, The Netherlands, 2020; pp. 1–49.
36. Fang, Y. Economic welfare impacts from renewable energy consumption: The China experience. *Renew. Sustain. Energy Rev.* **2011**, *15*, 5120–5128. [CrossRef]
37. Chetthamrongchai, P.; Jermsittiparsert, K.; Saengchai, S. How the nexus among the free trade, institutional quality and economic growth ffect the trade from ASEAN countries. *Entrep. Sustain. Issues* **2020**, *7*, 2079–2094. [CrossRef]
38. Narayan, P.K.; Narayan, S. Does environmental quality influence health expenditures? Empirical evidence from a panel of selected OECD countries. *Ecol. Econ.* **2008**, *65*, 367–374. [CrossRef]
39. Haseeb, M.; Kot, S.; Hussain, H.I.; Jermsittiparsert, K. Impact of economic growth, environmental pollution, and energy consumption on health expenditure and R&D expenditure of ASEAN countries. *Energies* **2019**, *12*, 3598.
40. Shahbaz, M.; Shahzad, S.J.H.; Mahalik, M.K.; Sadorsky, P. How strong is the causal relationship between globalization and energy consumption in developed economies? A country-specific time-series and panel analysis. *Appl. Econ.* **2018**, *50*, 1479–1494. [CrossRef]
41. Shahbaz, M.; Lahiani, A.; Abosedra, S.; Hammoudeh, S. The role of globalization in energy consumption: A quantile cointegrating regression approach. *Energy Econ.* **2018**, *71*, 161–170. [CrossRef]
42. Shahbaz, M.; Mallick, H.; Mahalik, M.K.; Sadorsky, P. The role of globalization on the recent evolution of energy demand in India: Implications for sustainable development. *Energy Econ.* **2016**, *55*, 52–68. [CrossRef]
43. Cole, M.A. Does trade liberalization increase national energy use? *Econ. Lett.* **2006**, *92*, 108–112. [CrossRef]
44. Dreher, A. Does globalization affect growth? Evidence from a new index of globalization. *Appl. Econ.* **2006**, *38*, 1091–1110. [CrossRef]
45. Antweiler, W.; Copeland, B.R.; Taylor, M.S. Is free trade good for the environment? *Am. Econ. Rev.* **2001**, *91*, 877–908. [CrossRef]
46. McAusland, C. Globalisation's direct and indirect effects on the environment. *Glob. Transp. Environ.* **2010**, *31*, 31–53.
47. Zaidi, S.A.H.; Zafar, M.W.; Shahbaz, M.; Hou, F. Dynamic linkages between globalization, financial development and carbon emissions: Evidence from Asia Pacific Economic Cooperation countries. *J. Clean. Prod.* **2019**, *228*, 533–543. [CrossRef]
48. Coxhead, I. A new resource curse? Impacts of China's boom on comparative advantage and resource dependence in Southeast Asia. *World Dev.* **2007**, *35*, 1099–1119. [CrossRef]
49. Ross, M.L. The political economy of the resource curse. *World Politics* **1999**, *51*, 297–322. [CrossRef]
50. Yanıkkaya, H.; Turan, T. Curse or Blessing? An Empirical Re-examination of Natural Resource-Growth Nexus. *J. Int. Dev.* **2018**, *30*, 1455–1473. [CrossRef]
51. Sharif, A.; Afshan, S.; Qureshi, M.A. Idolization and ramification between globalization and ecological footprints: Evidence from quantile-on-quantile approach. *Environ. Sci. Pollut. Res.* **2019**, *26*, 11191–11211. [CrossRef]
52. Holmberg, K.; Kivikytö-Reponen, P.; Härkisaari, P.; Valtonen, K.; Erdemir, A. Global energy consumption due to friction and wear in the mining industry. *Tribol. Int.* **2017**, *115*, 116–139. [CrossRef]
53. Apergis, N.; Payne, J.E.; Menyah, K.; Wolde-Rufael, Y. On the causal dynamics between emissions, nuclear energy, renewable energy, and economic growth. *Ecol. Econ.* **2010**, *69*, 2255–2260. [CrossRef]
54. Speirs, J.; McGlade, C.; Slade, R. Uncertainty in the availability of natural resources: Fossil fuels, critical metals and biomass. *Energy Policy* **2015**, *87*, 654–664. [CrossRef]

55. Saurin, J. Global environmental degradation, modernity and environmental knowledge. *Environ. Politics* **1993**, *2*, 46–64. [CrossRef]
56. Bilgili, F.; Koçak, E.; Bulut, Ü. The dynamic impact of renewable energy consumption on CO2 emissions: A revisited Environmental Kuznets Curve approach. *Renew. Sustain. Energy Rev.* **2016**, *54*, 838–845. [CrossRef]
57. Begum, R.A.; Sohag, K.; Abdullah, S.M.S.; Jaafar, M. CO_2 emissions, energy consumption, economic and population growth in Malaysia. *Renew. Sustain. Energy Rev.* **2015**, *41*, 594–601. [CrossRef]
58. Nasreen, S.; Anwar, S.; Ozturk, I. Financial stability, energy consumption and environmental quality: Evidence from South Asian economies. *Renew. Sustain. Energy Rev.* **2017**, *67*, 1105–1122. [CrossRef]
59. Pérez-Lombard, L.; Ortiz, J.; Pout, C. A review on buildings energy consumption information. *Energy Build.* **2008**, *40*, 394–398. [CrossRef]
60. Azam, M.; Khan, A.Q.; Zafeiriou, E.; Arabatzis, G. Socio-economic determinants of energy consumption: An empirical survey for Greece. *Renew. Sustain. Energy Rev.* **2016**, *57*, 1556–1567. [CrossRef]
61. Bhattacharjee, S.; Reichard, G. Socio-economic factors affecting individual household energy consumption: A systematic review. In Proceedings of the ASME 2011 5th International Conference on Energy Sustainability, Washington, DC, USA, 7–10 August 2011.
62. Keho, Y. What drives energy consumption in developing countries? The experience of selected African countries. *Energy Policy* **2016**, *91*, 233–246. [CrossRef]
63. Azam, M.; Khan, A.Q.; Zaman, K.; Ahmad, M. Factors determining energy consumption: Evidence from Indonesia, Malaysia and Thailand. *Renew. Sustain. Energy Rev.* **2015**, *42*, 1123–1131. [CrossRef]
64. Murshed, M.; Tul Jannat, F.; Amin, S. An Empirical Investigation of globalization and energy consumption: Evidence from Bangladesh. *World J. Soc. Sci.* **2018**, *8*, 54–68.
65. Shahbaz, M.; Solarin, S.A.; Sbia, R.; Bibi, S. Does energy intensity contribute to CO_2 emissions? A trivariate analysis in selected African countries. *Ecol. Indic.* **2015**, *50*, 215–224. [CrossRef]
66. Koengkan, M.; Poveda, Y.E.; Fuinhas, J.A. Globalisation as a motor of renewable energy development in Latin America countries. *GeoJournal* **2019**, 1–12. [CrossRef]
67. Shahbaz, M.; Mahalik, M.K.; Shahzad, S.J.H.; Hammoudeh, S. Does the environmental K uznets curve exist between globalization and energy consumption? G lobal evidence from the cross-correlation method. *Int. J. Financ. Econ.* **2019**, *24*, 540–557. [CrossRef]
68. Bekun, F.V.; Alola, A.A.; Sarkodie, S.A. Toward a sustainable environment: Nexus between CO_2 emissions, resource rent, renewable and nonrenewable energy in 16-EU countries. *Sci. Total Environ.* **2019**, *657*, 1023–1029. [CrossRef] [PubMed]
69. Jeong, K.; Härdle, W.K.; Song, S. A consistent nonparametric test for causality in quantile. *Econom. Theory* **2012**, *28*, 861–887. [CrossRef]
70. Nishiyama, Y.; Hitomi, K.; Kawasaki, Y.; Jeong, K. A consistent nonparametric test for nonlinear causality—Specification in time series regression. *J. Econom.* **2011**, *165*, 112–127. [CrossRef]
71. Hussain, H.I.; Slusarczyk, B.; Kamarudin, F.; Thaker, H.M.T.; Szczepańska-Woszczyna, K. An investigation of an adaptive neuro-fuzzy inference system to predict the relationship among energy intensity, globalization, and financial development in major ASEAN economies. *Energies* **2020**, *13*, 850. [CrossRef]
72. Arain, H.; Han, L.; Sharif, A.; Meo, M.S. Investigating the effect of inbound tourism on FDI: The importance of quantile estimations. *Tour. Econ.* **2019**. [CrossRef]
73. Shahzad, S.J.H.; Shahbaz, M.; Ferrer, R.; Kumar, R.R. Tourism-led growth hypothesis in the top ten tourist destinations: New evidence using the quantile-on-quantile approach. *Tour. Manag.* **2017**, *60*, 223–232. [CrossRef]
74. Mishra, S.; Sharif, A.; Khuntia, S.; Meo, S.A.; Khan, S.A.R. Does oil prices impede Islamic stock indices? Fresh insights from wavelet-based quantile-on-quantile approach. *Resour. Policy* **2019**, *62*, 292–304. [CrossRef]
75. Broock, W.A.; Scheinkman, J.A.; Dechert, W.D.; LeBaron, B. A test for independence based on the correlation dimension. *Econom. Rev.* **1996**, *15*, 197–235. [CrossRef]
76. Xiao, Z. Functional-coefficient cointegration models. *J. Econom.* **2009**, *152*, 81–92. [CrossRef]
77. Nawaz, K.; Lahiani, A.; Roubaud, D. Natural resources as blessings and finance-growth nexus: A bootstrap ARDL approach in an emerging economy. *Resour. Policy* **2019**, *60*, 277–287. [CrossRef]

78. Hassan, S.T.; Xia, E.; Huang, J.; Khan, N.H.; Iqbal, K. Natural resources, globalization, and economic growth: Evidence from Pakistan. *Environ. Sci. Pollut. Res.* **2019**, *26*, 15527–15534. [CrossRef] [PubMed]
79. Kandil, M.; Shahbaz, M.; Mahalik, M.K.; Nguyen, D.K. The drivers of economic growth in China and India: Globalization or financial development? *Int. J. Dev. Issues* **2017**, *16*, 54–81. [CrossRef]

energies

MDPI

Article

Power Market Formation for Clean Energy Production as the Prerequisite for the Country's Energy Security

Manuela Tvaronavičienė [1,*], Evgeny Lisin [2,*] and Vladimir Kindra [3]

1 Department of Business Economics and Management, Vilnius Gediminas Technical University, Sauletekio 11, 10223 Vilnius, Lithuania
2 Department of Economics in Power Engineering and Industry, National Research University "Moscow Power Engineering Institute", Krasnokazarmennaya st. 14, 111250 Moscow, Russia
3 Department of Thermal Power Plants, National Research University "Moscow Power Engineering Institute", Krasnokazarmennaya st. 14, Moscow 111250, Russia; kindravo@mpei.ru
* Correspondence: manuela.tvaronaviciene@vgtu.lt (M.T.); lisinym@mpei.ru (E.L.)

Received: 15 August 2020; Accepted: 18 September 2020; Published: 20 September 2020

Abstract: The paper analyzes the main issues of power market development for clean energy production within the broader framework of ensuring the country's energy security. In addition, special attention is paid to the technologies aimed at reducing emissions of toxic substances and greenhouse gases by the fossil-fired power plants. Even though the future electricity markets would most likely depend on the high shares of renewable energy sources (RES) in the electricity system, energy efficiency such as the one based on the near-zero emission technologies might also play a crucial role in the transition to the carbon-free energy future. In particular, there are the oxy-fuel combustion technologies that might help to reduce the proportion of unburned fuel and increase the efficiency of the power plant while reducing the emissions of flue gases. Our paper focuses on the role and the place of the near-zero emission technologies in the production of clean energy. We applied economic and mathematical models for assessing the prospects for applying oxy-fuel combustion technology in thermal power plants, taking into account the system of emission quotas and changes in the fuel cost. Our results demonstrate that at the current fuel prices, it is advisable to use economical combined cycle gas turbines (CCGT). At the same time, when quotas for greenhouse gas emissions are introduced and fuel costs increase by 1.3 times, it becomes economically feasible to use the oxy-fuel combustion technology which possesses significant economic advantages over CCGT with respect to the capture and storage of greenhouse gases.

Keywords: energy security; power market; near-zero emission technologies; oxy-fuel combustion cycles; emission rights sales mechanism; economic assessment

1. Introduction

The energy sector plays a key role in any country's economy in terms of maintaining its sustainability and ensuring its energy security. Moreover, it is the level of energy efficiency and reliability that defines the country's national security [1–4]. The development of the energy sector should take place at a faster pace than the development of other sectors of the economy which is explained by the need for satisfying the ever-growing demand of the economy for the electric energy.

Our analysis of the available data [5–7] on the construction of generating facilities of all possible types from various countries around the world allows us to conclude that there has been a significant increase in electricity consumption over the past 30 years. At the same time, despite the growth that was achieved over the past decade in the installed capacity of alternative and renewable energy sources, the consumption of classical fuel and energy resources (oil, gas, coal) is growing, and the role of traditional thermal power industry based on the chemical conversion of organic fuel still remains crucial [3,4].

The generating unit of the traditional thermal power industry is a thermal power plant (TPP). In Russia, TPPs account for more than 57% of the country's total electricity generation and 37% of heat production intended for the consumers' needs. At the same time, thermal power plants consume 38% of the extracted fossil fuel. The conversion of this fuel into energy becomes the main source of environmental pollution (forming about 57% of emissions among all processing industries).

The active use of fossil fuels at thermal power plants that is intended to meet the demand for electricity while this resource is limited is one of the main factors in the development of effective energy conversion technologies. Figure 1 presents a graph reflecting a gradual increase in efficiency on the example of coal TPP in countries with the most developed energy sector [6–10]. As it can be observed from the example of countries with the highest deficit of fossil fuels, the development of effective power generation technologies is faster.

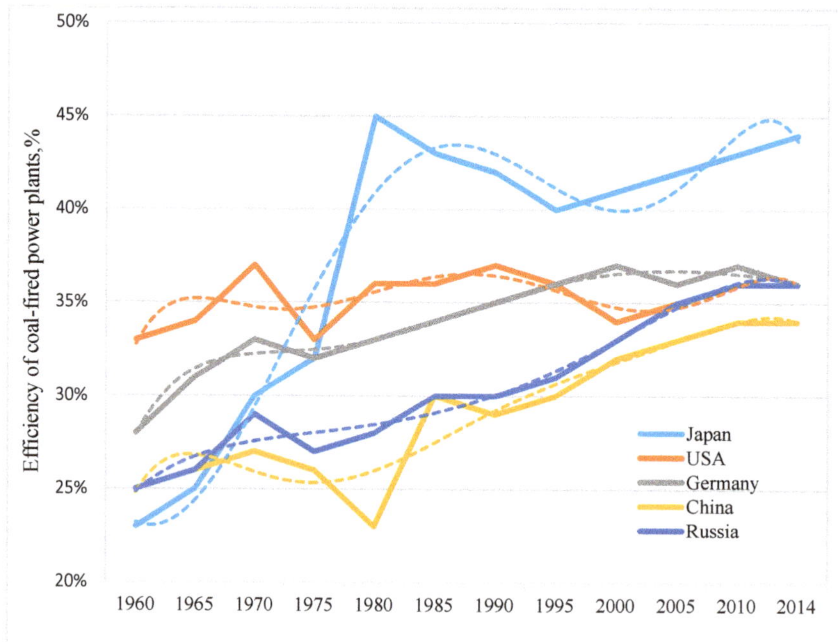

Figure 1. Development of coal thermal power plants (TPPs) in countries with the most developed energy sector.

Nowadays, the development of technologies for the production of energy products is moving towards the increase of the initial parameters of steam-pressure P_0 and temperature t_0. From a thermodynamic point of view, this is the only possible way to significantly increase the efficiency of turbine units and, hence, the efficiency of a power plant by surging its environmental friendliness in general. The most significant effect on the efficiency of a power unit is a change in the initial temperature of the steam t_0 in the combustion chamber. Thus, an increase in t_0 by 1% leads to an increase in the efficiency of thermal power plants by an average of 0.13% with a simultaneous increase in pressure P_0 by 0.0086% [11,12].

The existing modern technologies used for increasing the initial temperature of fuel combustion are based on its oxygen combustion which is a process when oxygen is served as the oxidizer instead of atmospheric air in the combustion chamber. In turn, oxygen is produced at a power plant or purchased from an external supplier. The increased oxygen content leads to an increase in the combustion temperature and the amount of heat transferred to the technological process. Thus, the share of unburned fuel is reduced and the efficiency of the power unit of the TPP is increased while the nitrogen oxide emissions are reduced (NO_X) [2,13–16].

Since the atmospheric air contains 80% of nitrogen, the use of the oxygen combustion leads to a significant reduction in flue gas consumption. At the same time, an increased concentration of greenhouse gases is formed in the volume of exhaust gases which creates favorable conditions for capture and storage CO_2 [16–19]. Therefore, in the research literature, this technology is typically characterized as a near-zero emission technology [20–22].

The implementation and dissemination of the near-zero emission technologies for the production of electricity in power plants is largely due to their economic efficiency which in this case is determined by the environmental policy pursued by the state, namely, the established standards for the maximum permissible concentration of harmful substances in the atmospheric air and the market mechanisms for emissions trading [22–24]. The paper studies the issue of near-zero emission technologies and draws conclusions about the prospects for the application of these technologies in Russia under various scenarios of tightening the norms of the maximum permissible concentration of harmful substances in the air and developing the market for flue gas emission quotas.

2. Analysis of the Oxy-Fuel Combustion Technology and Its Environmental Benefits

Nowadays, most of the energy produced in TPPs is produced by the combustion of fossil fuels. The oxygen contained in the atmospheric air is usually used as the oxidizing agent. The combustion products contain greenhouse gases and toxic substances such as the nitrogen oxides. A significant reduction in the emissions of these substances into the atmosphere is cumbersome for technical and economic reasons. In particular, the emission of carbon dioxide from flue gases is inefficient due to its low partial pressure. Moreover, the insufficient combustion temperature leads to the formation of unburned fuel and its emission into the atmosphere along with the flue gases [25–27]. With regard to the above, oxygen-fuel technologies for energy production have a special meaning. They allow to almost completely reduce emissions of harmful substances into the atmosphere.

According to the technology of oxy-fuel combustion, three flows enter the combustion chamber:

- fuel (gaseous, including based on coal gasification),
- oxygen,
- carbon dioxide flow limiting the maximum temperature in the combustion chamber.

As a result of the combustion reaction, a mixture of carbon dioxide and water steam is formed. The predominantly two-component medium at a temperature of 1000 to 1700 °C is sent to a cooled turbine, expands in it, and then enters a surface heat exchanger which can be a waste heat boiler or regenerative heat exchanger. Having given off most of the thermal energy, the flow is directed to a cooler-separator. There, the cooling of the working medium is accompanied by the formation of condensate of water steam removed from the cycle. After that, a stream rich in carbon dioxide is sent to the compressor, increases its pressure and is fed to the combustion chamber for recycling. Thus, the thermodynamic cycle is reserved [27–29].

In order to replenish the material balance, part of the working environment is removed for further burial. The storage tanks for carbon dioxide can be both natural and artificial.

The schematic diagram of the oxygen-fuel energy complex is shown in Figure 2 that follows.

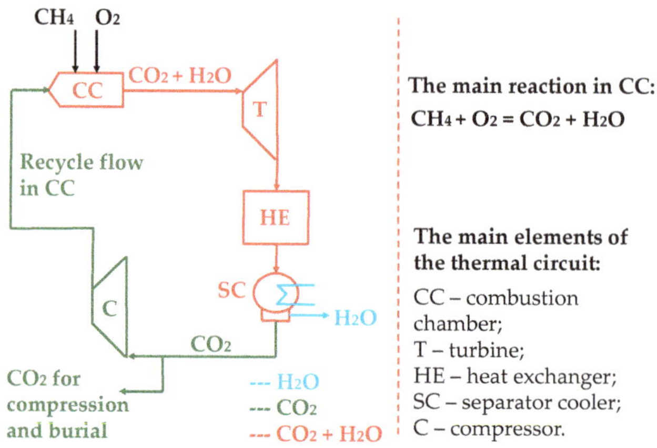

Figure 2. Schematic diagram of the oxygen-fuel energy complex.

The competitive technology for oxy-fuel combustion is the thermodynamic cycles used in the combined cycle power plants which are characterized by high efficiency and can significantly reduce thermal emissions into the atmosphere due to the long stay of the fuel–air mixture at high temperatures [30–33]. Unlike a combined cycle power plant, the oxygen-fuel energy complex is semi-closed. Most of the working medium circulates in a circle, and its mass and chemical composition only changes in the combustion chambers and cooler-separators. The maximum and minimum temperatures of the working medium are approximately at the same level as those of a combined-cycle power plant, and the maximum pressure is much higher: from 2 to 30 MPa, depending on the cycle [25,26].

These features of oxygen-fuel cycles lead to a number of advantages in comparison with the traditional combined-cycle power plants using air as an oxidizing agent:

1. Due to the lack of nitrogen among the components of the combustion reactions, the formation of high temperature nitrogen oxides is prevented. This allows to shift the main emphasis in the development of the combustion chamber to achieve high rates of efficiency and combustion stability.
2. The two-component composition of the working medium makes it possible to organize the simplest principle of thermodynamic separation of carbon dioxide and water steam in a cooler-separator by condensing the latter. As a result, there are no additional costs associated with respect to the capture of carbon dioxide from the flue gases.
3. Carbon dioxide is an inert gas that does not corrode power equipment, both at high and low temperatures.

The disadvantages of oxygen-fuel cycles include:

1. In order to avoid high energy costs for the production of oxygen, it must be supplied to the combustion chamber with a small excess coefficient relative to the stoichiometric ratio. The occurrence of the combustion reaction in the combustion chamber with a stoichiometric ratio of fuel to oxidizing agent necessitates the prevention of burning the fuel.
2. The need for high purity oxygen production causes higher costs of energy own use.

Table 1, which follows, shows a comparative analysis of the technical, economic, and environmental characteristics of oxygen combustion technologies and combined steam-gas cycle [25,34–41].

Table 1. Technical and economic characteristics of oxy-fuel combustion technologies and combined steam-gas cycles.

Oxy-Fuel Combustion Cycles	Fuel	Oxidizer	Net Efficiency, %	Specific Amount of Produced CO_2, g/kWh	CO_2 Capture Rate, %	Specific Amount of Captured CO_2, g/kWh	Specific Amount of CO_2 Emitted to the Atmosphere, g/kWh
				oxy-fuel combustion cycle			
SCOC-CC	CH_4	O_2	47.7	406	98.9	402	4
MATIANT	CH_4	O_2	43	451	98.9	446	5
E-MATIANT	CH_4	O_2	44	440	98.9	436	5
CC-MATIANT	CH_4	O_2	46	421	98.9	417	5
Allam cycle	CH_4	O_2	56.5	343	98.9	339	4
S-Graz cycle	CH_4	O_2	54	359	98.9	355	4
CES cycle	CH_4	O_2	48	404	98.9	399	4
AZEP	CH_4	O_2	50	388	98.9	383	4
ZEITMOP	CH_4	O_2	51	380	98.9	376	4
				combined steam-gas cycle			
Combined cycle gas turbine with CCS	CH_4	Air	48	404	89	359	44
Combined cycle gas turbine without CCS	CH_4	Air	60	323	0	0	323

As one can see from Table 1, the oxygen-fuel cycles have significant environmental advantages in comparison with the combined steam-gas cycle (in particular, expressed in the reduction of greenhouse gas emissions) with comparable values of the efficiency of power units. To date, the use of oxy-fuel combustion is restrained in the first place by the economic factors. However, due to the tightening of standards for emissions of harmful substances and the development of trade with quotas for greenhouse gas emissions, oxygen-fuel cycles of electricity production can become quite attractive.

3. Analysis of the Market Formation for Emissions Quotas of the Harmful Substances

The industrial and economic activities of thermal power plants are associated with direct environmental impacts in the form of emissions of the harmful substances. The criterion of environmental cleanliness of the generating facility is the condition that the concentration of harmful substances does not exceed the threshold value of the maximum permissible concentration (MPC). Current values of MPC of pollutants in the air are presented in the state regulations of countries. Some excerpts from the state standard of Russia GN 2.1.6.1338-03 are shown in Table 2 below [42].

Table 2. Maximum permissible concentrations of harmful substances in the air.

No.	Substance	Hazard Class	One-Time MPC, mg/m^3	Daily Average MPC, mg/m^3
1	Carbonic oxide	4	5	3
2	Nitrogen dioxide	2	0.2	0.04
3	Nitrogen oxide	3	0.4	0.06
4	Sulfur dioxide	3	0.5	0.05
5	Ammonia	4	0.2	0.04
6	Hydrogen sulfide	2	0.008	-

From Table 2, it becomes apparent that the existing system for regulating the amount of harmful substances used in Russia is rather soft and does not stimulate the development and implementation of the new and more environmentally friendly technologies for the production of electricity, in particular, oxygen combustion technology [43–46].

The above situation can be corrected by the development of the market mechanisms for the sale of emissions of harmful substances by the power plants [47–49]. Based on this market, the initially allocated quotas for pollutants and greenhouse gases will be redistributed from a position of minimizing total costs. In this case, the market will act as an optimizer and better cope with the task of reducing emissions than the introduction of a pollution tax [48,50]. The fact is that the determination of the required tax rate by regulatory authorities is very difficult, while the market usually copes with the optimization task much better.

The emissions trading scheme is shown in Figure 3.

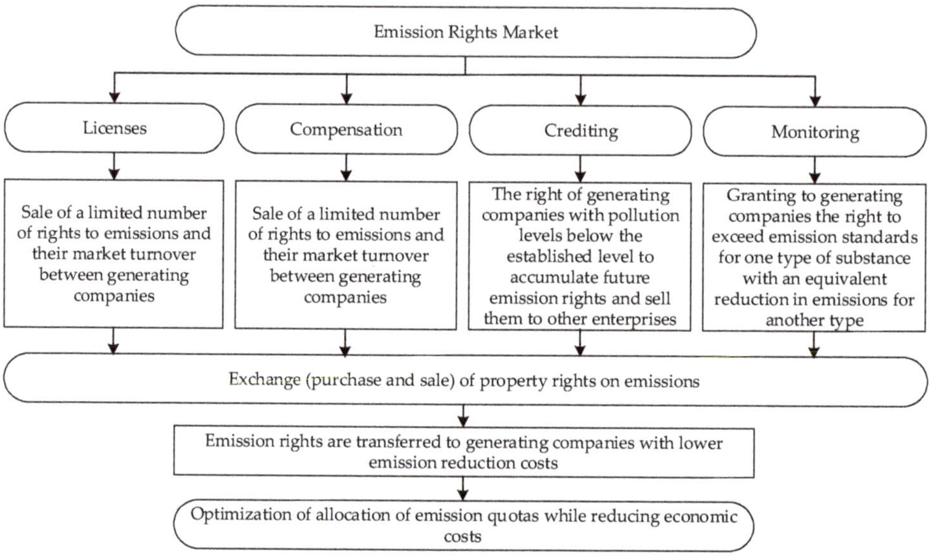

Figure 3. The emissions trading scheme of harmful substances.

A limit on a certain amount of pollutant emissions is introduced within a limited area. Emission permits are distributed among individual generating companies. The generating company is required to either meet the limits by investing in the appropriate technology, or, if this appears to be too expensive, purchase permits for additional emissions from those companies that are more profitable to reduce their emissions above the established limits [51,52].

A special role is played by the creation of a market mechanism for the sale of pollution rights at the global level [53,54]. The existing carbon tax creates a lot of problems for the industries of the developed countries which have almost exhausted all cheap ways for reducing their emissions. In these conditions, Russia can make itself significantly better off. The country possesses more than 75% of the world's carbon reserves. The creation of the international transfer mechanisms for offsetting the carbon balance is going to bring significant benefits and attract additional foreign investment.

In 2010, Russia transferred quotas for the emission of harmful substances and greenhouse gases into the atmosphere to the foreign companies for the first time. Japanese enterprises Mitsubishi and Nippon Oil bought quotas for the emission of 290,000 t of greenhouse gases from a Russian company PJSC Gazprom Neft. The cost of emissions is estimated to be at a total of 3.3 million euros. The transfer of quotas was carried out as part of joint implementation projects (JIP), the mechanism of which is prescribed in the Kyoto Protocol. The Kyoto Protocol obliges developed countries and transition countries to reduce or stabilize greenhouse gas emissions [55–57]. As a result of the JIP implementation, Russian enterprises received foreign investment in energy-efficient technologies, harmful emissions into the atmosphere were reduced, and the emission reduction volume could be implemented on the market.

In the case of the introduction of stringent standards for maximum permissible concentrations of harmful substances in the air and tax rates for flue gas emissions, one can expect the widespread usage of pollution rights sold for market purposes between different countries in the world. This will be conducive to environmental protection at the global level as well as the introduction of environmentally friendly technologies for the production of energy products in developing countries, in particular, based on oxy-fuel combustion, due to the influx of financial resources from developed countries, where the control of harmful substances is much more expensive.

At the local level, the development of the market for the emission quotas is facilitated by the development of the distributed electricity generation which complements the existing organization of energy supply to consumers on the basis of the centralized energy system of the country. The distributed power supply system is based on the idea of involving small distributed generation and consumer resources in the management of electric power systems with the achievement of the effect of increasing the total efficiency of work.

At the same time, there are many transaction costs and barriers arising on the way to the development of distributed energy such as the large number of participants in the system and economic relations between them, the costs of their information integration into control loops, as well as the costs of integrating their equipment into electrical networks which would prevent the power system from losing in reliability and stability of functioning. In order to remove these costs, it is necessary to develop a new system architecture that allows to support numerous fast transactions between participants and provides flexible connection to the electric networks of producers and consumers. These requirements are met by the architecture of the Internet of Energy (IoE) which represents a decentralized electric power system implementing an intelligent distributed control based on the energy transactions between its users [58,59]. The formation of the Internet of Energy will facilitate the transition of small generation to environmentally friendly energy supply to consumers based on the market principles.

4. Development of Economic Effect Assessing Model for the Application of the Oxygen Combustion Technology, Taking into Account the Environmental Aspect

The oxygen-fuel cycles increase the initial parameters of steam power plants which are currently one of the main ways to increase the efficiency and environmental friendliness of thermal power plants. Let us consider the social and environmental effectiveness of the use of oxy-fuel combustion from the perspective of economic criteria.

As an economic criterion which reflects the social significance of technology, we chose the minimum specific cost of electricity production. A reduction in the operating costs for electricity production will limit the increase in the electricity tariffs. The transition from the technology of one generation to a more efficient one should occur when the specific cost of electricity production using a more advanced technology becomes lower than the technology already used in the generation of energy. In many ways, the change in the cost of electricity production is determined by the change in the price of fossil fuels which grows as the available resource is exhausted.

In order to assess the level of specific costs for the production of electric energy for various generation technologies, we composed the following mathematical expression:

$$C = C_f + C_w + C_d + C_r + C_o \tag{1}$$

where C—annual cost of the electricity production; C_f—fuel costs of TPP; C_w—labor costs; C_r—repair costs; C_o—other expenses.

The most significant component is the fuel costs. The amount of the fuel costs depends on many factors. Among them, there are price of fuel, the efficiency of electric energy generation, determined by the level of generation technology, and the installed capacity utilization factor (ICUF), which sets the load on the generating facility during the year.

The fuel costs can be represented as the following mathematical expression:

$$C_f = N_p \cdot T_y \cdot K_{cf} \frac{1}{\eta \cdot Q_f} P_f \tag{2}$$

where N_p—installed capacity of TPP, kw; T_y—number of hours of operation of the power plant in a year, h; K_{cf}—installed capacity utilization factor; η—TPPs efficiency of the electricity production; Q_f—fuel calorific value, kJ/kg; P_f—fossil fuel price.

Labor costs can be estimated according to the following expression:

$$C_w = n\overline{W}(1 + \alpha) \tag{3}$$

where n—number of personnel at TPP, determined in accordance with the standard depending on the installed capacity of power units of the power plant; $\overline{W}$—industry average wage; α—the share of the mandatory insurance contributions.

The number of personnel is determined by the installed capacity of the power plant and the number of power units. In order to ensure the comparability of simulation results for each individual technology, we will calculate for one power unit. In this case, the electric power of the power unit will vary depending on the level of technology.

Depreciation can be calculated linearly. Then the share of depreciation charges can be determined by the life of the TPP power unit. For units at different steam parameters, let us assume the service life to last 30 years. Although the increase of the initial steam parameters is accompanied by the technological difficulties such as metal deformation and high temperature corrosion, the design life is maintained due to economic reasons, such as maintaining the investment attractiveness of power plant construction projects.

Repair costs are estimated in fractions of the amount of capital costs. The higher the steam parameters, the more expensive the steels and alloys used to create power units; therefore, repairs will cost more. The repair costs can be determined according to the following expression:

$$C_r = \beta \cdot I_s \cdot N_p \tag{4}$$

where β—share of contributions to the repair fund; I_s—specific investment, \$/kW; N_p—installed power capacity.

By other costs, we mean those costs that are not included in the above calculation, in particular, general expenses. Other costs depend on the power of the unit and can be estimated as part of the fixed costs:

$$C_o = \gamma(C_r + C_w + C_d) \tag{5}$$

where γ—share of other expenses.

The compiled functional model (1) allows us to calculate the specific cost of electricity production for technologies with different initial parameters of steam (including supercritical parameters achieved by oxygen combustion of fuel) at different prices for fossil fuels. Thus, it is possible to determine the cost of fossil fuels at which the use of oxygen-fuel cycles becomes economically feasible.

The introduction of a market-based mechanism for trading in CO_2 emissions quotas leads to an environmental factor affecting the cost of electricity production. In this case, the technology of oxygen combustion of fuel receives additional economic benefits. Incorporating environmental considerations leads to the following changes in the cost of electricity production model:

$$C = C_f + C_w + C_d + C_r + C_o + C_e \tag{6}$$

where C_e—costs for payment of quotas for CO_2 emissions.

In 2018–2019, CO_2 quotas were traded on the London Stock Exchange ICE Futures Europe at an average price of 20.4 €/t [60]. In this case, the price of quotas is set not by a decision of the authority, but by market mechanisms; therefore, the indicated value can also be used in calculating the economic efficiency of oxy-fuel combustion technology for Russian power plants.

5. Results and Discussions

The initial data for calculating the specific cost of production of energy products at various prices for fossil fuels and the formation of quotas for greenhouse gas emissions are provided in Table 3, which follows. The table contains model parameters of the existing and future environmentally friendly power generation plants at thermal power plants: gas turbine plants (GTU TPP without CO_2 capture) [61], combined cycle plants (CCGTU TPP without CO_2 capture and storage) [62], combined cycle plants with CO_2 capture and storage (CCGTU TPP with CO_2 capture and storage) [63], and oxygen-fuel plants (TPP with the oxygen-fuel cycle) (which are discussed in this paper).

Table 3. The calculated parameters of the model for determining the cost of electricity at different costs of fossil fuels and the availability of quotas for greenhouse gas emissions.

Model Parameter	GTU TPP Without CO_2 Capture	CCGTU TPP Without CO_2 Capture and Storage	CCGTU TPP with CO_2 Capture and Storage	TPP with Oxygen-Fuel Cycle
Installed capacity, MW	100	300	300	300
Net efficiency for the generation of electric energy, %	40	58	48	50
Specific CO_2 emissions, g/kWh	485	323	44	4
Specific capital investment, $/kW	1200	1700	2200	1900
Period of operation, years	30	30	30	30
Installed capacity utilization factor	0.7	0.7	0.7	0.7
Calorific value, kJ/m^3	36,000	36,000	36,000	36,000
The number of staff, persons	120	250	250	250
Average salary level in the industry, rubles	47,000	47,000	47,000	47,000
The coefficient of social contributions, %	30.2	30.2	30.2	30.2
The coefficient of contributions to the repair fund, %	5	5	5	5
The ratio of other costs, %	25	25	25	25

According to the functional model (1), the dependences of the unit cost of the electricity production on the price of fossil fuels (coal gasification) were obtained with an increase in the initial steam parameters by supplying the purified oxygen to the combustion chamber. The simulation results are presented in Figure 4.

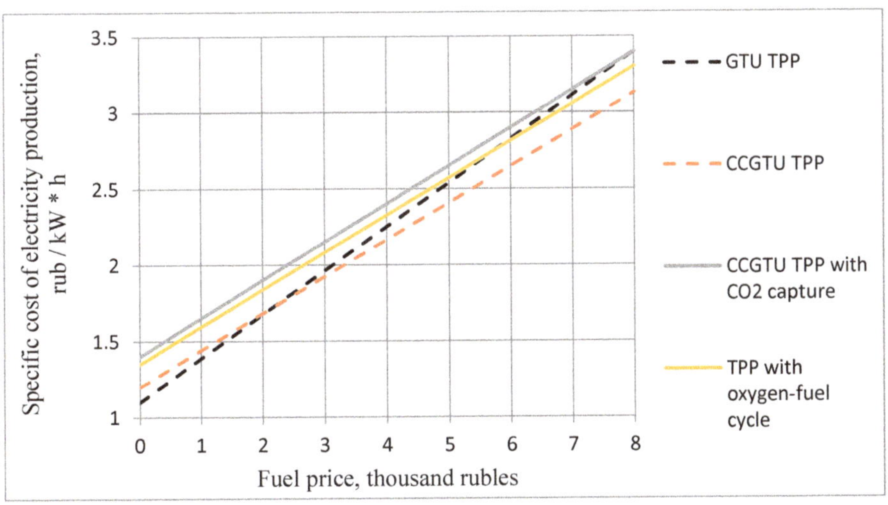

Figure 4. The dependence of the unit cost of electricity production on the price of fuel during the conventional and oxygen combustion.

In 2019, the average price of gasified fuel in Russia amounted to 2600 rubles/ton. At this level of fuel prices, according to the results of thermal power plants, CCGTU TPP units provide the lowest specific cost (1.78 rubles/kWh) of electricity production, which explains the increased interest of generating companies in their construction and commissioning. It also helps to reduce the burden on the environment.

The use of oxy-fuel combustion at TPPs will be economically feasible at a gasified fuel price of 5700 rubles/ton, which is 2.2 times the current cost of fuel. While maintaining the current dynamics of coal price growth, it will be economically feasible to build TPPs with oxygen-fuel cycles after 30 years.

Consider the simulation results obtained using model (2). This model of the specific cost of electricity production includes a quota system that provides for the purchase of rights for carbon dioxide emissions by generating companies. Accounting for the costs of ensuring environmental standards leads to a decrease in the threshold value of the fuel cost, which determines the economic feasibility of switching to the oxygen-fuel cycles. This is explained by the environmental benefits of TPPs operating at higher steam parameters and the possibility of almost complete capture of greenhouse gases during oxygen combustion of fuel. The simulation results are shown in Figure 5.

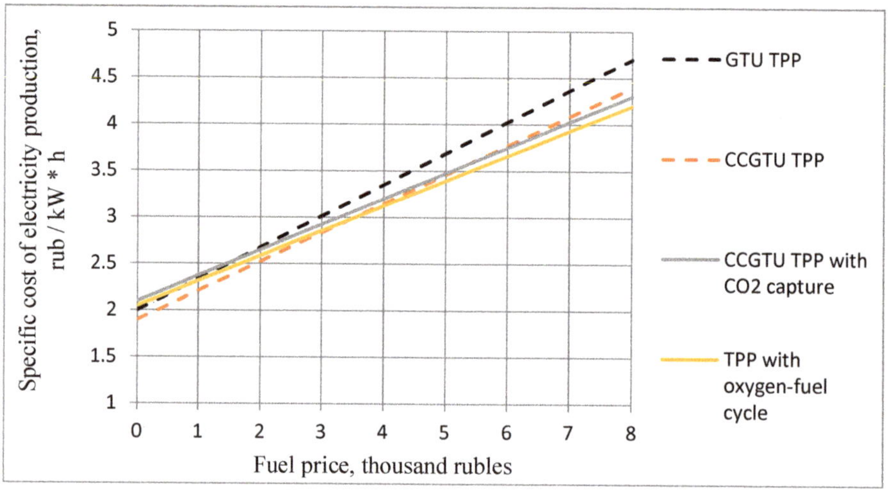

Figure 5. Dependence of the unit cost of electricity production on the price of fuel for the conventional and oxygen combustion, taking into account quotas for the greenhouse gas emissions.

From the analysis of the results presented in Figure 5, it is possible to determine the new threshold values of the fuel price at which it is economically feasible to build TPPs using oxy-fuel combustion technology. Table 4 presents the results of comparing threshold values for coal cost, which determines the economic feasibility of using oxy-fuel combustion technology without taking into account (model (1)) and taking into account (model (2)) the system of quotas for the greenhouse gas emissions.

Table 4. Comparison of threshold values for gasified coal prices that determine the economic feasibility of using oxy-fuel combustion technology.

Electricity Production Technology	Fuel Price Range Providing a Minimum of Unit Cost of Electricity Production, Rub/t	
	Without Emission Quota	With Emission Quotas
GTU TPP	0–1910	-
CCGTU TPP	>1910	0–3461
CCGTU TPP with CO_2 capture	-	-
TPP with the oxygen-fuel cycle	-	>3461

The results presented in Table 4 allow us to conclude that the introduction of a system of quotas for the greenhouse gas emissions would lead to the formation of market signals for owners of TPPs that determine the commissioning of the more advanced technologies for the fuel combustion and power units. The introduction of the oxy-fuel combustion technology provides a reduction in emissions of harmful substances into the atmosphere and also allows for the more careful use of limited fossil fuels which creates the basis for the sustainable development of the national economy.

It is worth noting that the introduction of quotas for the CO_2 emissions would inevitably lead to the increase of production costs and electricity prices for end consumers. However, a one-time increase in the prices for energy products in order to build a mechanism for the functioning of the energy industry that meets the principles of sustainable development appears to be an acceptable measure.

6. Conclusions

Our paper analyzed the prospects for the oxy-fuel combustion technologies from the position of ensuring the sustainable economic development of the country. Moreover, in our analysis, we also took into account the introduction of quotas for the greenhouse gas emissions.

Currently, the development of the technologies aimed at the production of electricity at traditional TPPs is associated with an increase in the temperature of fuel combustion in the combustion chamber and the beneficial use of flue gases. Thus, the initial steam parameters at the turbine inlet are significantly increased, which leads to an increase in the efficiency and environmental friendliness of the power plant due to a decrease in the formation of unburned fuel and its emission into the atmosphere. At the same time, capital costs are rising and have an impact on the cost of electricity through depreciation.

In connection with the tightening of quotas for emissions of harmful substances, oxygen-fuel technologies for energy production which allow to reduce emissions of harmful substances into the atmosphere almost completely, are of particular relevance. At the same time, their widespread use is hampered by the spread in the industry of competitive technology for energy production based on CCGT TPPs, the thermodynamic cycle of which is characterized by high efficiency and can significantly reduce thermal emissions into the atmosphere in comparison with the classical steam-powered power units.

The economic and mathematical model which was developed in this paper made it possible to formulate the dependences of the specific cost of electricity production on the price of gasified coal for the various levels of electricity production technologies, taking into account the possibility of the CO_2 capture and storage at CCGT and oxygen-fuel combustion. It is shown that at the current fuel prices it is advisable to use typical CCGT. At the same time, when quotas for greenhouse gas emissions are introduced and the cost of fuel rises by 1.3 times, it becomes economically feasible to apply the oxy-fuel combustion technology which has significant economic advantages over CCGT units with respect to the capture and storage of greenhouse gases.

Author Contributions: Conceptualization, M.T., V.K. and E.L.; methodology, V.K., E.L., and M.T.; formal analysis, E.L.; investigation, V.K., M.T. and E.L.; resources, V.K. and M.T.; data curation, V.K. and E.L.; writing—original draft preparation, E.L. and M.T.; writing—review & editing, V.K., M.T. and E.L.; visualization, E.L. All authors have read and agreed to the published version of the manuscript.

Funding: This research was funded by the RF President Council on Grants, grant number MD-914.2020.6".

Acknowledgments: This work described in this paper was performed thanks to the support of the grant of the RF President for the governmental support of young scientists PhD and DSc (MD-914.2020.6).

Conflicts of Interest: The authors declare no conflict of interest.

References

1. Chehabeddine, M.; Tvaronavičienė, M. Securing regional development. *Insights Region. Dev.* **2020**, *2*, 430–442. [CrossRef]
2. El Iysaouy, L.; El Idrissi, N.E.; Tvaronavičienė, M.; Lahbabi, M.; Oumnad, A. Towards energy efficiency: Case of Morocco. *Insights Region. Dev.* **2019**, *1*, 259–271. [CrossRef]
3. Dudin, M.N.; Frolova, E.E.; Protopopova, O.V.; Mamedov, A.A.; Odintsov, S.V. Study of innovative technologies in the energy industry: Nontraditional and renewable energy sources. *Entrep. Sustain. Issues* **2019**, *6*, 1704–1713. [CrossRef]

4. Tvaronavičienė, M.; Ślusarczyk, B. *Energy Transformation towards Sustainability*; Elsevier: Amsterdam, The Netherlands, 2020; p. 333. [CrossRef]
5. International Energy Statistics. Total Primary Energy Production. Available online: http://www.eia.gov/cfapps/ipdbproject/IEDIndex3.cfm (accessed on 25 July 2020).
6. International Energy Agency. Total Primary Energy Supply. Available online: http://www.iea.org/statistics/ (accessed on 25 July 2020).
7. Global Energy Statistical Yearbook. 2019. Available online: https://yearbook.enerdata.net/electricity/electricity-domestic-consumption-data.html (accessed on 25 July 2020).
8. Lisin, E.; Shuvalova, D.; Volkova, I.; Strielkowski, W. Sustainable development of regional power systems and the consumption of electric energy. *Sustainability* **2018**, *10*, 1111. [CrossRef]
9. International Energy Agency. Market Report Series: Energy Efficiency. 2018. Available online: https://webstore.iea.org/market-report-series-energy-efficiency-2018 (accessed on 25 July 2020).
10. Chen, T.Y.; Yeh, T.L.; Lee, Y.T. Comparison of power plants efficiency among 73 countries. *J. Energy* **2013**, *2013*, 916413. [CrossRef]
11. Lisin, E.; Strielkowski, W.; Krivokora, E. Economic Analysis of Industrial Development: A Case of Russian Coal Industry. *Monten. J. Econ.* **2016**, *12*, 129–139. [CrossRef]
12. Lisin, E.; Rogalev, A.; Strielkowski, W.; Komarov, I. Sustainable modernization of the Russian power utilities industry. *Sustainability* **2015**, *7*, 11378–11400. [CrossRef]
13. Lisin, E.; Sobolev, A.; Strielkowski, W.; Garanin, I. Thermal efficiency of cogeneration units with multi-stage reheating for Russian municipal heating systems. *Energies* **2016**, *9*, 269. [CrossRef]
14. Buhre, B.J.; Elliott, L.K.; Sheng, C.D.; Gupta, R.P.; Wall, T.F. Oxy-fuel combustion technology for coal-fired power generation. *Prog. Energy Combust. Sci.* **2005**, *31*, 283–307. [CrossRef]
15. Toftegaard, M.B.; Brix, J.; Jensen, P.A.; Glarborg, P.; Jensen, A.D. Oxy-fuel combustion of solid fuels. *Prog. Energy Combust. Sci.* **2010**, *36*, 581–625. [CrossRef]
16. Glarborg, P.; Bentzen, L.L. Chemical effects of a high CO_2 concentration in oxy-fuel combustion of methane. *Energy Fuels* **2007**, *22*, 291–296. [CrossRef]
17. Li, H.; Yan, J.; Anheden, M. Impurity impacts on the purification process in oxy-fuel combustion based CO_2 capture and storage system. *Appl. Energy* **2009**, *86*, 202–213. [CrossRef]
18. Ekström, C.; Schwendig, F.; Biede, O.; Franco, F.; Haupt, G.; de Koeijer, G.; Røkke, P.E. Techno-economic evaluations and benchmarking of pre-combustion CO_2 capture and oxy-fuel processes developed in the European ENCAP project. *Energy Procedia* **2009**, *1*, 4233–4240. [CrossRef]
19. Park, S.K.; Kim, T.S.; Sohn, J.L.; Lee, Y.D. An integrated power generation system combining solid oxide fuel cell and oxy-fuel combustion for high performance and CO_2 capture. *Appl. Energy* **2011**, *88*, 1187–1196. [CrossRef]
20. Wu, F.; Argyle, M.D.; Dellenback, P.A.; Fan, M. Progress in O_2 separation for oxy-fuel combustion–a promising way for cost-effective CO_2 capture: A review. *Prog. Energy Combust. Sci.* **2018**, *67*, 188–205. [CrossRef]
21. Ding, G.; He, B.; Cao, Y.; Wang, C.; Su, L.; Duan, Z.; Li, X. Process simulation and optimization of municipal solid waste fired power plant with oxygen/carbon dioxide combustion for near zero carbon dioxide emission. *Energy Convers. Manag.* **2018**, *157*, 157–168. [CrossRef]
22. Strielkowski, W.; Štreimikienė, D.; Bilan, Y. Network charging and residential tariffs: A case of household photovoltaics in the United Kingdom. *Renew. Sustain. Energy Rev.* **2017**, *77*, 461–473. [CrossRef]
23. Rogalev, A.; Grigoriev, E.; Kindra, V.; Rogalev, N. Thermodynamic optimization and equipment development for a high efficient fossil fuel power plant with zero emissions. *J. Clean. Prod.* **2019**, *236*, 117592. [CrossRef]
24. Igaliyeva, L.; Niyazbekova, S.; Serikova, M.; Kenzhegaliyeva, Z.; Mussirov, G.; Zueva, A.; Tyurina, Y.; Maisigova, L. Towards environmental security via energy efficiency: A case study. *Entrep. Sustain. Issues* **2020**, *7*, 3488–3499. [CrossRef]
25. Huang, X.; Guo, J.; Liu, Z.; Zheng, C. Opportunities and Challenges of Oxy-fuel Combustion. In *Oxy-Fuel Combustion*; Academic Press: New York, NY, USA, 2018; pp. 1–12. [CrossRef]
26. Balitskiy, S.; Bilan, Y.; Strielkowski, W.; Štreimikienė, D. Energy efficiency and natural gas consumption in the context of economic development in the European Union. *Renew. Sustain. Energy Rev.* **2016**, *55*, 156–168. [CrossRef]
27. Rogalev, A.N.; Kindra, V.O.; Rogalev, N.D.; Sokolov, V.P.; Milukov, I.A. Methods for efficiency improvement of the semi-closed oxy-fuel combustion combined cycle. *J. Phys. Conf. Ser.* **2018**, *1111*, 012003. [CrossRef]

28. Cau, G.; Tola, V.; Ferrara, F.; Porcu, A.; Pettinau, A. CO_2-free coal-fired power generation by partial oxy-fuel and post-combustion CO_2 capture: Techno-economic analysis. *Fuel* **2018**, *214*, 423–435. [CrossRef]
29. Sarma, U.; Karnitis, G.; Zuters, J.; Karnitis, E. District heating networks: Enhancement of the efficiency. *Insights Region. Dev.* **2019**, *1*, 200–213. [CrossRef]
30. Song, Y.; Zou, C.; He, Y.; Zheng, C. The chemical mechanism of the effect of CO_2 on the temperature in methane oxy-fuel combustion. *Int. J. Heat Mass Transf.* **2015**, *86*, 622–628. [CrossRef]
31. Lisin, E.; Strielkowski, W.; Garanin, I. Economic efficiency and transformation of the Russian energy sector. *Econ. Res.-Ekon. Istraž.* **2015**, *28*, 620–630. [CrossRef]
32. Park, H.; Baldick, R. Stochastic generation capacity expansion planning reducing greenhouse gas emissions. *IEEE Trans. Power Syst.* **2015**, *30*, 1026–1034. [CrossRef]
33. McLellan, B.; Florin, N.; Giurco, D.; Kishita, Y.; Itaoka, K.; Tezuka, T. Decentralised energy futures: The changing emissions reduction landscape. *Procedia Cirp* **2015**, *29*, 138–143. [CrossRef]
34. Colmenar-Santos, A.; Gómez-Camazón, D.; Rosales-Asensio, E.; Blanes-Peiró, J.J. Technological improvements in energetic efficiency and sustainability in existing combined-cycle gas turbine (CCGT) power plants. *Appl. Energy* **2018**, *223*, 30–51. [CrossRef]
35. Rogalev, A.; Kindra, V.; Zonov, A.; Rogalev, N.; Agamirov, L. Evaluation of bleed flow precooling influence on the efficiency of the E-MATIANT cycle. *Mech. Mech. Eng.* **2018**, *22*, 593–602. [CrossRef]
36. Zhao, Y.; Chi, J.; Zhang, S.; Xiao, Y. Thermodynamic study of an improved MATIANT cycle with stream split and recompression. *Appl. Therm. Eng.* **2017**, *125*, 452–469. [CrossRef]
37. Rogalev, A.; Kindra, V.; Osipov, S. Modeling methods for oxy-fuel combustion cycles with multicomponent working fluid. *AIP Conf. Proc.* **2018**, *2047*, 020020.
38. Allam, R.; Martin, S.; Forrest, B.; Fetvedt, J.; Lu, X.; Freed, D.; Manning, J. Demonstration of the Allam Cycle: An update on the development status of a high efficiency supercritical carbon dioxide power process employing full carbon capture. *Energy Procedia* **2017**, *114*, 5948–5966. [CrossRef]
39. Nami, H.; Ranjbar, F.; Yari, M.; Saeidi, S. Thermodynamic analysis of a modified oxy-fuel cycle, high steam content Graz cycle with a dual-pressure heat recovery steam generator. *Int. J. Exergy* **2016**, *21*, 331–346. [CrossRef]
40. Ferrari, N.; Mancuso, L.; Davison, J.; Chiesa, P.; Martelli, E.; Romano, M.C. Oxy-turbine for Power Plant with CO_2 capture. *Energy Procedia* **2017**, *114*, 471–480. [CrossRef]
41. Siegelman, R.L.; Milner, P.J.; Kim, E.J.; Weston, S.C.; Long, J.R. Challenges and opportunities for adsorption-based CO_2 capture from natural gas combined cycle emissions. *Energy Environ. Sci.* **2019**, *12*, 2161–2173. [CrossRef]
42. Garcia, S.I.; Garcia, R.F.; Carril, J.C.; Garcia, D.I. Critical review of the first-law efficiency in different power combined cycle architectures. *Energy Convers. Manag.* **2017**, *148*, 844–859. [CrossRef]
43. Electronic Fund of Legal and Scientific-Technical Documentation of Russia. Maximum Allowable Concentrations of Pollutants GN 2.1.6.1338-03. Available online: http://docs.cntd.ru/document/901865554 (accessed on 25 July 2020).
44. Kondrat'eva, O.E.; Roslyakov, P.V. Comparative study of gas-analyzing systems designed for continuous monitoring of TPP emissions. *Therm. Eng.* **2017**, *64*, 437–449. [CrossRef]
45. Fedorov, M.P.; Elistratov, V.V.; Maslikov, V.I.; Sidorenko, G.I.; Chusov, A.N.; Atrashenok, V.P.; Zinchenko, A.V. Reservoir greenhouse gas emissions at Russian TPP. *Power Technol. Eng.* **2015**, *49*, 33–39. [CrossRef]
46. Tumanovskii, A.G. Prospects for the development of coal-steam plants in Russia. *Therm. Eng.* **2017**, *64*, 399–407. [CrossRef]
47. Kondrateva, O.E.; Roslyakov, P.V.; Borovkova, A.M.; Loktionov, O.A. Problems of requirements implementation for the new environmental legislation in energy sector. *J. Phys. Conf. Ser.* **2018**, *1128*, 012130. [CrossRef]
48. Ang, B.W.; Su, B. Carbon emission intensity in electricity production: A global analysis. *Energy Policy* **2016**, *94*, 56–63. [CrossRef]
49. Meng, M.; Wang, L.; Chen, Q. Quota Allocation for Carbon Emissions in China's Electric Power Industry Based Upon the Fairness Principle. *Energies* **2018**, *11*, 2256. [CrossRef]
50. Mardones, C.; Flores, B. Effectiveness of a CO_2 tax on industrial emissions. *Energy Econ.* **2018**, *71*, 370–382. [CrossRef]

51. Kolster, C.; Mechleri, E.; Krevor, S.; Mac Dowell, N. The role of CO_2 purification and transport networks in carbon capture and storage cost reduction. *Int. J. Greenh. Gas Control* **2017**, *58*, 127–141. [CrossRef]
52. Rogge, K.S.; Schneider, M.; Hoffmann, V.H. The innovation impact of the EU Emission Trading System—Findings of company case studies in the German power sector. *Ecol. Econ.* **2011**, *70*, 513–523. [CrossRef]
53. Ren, S.; Yuan, B.; Ma, X.; Chen, X. International trade, FDI (foreign direct investment) and embodied CO_2 emissions: A case study of Chinas industrial sectors. *China Econ. Rev.* **2014**, *28*, 123–134. [CrossRef]
54. Ellerman, A.D.; Buchner, B.K. The European Union emissions trading scheme: Origins, allocation, and early results. *Rev. Environ. Econ. Policy* **2007**, *1*, 66–87. [CrossRef]
55. Kander, A.; Jiborn, M.; Moran, D.D.; Wiedmann, T.O. National greenhouse-gas accounting for effective climate policy on international trade. *Nat. Clim. Chang.* **2015**, *5*, 431. [CrossRef]
56. Villoria-Sáez, P.; Tam, V.W.; del Río Merino, M.; Arrebola, C.V.; Wang, X. Effectiveness of greenhouse-gas Emission Trading Schemes implementation: A review on legislations. *J. Clean. Prod.* **2016**, *127*, 49–58. [CrossRef]
57. Hickmann, T. Voluntary global business initiatives and the international climate negotiations: A case study of the Greenhouse Gas Protocol. *J. Clean. Prod.* **2017**, *169*, 94–104. [CrossRef]
58. Zhang, H.; Li, Y.; Gao, D.W.; Zhou, J. Distributed optimal energy management for energy internet. *IEEE Trans. Ind. Inform.* **2017**, *13*, 3081–3097. [CrossRef]
59. Li, Y.; Gao, D.W.; Gao, W.; Zhang, H.; Zhou, J. Double-Mode Energy Management for Multi-Energy System via Distributed Dynamic Event-Triggered Newton-Raphson Algorithm. *IEEE Trans. Smart Grid* **2020**. [CrossRef]
60. ICE Futures Europe. Available online: https://www.theice.com/products/Futures-Options/Energy/Emissions (accessed on 25 July 2020).
61. Shavochkin, I.A.; Grekhov, L.L.; Dubov, E.A. Development of a process control system for the gas-turbine units operating at the thermal power plant of the Tuapse oil refinery plant. *Therm. Eng.* **2013**, *60*, 722–729. [CrossRef]
62. Kudinov, A.A.; Ziganshina, S.K.; Khusainov, K.R.; Demina, Y.E. Development of technologies to increase efficiency and reliability of combined cycle power plant with double-pressure heat recovery steam generator. *Mater. Sci. Eng.* **2020**, *791*, 012014. [CrossRef]
63. Davison, J. Performance and costs of power plants with capture and storage of CO_2. *Energy* **2007**, *32*, 1163–1176. [CrossRef]

energies

MDPI

Article

Problems of Innovative Development of Oil Companies: Actual State, Forecast and Directions for Overcoming the Prolonged Innovation Pause

Yana S. Matkovskaya [1,*], Elena Vechkinzova [2], Yelena Petrenko [3] and Larissa Steblyakova [4]

1 Department of Management and Innovation, Higher School of Management, Financial University under the Government of the Russian Federation, 49 Leningradsky Prospekt, 125993 Moscow, Russia
2 Department of Business Administration, Higher School of Management, Financial University under the Government of the Russian Federation, 49 Leningradsky Prospekt, 125993 Moscow, Russia; kvin07@list.ru
3 Academic Department of Management and Business Technologies, Faculty of Management, Plekhanov Russian University of Economics, 36 Stremyanny Lane, 117997 Moscow, Russia; petrenko_yelena@bk.ru
4 Department of Engineering Entrepreneurship and Marketing, Faculty of Engineering Economics and Management, Karaganda State Technical University, Nursultan Nazarbayev Avenue, 56, Karaganda 100027, Kazakhstan; larissastkaz@mail.ru
* Correspondence: yana.s.matkovskaya@gmail.com; Tel.: +7-925-259-2674

Abstract: The study of the rates of innovative development of various sectors of the modern economy makes it possible to determine the existence of a scientific and practical problem, eliciting the need for urgent identification of the reasons for non-innovative development of Oil and Gas Companies and development of the directions for innovation development. Based on a number of methods, including methods of graphical analysis, time series forecasting, construction of linear trends, correlation analysis and scenario forecasting, the authors stated the fact of the serious depth of the problem of innovative insufficiency in the oil sector in comparison with other sectors and they built six scenarios for the development of these companies. The applied methods made it possible to not only come to the conclusion that with the current level of investment in R&D in the oil and gas sector, Oil Companies may find themselves in difficult conditions, especially if breakthrough technologies show themselves in the non-hydrocarbon energy of the future, but also made it possible to determine the most important directions for the development of Oil Companies, including the formation and development of the oil and gas industry 4.0, marketing strategic management of the activities of these companies.

Keywords: Oil Companies; innovations; investments in R&D; forecasting the Innovation activities; industry 4.0

Citation: Matkovskaya, Y.S.; Vechkinzova, E.; Petrenko, Y.; Steblyakova, L. Problems of Innovative Development of Oil Companies: Actual State, Forecast and Directions for Overcoming the Prolonged Innovation Pause. *Energies* **2021**, *14*, 837. https://doi.org/10.3390/en14040837

Academic Editor: Manuela Tvaronavičienė

Received: 28 December 2020
Accepted: 1 February 2021
Published: 5 February 2021

1. Introduction

The modern energy sector is experiencing a number of serious problems. On the one hand, there is a number of common problems, experienced by most companies, related, for example, to internal management peculiarities [1–4]; networking strategy [5]; concentration and diversification issues [6] and so forth. On the other hand, the energy sector and specifically Oil Companies are affected by specific external factors, such as changes in OPEC (The Organization of the Petroleum Exporting Countries) policy, the "shale revolution," green course [7,8]. The pandemic and lockdowns in their course had a profound financial impact, influenced the decrease in demand for the products of the Energy Industry and thus reduced the investment potential necessary for the development of companies of this industry [9].

The consequences of the pandemic and lockdowns are not only not yet overcome but have also not yet been determined, since the pandemic is not over yet, the crisis phenomena are growing, their duration and timeline are unknown. But these are common problems, they apply to practically all sectors and spheres of the Modern Economy. However, companies in the Energy Sector are influenced not only by common but also specific problems due to the fact that production in this Industry, as a rule, has a continuous nature; the suspension of a number of technological processes can lead to the complete interruption of production. In addition, a significant part of the sub-sectors of the Energy sector does not have the technical and technological capabilities of warehousing and storage of products in principle. Sometimes this is technically impossible (as in the Electric Power Industry) and in the oil and gas industry, for example, these possibilities (storage and warehousing) are extremely limited. Speaking about the problems that relate to the energy sector and, in particular, oil and gas companies, it is worth recalling the collapse in oil prices, as well as the procedural and positional difficulties in making OPEC decisions in the spring of 2020 [10,11].

At the same time, there is still a number of fundamental problems related to the energy sector. Thus, it is the Energy Companies that are primarily charged with the burden of environmental responsibility before society. But this accusation is not entirely fair, because they are only providing energy resources for other industries of the economy. If the Economy had a demand for environmentally friendly energy resources, the Energy Companies would change the structure of their supply and begin to provide energy resources that meet such demand. But for the sake of fairness, it must be said that most energy companies do not show the necessary activity to produce environmentally friendly energy resources, they are the least innovatively active and negligibly little is invested in R&D, compared to companies in other industries. And all this despite the fact that it is they who have significant investment opportunities for the implementation of innovative processes. Of course, this applies, first of all, to Oil Companies (hereinafter, under Oil Companies we mean companies engaged in the production, transportation and processing of oil and gas (Upstream, Midstream, Downstream) as a Business model, mainly it is large companies that are often called supermajors, including state-owned). It should be noted here that the existence of the problem of insufficient investment in R&D by Oil Companies, on the one hand, cannot remain unnoticed for those researchers who analyze data on the structure and growth rates of all sectors of the modern world economy, as well as for those who have researched the regional profile by conducting a comparative analysis between the scale of companies' activities, their Profitability, Innovation and Investment in R&D. But this problem, fraught with a threat to the Oil Business, is clearly ignored by the Oil Companies management. They prefer not only not to increase the rate of innovation and investment in R&D but even reduce their volumes both in relative and absolute indicators. The reasons for this—as a rule consider the fact that these companies are too profitable and have stable positions—have significant market power, in order to think about the importance of increasing the pace of their innovative and technological development. Innovative development, corresponding to the pace of technological development typical for other sectors of the economy, does not represent to Oil Companies the relevance and importance that it represents for heads of companies from other sectors of the Economy. In general, one gets the impression that this problem of development by Oil Companies (it should be emphasized that it is precisely the Oil Companies and not the oil and gas or energy sectors as such, operating according to these familiar extensive schemes corresponding to the industrial era) are not considered as posing a threat; it is veiled. This situation apparently suits the management of the Oil Companies. But this is a very serious problem, the manifestation of which, taking into account the development of energy technologies, can create conditions for the impossibility of the functioning of modern Oil Companies even in the foreseeable future, regardless of the time of depletion of hydrocarbon energy sources.

If the management of Oil Companies were to compare their capabilities and threats to their business based not on an intra-industry analysis but taking into account the pace of innovation in other areas of the economy, where success is achieved through systematic, effective innovation, then this threat would be understood by management and they would begin to revise strategies for their development. Therefore, we can say that the reason for the inattention and low rates of innovative development of Oil Companies is that the threat of the emergence and impact on them of the fifth force of competition M. Porter [12–14] is not taken seriously by them. The importance of changes in the development strategy in accordance with the changes taking place in the modern economy is not adequately considered by them, they have not drawn conclusions that the general drop in demand from production consumers realizing strategies to increasing energy efficiency also contribute to the reduction in oil prices. As a consequence, the competitiveness of hydrocarbon-oriented Oil Companies is under threat.

It must be said that it is not only these factors and factors of global oil prices that determine the problems of increasing the vulnerability of the oil industry. Compared to high-tech industries (ICT, Pharmaceuticals and Automobiles), Oil companies are characterized by low operational flexibility and insufficient marketing activity. But, given the persisting internal resource potential, high barriers to entry into the industry and the strong dependence of the world economy on oil and gas, their position cannot yet be called critical. However, the changes in the global market require Oil Companies to rethink their corporate strategies. For example, due to a man-made accident in the Gulf of Mexico, the US government took a tough stance against Oil Companies [15,16]. It banned upstream in this region for an extended period and blamed the oil company (BP) for this technological accident. Despite the fact that as a result of the elimination of the consequences of this catastrophe, a number of innovations have arisen; nevertheless, the presence of a more developed scientific and technical base in the field of ecology, as well as more attention to R&D in the field of environmental safety could, if not prevent the catastrophe, then reduce the scale of consequences for the environment and, accordingly, for the economic position of the oil company—the perpetrator of the technogenic accident in the Gulf of Mexico.

Thus, the problem of the non-innovative development of modern Oil Companies can also be viewed as a problem of the short-sightedness of their development strategies.

Strategies of modern companies must be innovation-oriented. This is where one should look for reasons and not condone Oil Companies in their choice to follow a formal approach to innovative development.

These premises define the aim of this manuscript. It consists of the determination of the causes and identifies the expected consequences of low innovation activity of Oil Companies, their relatively low investment in R&D and design of possible scenarios for the development of Oil Companies and innovative development of Energy generally. The main research tasks of this article are, first, the task to identify the reasons for the low innovative activity of Oil Companies and, based on a comparative analysis, to determine the depth of the problem; secondly, the task to form a scientific and methodological basis for solving the problem under study, to choose a set of methods necessary to solve the problem; third, the task to review the current situation and the determination of the nature of innovative development of Oil Companies; fourth, the task of making estimates (forecast) of the development prospects of Oil Companies; fifth, the task of working out scenarios for the development of Oil Companies; sixth, the task of proposing the most important directions for the development of Oil Companies.

The hypothesis of this study is that if Oil Companies maintain their existing low investment in R&D and low innovation activity, in a few years they will find themselves in a difficult situation, which will mark a collapse for those of them that will not revise their development strategies, having seriously increased attention to the issues of innovative and technological development, taking into account the fact that the return on investment in R&D takes a rather long period. The expanded basic hypothesis has a general character and implies the formulation in the study of a number of intermediate hypotheses that correlate with the research tasks indicated above and the general hypothesis itself is supported by the results of the research given this article. This study is based on a comparative cross-sectoral analysis, with the construction of development trends and possible scenarios for the development of Oil Companies. As conclusions, a change the strategic guidelines of these companies was proposed, to transform them into more innovatively active ones; the authors highlight the areas of innovative development and emphasize the importance of diversifying their activities to create energy products that meet the requirements of environmental friendliness and innovation, also involving the widespread introduction of digital technologies and development innovative marketing culture.

Section 1—"Introduction"—of the manuscript provides its actuality, purposes, tasks and the significance of studying the current state of the research field. Section 2—"Materials and Methods"—of the manuscript provides the sources of used data, the basic approaches used in the article and the methods that make up the methodological basis for studying the problems studied in the article. Section 3—"Results"—contains the substantive part of the results of the research, the problem and the results of its solution, providing ways of solving the problems of development of Oil Companies, are described. In addition to analyzing the current state, it provides forecasts and scenarios for the development of Oil Companies and the oil and gas industry as a whole. In Section 4—"Discussion"—the authors discuss the results, the correctness of the working hypotheses and highlight future research directions. Section 5—"Conclusions"—presents the conclusions of the article.

This structure of article is determined by the need to achieve confirmation of the basic hypothesis of the study and its accompanying hypotheses (including those related to determining the degree of probability of the implementation of positive scenarios for the development of Oil Companies and their qualitative transformation) and it also allows the removal of the gaps in the work of other authors. In particular, these research gaps are associated, first of all, with theoretical gaps that dominate in a number of publications on this topic and reflect an absence of interdisciplinary approaches to studying the problem of the development of this industry. Thus, research on the development issues of Oil and Gas Companies is carried out within the framework of separate research areas—either current technological problems of the industry or issues of financial and economic development or within the framework of management science or in other rather narrow areas. As a result, the general situation and development opportunities of companies of this industry are ignored. In turn, the proposed approach in this article, which is the result of the conceptual integration of the research results of various authors, starting with the ideas presented in the works of the classics of economic science and taking into account the contribution of specialized research in this industry (see below, Sections 2, 3.1 and 3.2), conceptualizes the formation of a special intersectoral approach that allows study of the current and future problems of the development of the industry. It forms the basis for the development of a new institutional approach, which allows the formation of a system of ideas about the importance of developing a non-autonomous approach to the analysis of the industry, based on an understanding of the consistency of relationships that create opportunities and threats to development for individual companies in a globalizing world, increasing the impact of intersectoral competition on the development of industry enterprises in the context of growing uncertainty and acceleration of the pace of scientific and technological progress and uneven distribution of its results. With regard to the object of research, a significant theoretical gap also reflects the need to develop special marketing approaches that form strategies for the long-term development of companies at the corporate level and the development of Industry 4.0. in the field of Oil and Gas companies.

In addition, there are serious empirical problems in the research of Oil and Gas companies, which are caused by the dominance of studying and comparing only intra-industry development trends, which is closely interconnected and influenced by the development of the methodology for studying the activities of Oil and Gas companies and allows us to identify a number of serious methodological gaps in the study of the activities of companies in the industry (see below, Sections 3.1 and 3.2). They are predetermined by the existing methodological disadvantage of the domination of intra-industry analysis, which does not allow realistic evaluation of the likelihood of the realization of a negative development scenario for existing oil and gas companies, narrowing the horizon of their research, which does not allow assessment of the depth and scale of the problem of innovation insufficiency of the development of the industry and strengthens a non-systemic representation of the existence of Oil Companies and problems in the development of this industry, taking a nihilistic short-sighted traditionalist position. The gap in the dominance of the prevalent methodology, is based, as a rule, on the study of Oil Companies as objects isolated from the economic system, detached from the processes occurring in it, ineffectively abstracted from the development of the entire system, seems impermissible today, either from the point of view of the development of science, methodology and practices and is destructive in the modern world.

This methodological problem is solved in the article by decomposing the groups of companies under study and applying a comparative intersectoral analysis of the innovative activities of modern companies in the modern competitive world, based on the use of various forecasting methods and selection from among those that objectively reflect the depth of the problem, which is also applicable when analyzing other industries and areas of activity. These are general, conceptual gaps in the research. The Results Section (below) describes the private research gaps in Oil Companies and the oil and gas industry.

2. Materials and Methods

As materials in this article, we used data from open sources: data from international economic organizations, authoritative consulting and rating agencies, statistical information, as well as data published in scientific publications (books, articles).

At the same time, the basic approaches used in the article are based in a broad sense on the classical political economy, neoclassical economic school presented in the works of A. Marshall [17], P. Samuelson, P. Nordhaus [18], G. Mensh [19], F.A. Hayek [20], in the works of scientists who investigated the patterns of innovation and technological development N. D. Kondratiev [21], J. Schumpeter [22–24], R. Nelson and S. Winter [25], A. Toffler [26], D. Bell [27], J. Galbraith [28,29], B. Santo [30], M. Porter [12–14] and on the works of Russian scientists: A. Anchishkin [31], S. Glazyev [32,33], Yu. Yakovets [34] and others.

In a narrower sense, of interest are the results of empirical studies of the oil industry and Oil Companies of famous scientists, such as R. Oligny, A. Izquierdo, M. Economides [35], M. Kamien, N. Schwartz [36], H. G. Grabowski, N. Baxter [37,38], E. Mansfield et al. [39], W.S. Commanor, F.M.Sherer [40], A. Phillips [41], J. Shmookler [42], contributed to the understanding of the reasons explaining why Oil Companies are in a state of low innovation activity A. Mastepanov [43], M. Poleshchuk [44], M. Cherkasov [45,46] and others.

An important role is played by the methodological aspects from the works of M. Blaug [47], as well as the methods for developing forecasts and scenario forecasting—the works of K. Abt, R. Foster, R. Ri [48].

The main methods that make up the methodological basis for studying the problems studied in this article are a set of general scientific methods (analysis and synthesis, the method of scientific abstraction, generalization, analogies), methods of economic analysis, classification, ranking and structuring of data, rating analysis, statistical analysis, comparative analysis, as well as graphical analysis, analysis and forecasting of time series, method of constructing linear trends (extrapolation of trends), ETS forecasting (Exponential Triple Smoothing Forecasting), correlation analysis, scenario planning and forecasting.

Based on this methodology and on the mentioned theoretical and practical platforms, at the article have formulated the problem of the innovative insufficiency of the oil sector in comparison with other sectors, assesses the prospects for the development of Oil Companies, provided that while maintaining modern technological trends (the absence of breakthrough technologies in the field of production, processing oil and gas) and with the current level of their investments in R&D in the oil and gas sector, Oil Companies may find themselves in difficult conditions, especially in a situation where the disruptive energetic technologies will be commercializing by non-hydrocarbon energy companies.

3. Results

3.1. Review of the Current Situation—Determination of the Nature of Innovative Development of Oil Companies

The ICT sphere has demonstrated the highest rates of innovative development not the first decade. Of course, ICT is not as capital intensive as the oil and gas sector and has a much faster return on investment. However, in recent years, some Oil Companies have also started to appear in the global innovation ratings [49–52]. But their number is extremely small. There is no encouraging data from the WIPO (World Intellectual Property Organization) 2020—The Global Innovation Index 2020. In 2018–2019, the entire oil and gas sector accounted for less than 1% of all R&D spending in the modern economy [52]. It is not surprising that none of the Oil Companies ranked among the leaders in R&D investments in the GII 2020 rating. Moreover, the data even indicate a decrease in the level of innovative activity of oil and gas companies in 2018–2020, even without taking into account the impact of the pandemic (the impact of the pandemic on energy development is presented in detail in [53]). The data of the rating EU R&D Scoreboard: The 2017 EU Industrial R&D Investment Scoreboard [54,55] also testifies to the incomparably low innovative activity of Oil Companies, compared with companies from other industries. The intensity of Oil and Gas Companies innovation activity in comparison with companies from other industries is shown in the diagram (Figure 1).

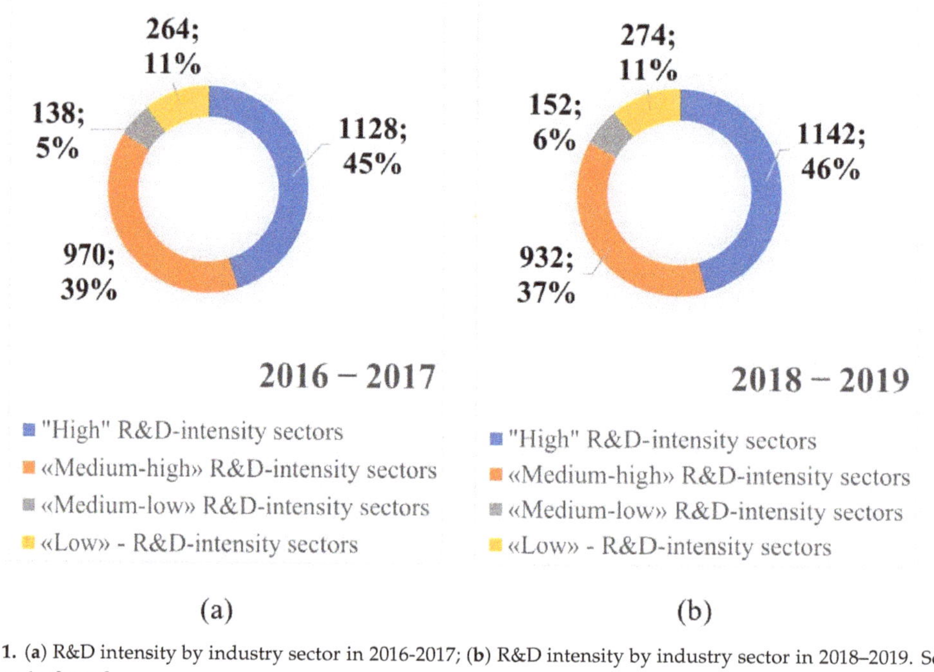

Figure 1. (**a**) R&D intensity by industry sector in 2016-2017; (**b**) R&D intensity by industry sector in 2018–2019. Source: drawn up by [54,55].

It is necessary to clarify the diagram in which the gradation of the analytical review of the European Commission is preserved. In it, the first group (high) of companies with a high intensity of R&D (>5%) included companies specializing in aerospace technologies, computer equipment, protection and security, the production of office equipment, leisure goods, medical equipment, pharmaceuticals, semiconductors, software, telecommunications equipment, as well as providing medical and Internet services.

The second group ("medium-high")—medium-high intensity of R&D (2–5%) included companies specializing in the production of auto components, automobiles, trucks, chemicals, packaging, electrical equipment, electronic equipment, household goods, industrial equipment, manufacture of automobile tires and financial and travel services.

The third group "medium-low"—groups companies with a medium-low level of R&D intensity (1–2%). It included only one segment of companies related to the oil sector, namely, oil equipment manufacturers. The same group includes companies specializing in alternative energy, beverage production, food, retail, media, tobacco production and distribution services.

Finally, the fourth group ("low") is characterized by a low degree of R&D intensity (up to 1%). It includes mining companies, companies specializing in the production of aluminum, precious metals, gas, steel, water supplies, timber processing, water suppliers, real estate, as well as the provision of services in the field of insurance, mobile telecommunications and transport [54,55].

The above data once again prove that Oil Companies are seriously lagging behind in terms of R&D intensity relative to companies from other sectors of the economy. It should be added that alternative energy also occupies a low position.

In general, the oil and gas sector accounted for 1.12% of all R&D expenditures in the world in 2018 (€9.3 billion out of €823.4 billion), while the Net Sales of Oil Companies amounted to €2812.5 billion (13.8% of the global volume of all industries). Their R&D intensity is 0.3% (the penultimate sector in the world), provided that in terms of profitability Oil Companies rank second in the world (14%), yielding first place to the banking sector (26.6%).

A detailed review of the positions of Oil Companies in 2019 shows that PetroChina, which received 81st place in the overall rating, ranks first among Oil Companies with an intensity of 0.6%. As noted in the 2019 EU Industrial R&D Investment Scoreboard: «In contrast, companies in the biotechnology & pharmaceuticals, software and technology hardware sectors have R&D intensities well into double figures and R&D is a key success factor for them» [55], p. 59.

To demonstrate the extent of the problem, let us compare the indicators of revenue (total income), R&D expenditures and R&D intensity of the three absolute leaders studied by PWC in 2016–2018. and leaders among Oil Companies in the same ratings (Table 1).

Table 1. Comparison of revenue, R&D Expenditures and R&D Intensity of innovations leaders of the world and the leaders of the Oil Companies, 2016–2018.

Company	2016			2017			2018		
	Revenue ($ Billions)	R&D Expenditures ($ Billions)	R&D Intensity, %	Revenue ($ Billions)	R&D Expenditures ($ Billions)	R&D Intensity, %	Revenue ($ Billions)	R&D Expenditures ($ Billions)	R&D Intensity, %
Amazon	107.01	12.50	11.7	135.99	16.10	11.8	177.9	22.6	12.7
Samsung	166.67	11.95	7.2	167.68	12.72	7.6	224.3	15.3	6.8
Volkswagen	225.16	12.51	5.6	229.35	12.15	5.3	277	15.8	5.7
PetroChina Company Ltd. (84 position, 2018)	265.2	1.8	0.7	248.5	1.7	0.7	309.8	1.9	0.6
Exxon Mobil Corporation (152 position, 2018)	241.1	1	0.7	201.6	1.1	0.5	238.9	1.1	0.4

Source: [56,57].

Table 1 shows that investments in R&D by Oil Companies and investments in R&D by leading companies are not comparable. The highest position (84th place) in 2018 is occupied by PetroChina (84th in 2017), followed by Exxon Mobil—152nd (129th in 2017), [56,57].

It should be noted that PetroChina's high positions in comparison with other Oil Companies could be explained only due to a large number of patents and not to commercialized innovations, which is confirmed by other sources, for example, data from the World Intellectual Property Organization (2020 and earlier) [52].

Regarding the patent activity of Oil Companies, note that the main areas of patent activity in the oil industry are Upstream (64% of patents), Downstream is account for 32% of patents, oil refining—1%, Midstream have 3% [50]. The number of patents by Oil Companies grew fivefold over the period 2000–2014. But the reason for this growth is only "shale revolution." After a slight decline in 2015, growth resumed the following year. So, Upstream accounts for 18,358 patents (in 2015—18,086), production of fuels and other oil products—9861 (against 9224 in 2015), Midstream has—1321 (+52%). On the contrary, in the Downstream sector there was a significant decrease—by 15%, despite the fact that its share in the structure of patenting in the industry is only 1% [51].

Meanwhile, the patenting indicator, although it reflects the presence of innovative ideas in companies, cannot fully reflect the fact of the presence of innovative activity, this indicator does not renew significantly worn-out production assets of a significant number of Oil Companies. Against the background of a decrease in the efficiency of the main activity due to the consequences of the financial and economic crisis and a decrease in the profitability of Oil Companies due to the pandemic, these indicators indicate not only low rates of innovative development but also characterize the presence of serious threats due to the growing likelihood of technogenic accidents. In general, the low innovative activity of Oil Companies cannot be explained only by those rather comfortable conditions in which modern Oil Companies exist, the rather high profitability of which, in general, does not stimulate them to invest enough funds in R&D (relative to other companies). This situation characterizes the presence of a scientific problem, the solution to which can also solve a number of practical problems.

3.2. Estimates of the Development Prospects of Oil Companies (Forecast)

Based on the hypothesis defined in this article we need to get an idea of the significance of this trend and the extent of its impact (the trend that if the current pace of R&D and investment in R&D of Oil Companies is maintained, they (and the oil and gas industry as a whole) will experience serious problems of their development in the future).

Unfortunately, the analytical data and forecasts below evidence just that.

To evaluate the development prospects of Oil Companies, it is necessary to identify future trends in terms of R&D intensity, R&D investment volumes and revenue from their activities based on forecasting time series in comparison with other industries (comparative analysis).

The forecasting methodology used by building linear trends has advantages and disadvantages. But for the tasks solved in this article, that methodology is applicable and its use is reasonable. This is due to the fact that it is necessary to first determine how the oil market will develop and what will be the main economic indicators of companies' development while maintaining the current level of R&D costs in comparison with other industries, when all other things are equal (macroeconomic stability and the absence of breakthrough commercialized technologies in energy sphere in general (in hydrocarbon and non-hydrocarbon energy)). Thus, the task is to identify a trend that will make it possible to assess what will happen to Oil Companies if they do not take appropriate measures to increase investment in innovation and in-crease the intensity and efficiency of their innovation activities (including based on the fact that Oil Companies also there are opportunities for the development of non-hydrocarbon technologies). Based on these data, a scenario forecast will be built.

To build forecast data, information was taken from open sources PWC (PricewaterhouseCoopers) for 2011–2017 [56] and the comparative approach allowed for cross-sectoral comparisons. The period 2011–2017 was chosen due to the fact that the dynamics during this period demonstrates the situation quite well, during this period there are no sharp declines and rises. This allows us to come to the most objective conclusions.

Forecasting objects. To forecasting was created the nonrandom (nonprobability) quota sample. The sample included companies grouped by industry: Software & Service & Semiconductors companies, Pharmaceuticals & Biotechnology companies, Technology Hardware & Equipment, Automobiles & Components Companies, Oil & Gas companies (using PWC terms). The sample consists of 40 companies and it is representative sample. The forecasting time-frame is 8 years—2018–2025. Forecasting methods—building linear and exponential trend.

The first forecasted parameter is the volume of investment in R&D in absolute terms. The trend is built on the average data for each quota (a group of companies belonging to a particular industry, the leader of cluster). When forecasting this indicator, the following are additionally introduced indicators: the average indicator for the sample by periods (2011–2017). it reflects the general trend of changes in investments in R&D for the entire sample and the averaged indicator for the sample for 2011–2017. Obtained linear trend in forecasting investments in R&D in absolute terms is shown in Figure 2.

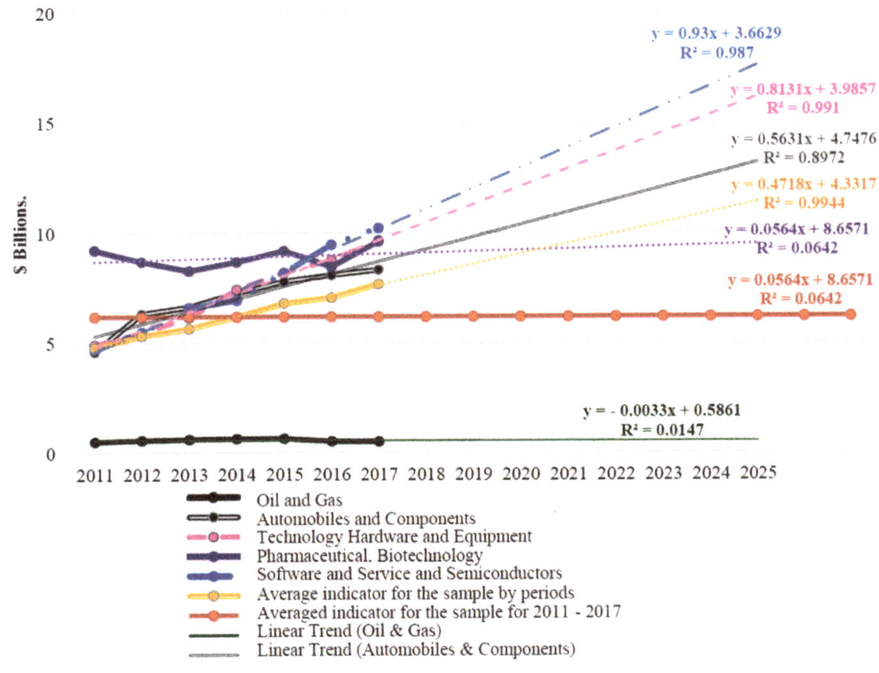

Figure 2. Forecast (linear trend) of investments in R&D of Oil Companies in comparison with companies in other industries until 2025, in billions of dollars. Source: calc. by author by data [56].

As can be seen from Figure 2, according to the forecast, the largest growth in investment in R&D since 2018, as in 2011–2017, belongs to Technology Hardware & Equipment Companies, similar positions belong to Pharmaceuticals & Biotechnology Companies. Investments in the Automobiles & Components and Software & Service & Semiconductors industries will grow above the average indicators of sample. Only Oil Companies, whose investments in R&D are in the range of $0.5–0.7 billion, will seriously lagging behind other companies from this sample. (It should be noted that in the calculations of the authors, both linear and exponential trends were built for all parameters but only linear trends could be recognized as reliable trend).

However, in general, the linear trend of the growth of investments in R&D in absolute terms is largely demonstrative. More significant conclusions can be drawn based on the analysis of relative indicators—the Intensity of R&D (Figure 3).

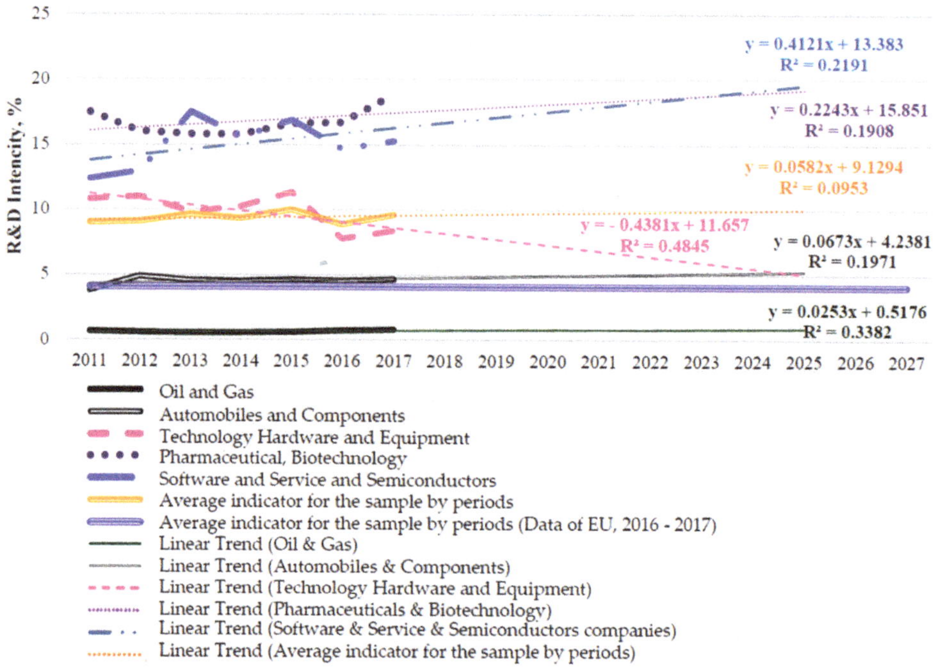

Figure 3. Forecast (building a linear trend) R&D intensity of Oil Companies in comparison with companies in other industries until 2025, %. Source: calc. by author by data [56].

Figure 3, in contrast to Figure 2, shows two linear trends in the average R&D Intensity indicators for the sample by periods. The first one was calculated by us on the basis of data for the entire period and the second one was calculated by the European Commission for 2016–2017. The indicator the European Commission calculated is based on an results of analysis of the R&D Intensity of 2500 R&D companies and its size is 4.1% [54], p. 36.

As can be seen from Figure 3, in this trend, the leadership was recorded by companies belonging to the Pharmaceuticals & Biotechnology and Software & Service & Semiconductors sectors. The rest of the groups of companies will maintain their positions within the average fluctuations in the trend. But this does not apply to Oil Companies, which will only be able to slightly intensify R&D and will remain lagging behind.

These trends (Figures 2 and 3), being linear, do not allow predicting possible changes in the socio-economic, scientific and technical development of the Economy, since they are implemented only with other things being equal. However, since the forecast horizon is of a medium-term nature and expert assessments indicate an almost unchanged nature of investments in R&D of Oil Companies, this forecast can be considered reliable as supported by additional data [54].

Under these conditions (according to the same estimates) Oil Companies will reduce or maintain the same level of investment in R&D. In this case, they should think about how effective such a policy is, because these companies need to intensify their innovation activity (even not within the framework of the industry's traditional lines of business). However, for investment in R&D, Oil Companies need funds, therefore, in order to identify the growth trends of the revenue (income) of the companies included in this sample, a linear trend in income was built within the same forecasting time-frame (Figure 4).

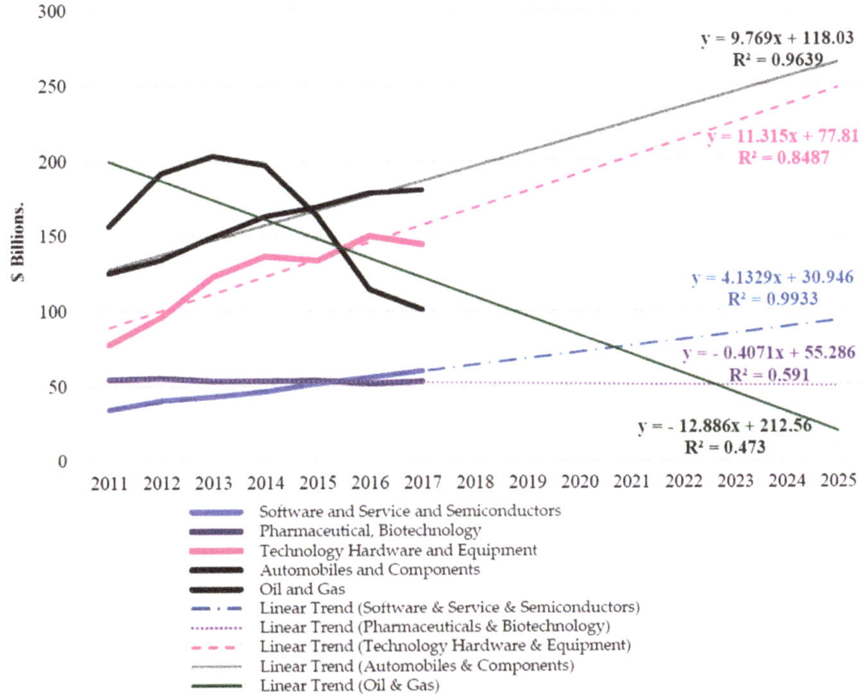

Figure 4. Forecast (Linear Trend) of revenue (income) of the Oil Companies in comparison with companies of other industries until 2025, in billions of dollars. Source: calc. by author by data [56].

The Linear Trend reflecting the forecast of changes in the company's revenues determines, firstly, that high positions are no longer held by Software & Service & Semiconductors companies and Pharmaceuticals & Biotechnology but by Oil & Gas and Automobiles & Components sectors, while maintaining the middle positions in Technology Hardware & Equipment. Second, revenues in all quotas will grow during the forecasting time-frame, while revenues in the Oil and Gas sector will seriously decline. So, according to the European Commission, Oil Companies form the majority of those companies whose sales volume decreased in 2016–2017. These are: Eni −18%, Total −11%, Royal Dutch Shell −12%, BP −18%; net sales decreased at Petroleos de Venezuela (−24%), Statoil (−23%), Petrochina (−6%), Chevron (−15%) and Exxon Mobil (−16%). It must be said that during this period Apple also experienced an 8% decrease in this indicator [56].

To clarify this fact, we note, first, that, as can be seen from Figure 2, in absolute terms, the investments of Oil Companies in R&D are also not large. Secondly, we note that the correlation coefficient we calculated indicates that there is a close and direct relationship for all groups of companies between the indicators of investment in R&D and the dynamics of income for the period under study. But this does not apply to Oil Companies, for which this relationship, due to low investment in R&D, is direct but not tight (close). And thirdly, perhaps the most important, perhaps, one should not explain the decline in the profitability of Oil Companies (moreover, diversified in a number of areas) only by the dynamics of oil prices and the tax burden. Indeed, at this stage of the analysis, it can be concluded that it is short-sighted of the explaining of the decline of company income only by market volatility. Perhaps it should be assumed (including based on the above data) that the problems of development of Oil Companies are contained in the insufficient investment in R&D that has been observed in the industry for several decades, as well as in their organizational management.

Uncertainty in this issue is removed when studying the results of empirical studies of the diffusion of innovations in the industry and investment in R&D by a number of well-known scientists. Thus, J. Schmookler back in the 60s of the twentieth century established that there is a direct relationship between the increase in industry investment and the invention of means of production in the oil refining industry [42]. Moreover, the same proportions are also characteristic of the construction sector; this fact fully explains why construction and Oil Companies belong to the 3rd and 4th groups in the study of the European Commission (Figure 1) and confirms the relevance of the forecast results (Figures 2–4). In addition, it also suggests that Oil Companies in the 21st century remain committed to the strategies that have brought them in successful in the 20th century.

E. Mansfield et al. came to conclusions that do not contradict J. Shmooklers position [39].Having studied 9 Oil Companies (along with 10 chemical companies and 11 steel companies), they found that the distinguishing quality of the oil industry is, firstly, a direct proportional relationship between the budget of these companies for R&D and the orderly distribution of their programs by the criterion of the quality and efficiency of their investments in R&D, and, secondly, by an inverse proportional relationship to the volume of their sales (while, for example, in the chemical industry, there is a direct proportional relationship between the increase in R&D costs and the results of their inventive activity).

The stimulus for the innovative development of Oil Companies was identified by H.G. Grabowski [37,38], who made an interesting conclusion (based on the study of oil refining, chemical and pharmaceutical companies). It consists in the fact that one of the most important factors ("determinants") of the intensity of R&D by companies is the "index of firm productivity before the start of research," measured by the ratio of the number of patents per scientist and engineer working in a given firm. He found, therefore, that companies in this industry are characterized by a long-term effect of R&D, characterized by the fact that the more patents accrued to scientists and engineers in the past, the higher the research intensity of these companies in relation to their competitors (all other things being equal) [37]. It should be noted that the evidence that the growth in the number of scientists and engineers in the company leads to the growth of patents was led by J. Schmookler [42] and W.S. Comanor with Sherer, F.M. found a positive effect and correlation between the volume of sales of new products within two years after their invention, the number of employees in R&D and the number of patents. This correlation effect has been found to be positive for a constant firm size [40]. M.I. Kamien and N.L. Schwartz adhere to a similar position [36].

So what conditions are needed for the industry to innovate? To answer, firstly, one must recall the warnings of J. Galbraith more than half a century ago, who believed that "the era of cheap innovations is a thing of the past and an era of diminishing returns has come to replace it" [28,29] and secondly, one should pay attention to the thoughts of A. Phillips, who argued that the special state of the industry can stimulate the development of innovations in it: "there is such a level of competition that occupies an intermediate position between perfect competition and monopoly and that it stimulates innovation as much as possible " [41]. Consequently, the expectation of an increase in innovative activity in the oil industry should be associated with changes in the macro environment and in the structure of the oil market.

There are opinions according to which a large monopoly firm is more inclined to innovative activity, since it has more opportunities for diversification [36]. For example, H. Grabowski and N. D. Baxter came to the conclusion that the more oligopolistic the industry is, the higher the R&D competition in it [38]. If we apply this conclusion to the situation on the oil market, which is characterized by an oligopolistic nature, then this sector of the economy is characterized by a high degree of competition in the field of R&D. However, the low degree of innovative activity of Oil Companies today suggests that this is not entirely true. There is that industry also no intense competition in the questions about innovation and their commercialization. But competition is happens going on in certain segments of production, such as such as vertical drilling technologies (shale gas upstream technologies), development and promotion of innovative fuels and oils and offshore production technologies.

In general, there are different points of view on the problem of the relationship between the intensity of R&D in the industry and the companies size. According to one of them, current profits are predetermined by future innovations and the most probable sources of technological progress are created by large companies, since they are in the best position in terms of profitability ("high current profits, which are a source of liquidity, are an essential condition for the application of significant efforts in the field of R&D") [36].

The above points of view of scientists and the conclusions made by the above authors are not typical for studies of the features and directions of development of companies in the oil and gas industry. Unlike the cited researchers, most analysts ignore the problem of the futility of maintaining the current extremely low rates and volumes of R&D and their technological lagging behind the general development trends of the modern economy, not to mention lagging behind the advanced industries. The main purpose of illustrating views of these scientists was the need to emphasize that the situation with the non-innovative nature of the development of Oil Companies has deep roots and requires a conceptual revision of the innovative and investment strategies for the development of Oil Companies. Therefore, the points of view are given not of opponents but of like-minded persons. That is why we did not dispute with these authors, although we see a number of disadvantages in their works. Such disadvantages and controversial points include the lack of study of the types of innovations being introduced and their impact on the quality of innovative development, the feedback between inventions and the development of an oil company is not studied [36,39,42], the focus on the study of patent activity and inattention to the organization of the process of commercialization of innovations [36,37,42], lack of attention to environmental factors, the use of digital technologies, organizational innovation and more.

Nevertheless, an important general conclusion can be made that there is still a direct relationship between the income of companies and their investments in R&D. But this relationship is differentiated depending on the characteristics and structure of the industry and, when certain proportions in the industry are reached, incentives for innovation are formed in it. (As M. Porter noted, "innovation is both a response to incentives created by the general structure of the industry and a powerful influence on this structure)" [58], p. 284).

The solution to the problem of low rates of innovative development of Oil Companies should be based on the concept of M. Porter including his industry life cycle concept. Since the oil industry is at a stage of maturity, which is predetermined by the objective reasons for its resource orientation, the following statement by M. Porter is true in relation to it: "as the industry moves to the stage of maturity, the product design changes more slowly and mass production techniques appear. Product innovation is giving way to organizational innovation ... the latter becomes the main form of technological activity in the industry, since the goal at this stage is to reduce the cost of producing a product ... Finally, in the later stages of industry maturity, the rate of innovation slows down and innovation gradually fizzles out: investments in technology in the industry reach the point, followed by a decline in profitability from further improvements" [58], p. 283.

However, the maturity of the industry at the moment is not a "verdict." Oil companies today have the opportunity to take a number of actions that will allow them in the future to apply the strategy of "rejuvenation," in conditions "when, due to major technological changes, the industry can be thrown back into a state of instability" [58], p. 283. We are talking about R&D and the commercialization of their results in the field of non-hydrocarbon energy and diversification (which is not yet active enough but a number of companies are already developing, see below).

The "state of instability," which is the result of technological evolution in interaction (according to M. Porter) with the life cycle, gives rise to five forces: first, "change in scale" due to the fact that as companies and the industry as a whole grow, they have more room for innovation. Secondly, "learning," which means that in the course of the life cycle of companies and the industry, they accumulate skills to improve this process. Third, "uncertainty reduction and borrowing" characterizes the process of "pushing for product standardization." Fourth, "the spread of technology," and fifth, "a fall in profits from technological innovations in various types of activities," when the limit of possibilities for further improvement of this technology comes [58], p. 284.

Therefore, Oil Companies should pay attention to the idea proposed by M. Porter about the need to form industry scenarios as a competitive strategy in conditions of uncertainty. The industry scenario is a "consistent and consistent system of views on the industry and its future structure" [58], p. 603. Uncertainty, in this case, can come from any of the five forces of competition and from the point of view of the oil market, the most "dangerous" for Oil Companies is the massive distribution of substitute goods (other types of energy). To protect business from them, M. Porter proposed a system of protection against substitute goods, which also includes an "attack on the industry" where substitute goods are produced [58], (pp. 428–430). Another "recipe" by M. Porter is to use market relationships, concerning the unification of marketing, logistics efforts, production of products, as well as—joint efforts in the field of R&D [58], (pp. 470–474), as it was implemented by corporations in the market electronics.

3.3. The Scenarios of Development of Oil Companies

Scenario planning is a promising method for the variable description of the position of an object in the future, taking into account the influence of various factors on the change in the object. Scenario planning and forecasting are often used to forecast the development of the oil and gas and energy sectors. The most famous in the energy sector are the scenarios of the International Energy Agency, OPEC scenarios and scenarios for individual countries and regions are being developed. Scenario forecasts are also made by Oil Companies such as BP, ExxonMobil, Shell. All these materials were studied before the development of the following scenarios for the development of the oil industry.

The peculiarity of the above scenarios is that they consider the problems caused by the development of the industry—these are intra-industry comparisons. Without questioning the significance of such forecasts and expressing respect for the authority of these studies, we note that, in general, intra-industry comparisons do not create conditions for the transition of Oil & Gas Industry companies to innovative development. Meanwhile, it is precisely the cross-sectoral comparisons that make it possible to assess the complexity of the position of Oil Companies, their future as such and their business models.

The starting point for making a decision on the formation of the scenarios outlined in this article was also the results obtained using the MS Excel toolkit (ETS forecast). The initial data was the information from PWC [56,59–61] and so forth on the incomes of companies used to build linear trends in the development of the industry but with a forecasting time-frame until 2032 (15 years) for Oil Companies (Figure 5).

As Figure 5 shows, if the current situation persists, the revenues of Oil Companies will take negative values by 2025 and by 2032 their losses will amount to more than $90 billion per year (all other things being equal). To establish whether the downward trend in Oil Companies' revenues is typical/atypical, a similar forecast was made for automobile component sectors. The choice of the automotive industry as a base for comparison was determined, firstly, by the relative similarity of the income values in both industries (see Figure 4), and, secondly, by the similarity both in the models of industry competition (oligopolistic) and factors influencing production and consumption in these industries. The forecast result is shown in Figure 6.

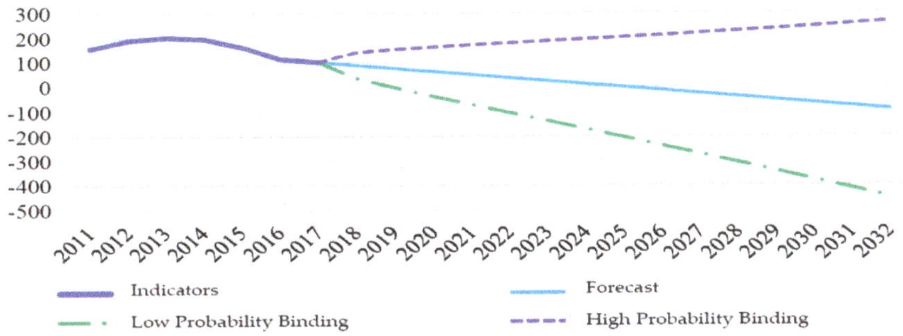

Figure 5. Forecast of revenues of Oil Companies 2018-2032, billions of dollars. Source: calc. by author by data [56].

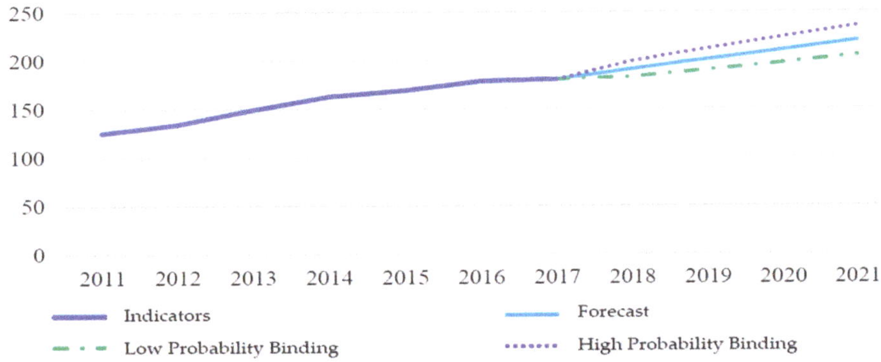

Figure 6. Forecast of revenues for the automotive industry 2018–2032, billions of dollars. Source: calc. by author by data [56].

From the forecasts (Figures 5 and 6), compiled by applying the ETS-forecast, it can be seen that with respect to the automotive industry (and other sectors of the economy), the situation with a decrease in income, which is characteristic of Oil Companies, is atypical. Of course, the decline in income of Oil Companies was also associated with a decrease in oil prices and the Excel forecast is based on the constancy of other factors. However, this is fair, since there are no guarantees that the price of oil will rise and this will create conditions for stabilization and growth of incomes of Oil Companies (in 2019 and 2020 there was a decline in prices for oil and oil products). Therefore, the probability of this forecast being realized is nonzero. This further reinforces the fact that Oil Companies must seek other sources of innovative growth. In particular, by diversifying and stepping up its innovative activities. An alternative forecast of the dynamics of income of Oil Companies for the same period can be presented using the method of analogies, assuming that the income of Oil Companies will increase if they are more active in innovation and bring the level of their investments in R&D to the level of other companies.

For forecasting, based on the correlation analysis, the degree of dependence of the income growth rates of the companies included in the sample on the level of their investments in R&D was established and the growth rate was revealed. In accordance with this, two forecasts were formed. In the first, a proportional relationship between these indicators is used for Oil Companies, by analogy with automobile concerns (Automobiles & Components) and in the second, a proportional relationship between these indicators is used to predict the revenue of Oil Companies, by analogy with the industries producing Technology Hardware & Equipment companies. The forecast is shown in Figure 7.

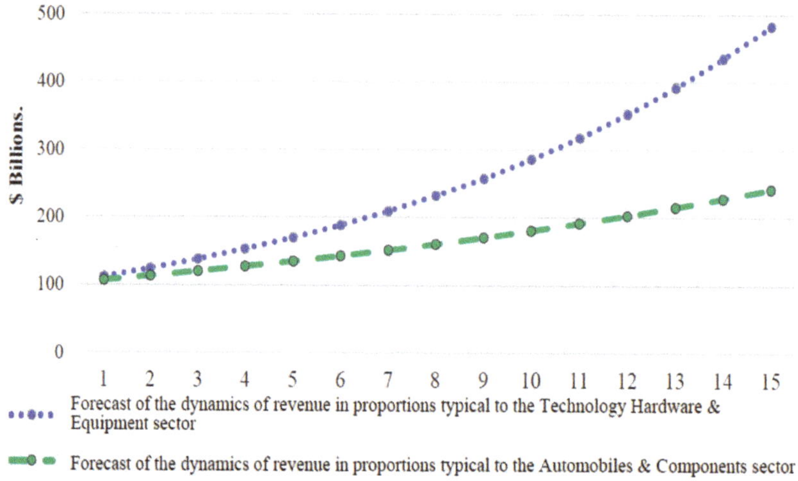

Figure 7. Forecast of the dynamics of revenue of Oil and Gas sector in proportions typical for the growth of the Automobiles & Components and Technology Hardware & Equipment sectors 2018–2032, billion dollars. Source: calc. by author by data [56].

As can be seen, increasing investments of Oil Companies in R&D can lead to the implementation of optimistic development scenarios. Based on the forecasting results, it is possible to build preliminary generalized scenarios for the development of the oil industry. They are shown in Table 2—scenarios for the development of Oil Companies (oil industry).

Here, it is necessary to add that the used forecasting methods are not multifactorial. They are not intended to reflect the entire set of factors in full, which serves as the key idea of the study—to study how events will develop, taking into account an important proviso for economists—"all other things being equal" (according to P. Samuelson [18], A. Marshall [17]). The indicated drawbacks of the method for constructing linear trends are compensated for by constructing a scenario forecast of the industry development and strategic initiatives of companies from other sectors of the Economy (below).

In addition to this remark, others should be added. First, the scenario is built based on traditional concepts of formation of the scenarios and can be identified as a contrasting type of scenario. This type of scenario allows the exploration of the conditions in which decisions will be made, assessing existing concepts and other factors, as well as to make more correct decisions. The create of this scenario aims to complete and concretize the forecasting process and determine the parameters of the transformation of the Oil Companies' market as a result of the implementation of one of the scenarios and the scale of the consequences for Oil Companies in their current state, as well as an assessment of the likelihood of acquisitions of Oil Companies by the non-Oil Companies.

Secondly, the scenario method presupposes a written description of problems and proposals for their solution, as a result of which a comprehensive summary of the problem is achieved. Third, the scenario shown in Table 2 was formed in accordance, as follows from the above study, with the principles, the most important of which was the principle of proportionality of the volumes of investments in R&D of Oil Companies in R&D, combined with the dynamics of their profitability/profitability with similar parameters of non-Oil Companies. Fourthly, this scenario makes it possible to understand that Oil Companies, provided that they maintain their current innovation, technological and investment development strategies may find themselves under the influence of a serious threat due to the growing risks of exposure to them from the fifth force of competition (according to M. Porter).

Table 2. The Scenarios of Development of the Oil Companies (Oil and Gas Industry).

Scenario	Macroeconomic Situation	Political Situation	Competition	R&D of Oil Companies	Revenue of the Oil Companies	R&D of other (Non-Oil) Companies	Revenue of Other (Non-Oil) Companies	Technological Situation	Other Factors: Threats	Other Factors: Opportunities	Scenario Probability (Heuristic Estimates)	the Type of the Scenario
1	2	3	4	5	6	7	8	9	10	11	12	13
Scenario 1. Extremely pessimistic	Deterioration of the macroenvironment of functioning (a decrease in oil prices, an increase in the tax burden, etc., technogenic disasters), a decrease in the index of confidence on the part of investors and consumers.	Rising political tensions in oil producing regions	Sharp intra-industry and cross-industry competition for resource and sales markets	Absent or reduced to a minimum due to lack of liquidity, low lending opportunities	Sharp decline	The efficiency of R&D, including in the energy sector, are growing (due to increased investment and intensification of innovation). Non-Oil Companies are expanding into the oil market.	They grow rapidly within the range of the excess of the income of Oil Companies by an order of magnitude or inversely proportional to the decrease in the income of Oil Companies	Lack of significant innovations in Oil Companies, emergence of breakthrough energy technologies and their successful commercialization by non-Oil Companies	Technogenic accidents caused by the fault of Oil Companies, deterioration of the environmental situation in the world, negative image of Oil Companies.	There are no opportunities and sources for the development of Oil Companies. The oil industry is in decline and its current companies are disappearing from the economic arena.	Low (until 2040–2050)	Catastrophic (for Oil Companies)
Scenario 2. Medium-pessimistic	Macroenvironment for the functioning of Oil Companies is stable without shocks but it does not create favorable conditions for their growth	Stable. But it is not conducive to the growth and development of Oil Companies	Sharp intra-industry competition for resource and sales markets	Decreasing or staying at the same low level (horizontal trend direction)	Decrease or remain at the same level (2016–2017 level)	Growing at an accelerating pace, increasing the efficiency of R&D, increasing the speed of commercializing effective innovations	Growing significantly, capital surpluses are formed.	Lack of significant innovations among Oil Companies against the background of growing innovations in other sectors. The probability of new energy technologies emerging, unavailable to Oil Companies is growing.	Accelerating development of non-hydrocarbon energy technologies, growth of companies specializing in non-hydrocarbon energy.	There are opportunities to increase investment in R&D of new types of energy.	High, provided that current R&D rates are maintained	Stagnation (for Oil Companies)
Scenario 3. Neutral	No shocks, stable, equally unfavorable for all spheres of the economy (global economic recession)	Stable. But it is not conducive to the growth and development of all spheres of the economy.	Medium intra-industry competition for resource markets.	Not-high rates of Growing	Growing at low or medium rates (relative to the level of 2016–2017)	Growing at an medium rate commensurate with the rate of growth in Oil Companies (in proportions 2016–2017)	Growing at a low rate (comparable to the rate of 2011–2018) or slightly growing	Stable. The Industry-specific improving innovation are dominates. Oil companies lack funds to accelerate innovation activity	Global threat of economic slowdown and financial crisis (for the entire economy)	Increase in R&D growth rates compared to other industries due to higher profitability of Oil Companies	High probability.	Inertial for the entire economy

Table 2. *Cont.*

Scenario	Macroeconomic Situation	Political Situation	Competition	R&D of Oil Companies	Revenue of the Oil Companies	R&D of other (Non-Oil) Companies	Revenue of Other (Non-Oil) Companies	Technological Situation	Other Factors: Threats	Other Factors: Opportunities	Scenario Probability (Heuristic Estimates)	the Type of the Scenario
Scenario 4 Medium-optimistic	No shocks, stable, favorable for all sectors of the economy (general economic recovery)	Stable, favorable for the growth and development of all sectors of the economy	Intra-industry and cross-industry competition in sales and technology markets.	Growing at an average or faster pace compared to the 2011–2017 level. (according to forecast 1, Figure 5)	Growing at a medium or faster rate pace compared to the 2011–2017 level (according to forecast 1, Figure 5).	Growing within the rate of 2011–2017.	Growing within the rate of 2011–2017	Fundamental research is developing, the rate of commercialization of innovations is increasing	Insufficient but not low, rates of innovative development remain, products and technologies are being improved, R&D in the field of new energy sources.	Investments in R&D are increasing, strategies for innovative development are being developed and implemented. Oil companies form inter-corporate integration alliances to create innovation.	Medium, given that baseline R&D growth rates are low for Oil Companies and high for non-Oil Companies	Progressive
Scenario 5 Optimistic	Especially favorable for Oil Companies	Stable, favorable for the growth and development of the oil business	Interindustry technology competition	They are growing at a significant rate relative to the level of 2011–2017 (according to forecast 2 or with rates exceeding relative to other industries Figure 6).	They are growing at a significant rate relative to the level of 2011–2017.) (according to forecast 2 or with rates exceeding relative to other industries Figure 6)	Reduced or minimized	Reduced	Oil companies commercialize effective innovative technologies, apply them outside the industry and intellectual property is protected	Reduced purchasing power due to downturns in other industries	Opportunities of high-quality transformation of Oil Companies; diversification, development of new technologies for the main and other types of activities. Qualitative growth of companies in economic, technological, social and environmental directions. Involvement of fundamental science.	The probability is medium, given that, in favorable conditions for Oil Companies, other companies do not use incentives for innovation or they are characterized by low innovation activity, lack of prerequisites for the transition to a new technological structure with subsequent economic growth	High-quality development of the industry Oil Companies are the leading driver of major technological changes or breakthrough innovations
Scenario 6. Extremely optimistic	Extremely favorable	Stable and favorable for the growth of Oil Companies and the entire economy	The competitive environment is conducive to economic and technological growth and development	Growing with rapidly rates				A major breakthrough was made in the energy sector, the transition to a new technological order	-	The transition of the world economy to a qualitatively new level—the transition to inexhaustible, environmentally friendly sources, "smart" energy, a radical increase in energy efficiency	Low, given the lack of prerequisites for the transition to a new technological order with subsequent economic growth	Oil companies are the driver of scientific and technological progress and the initiator of the transition to a new technological order

The proposed scenarios, in contrast to the development scenarios of the IEA (International Energy Agency) and another's scenarios are characterized by the fact that they are primarily assess the future of existing Oil Companies who continue to implement their low-innovation development strategies (or rather, regression strategies). The scenarios take into account the models of the future oil and gas business, their actual place in the structure of the world economy in situations of decreasing/maintaining the rates/growth of their R&D relative to other industries (including those that may arise in future, as a result of a technological shift). Indeed, the analysis of dynamics, comparison and forecasting of the rate of investment in R&D, the rate of commercialization, the efficiency of R&D and the profitability of Oil Companies and other companies indicated that the enterprises of this industry are faced with the need to increase their innovative activity. In this case and in the post-carbon era, companies related to oil today tomorrow will exist and develop, either diversifying their activities or increasing the pace of innovation or becoming at the forefront of innovative development. Otherwise, Oil Companies may not only become unprofitable but also be object of mergers and acquisitions by corporations from other sectors of the economy, which may have the competence of high rates of innovation activity and rapid rates of commercialization of innovations. There is also a high probability of a technological breakthrough in the field of energy production.

3.4. Important Directions for Development

When starting to consider the issue of important areas of development, it should be noted that, of course, one cannot say that companies are not doing any work—it is simply not enough. If we look at individual companies, it can be noted that, for example, traditionally leading places in the ratings in terms of absolute volume of funds among companies belonging to the oil and gas sector belonged to Exxon Mobil and Total, annually investing 700–800 million USD. But on the whole, as the ratings show, in comparison with other sectors, Oil Companies can hardly be called innovatively active. To the above indicators and facts, we add the fact that the average annual indicator of investments in R&D in the industry is 0.7%, however, specific values for countries vary significantly. For American and European companies, such expenses are generally equal to $1 per 1 ton of oil equivalent, while in China and Brazil they are 2.5–3.2 times higher. For Russian companies, the share of R&D in the same relative indicators is less than $0.2 per ton. Undoubtedly, such a spread in data is explained by the size of companies and the specific starting conditions for innovative development, determined by the history of investment in R&D. Western companies have been paying attention to this area for many years, while Chinese and Brazilian companies have turned to it relatively recently. For a long time, Russian companies have been operating production facilities created back in the Soviet period. A protracted innovation pause against the backdrop of low investment in R&D turns out to be a serious problem for them.

At the same time, given the major changes in the context of these article, it is not enough for oil and gas companies to simply increase R&D spending and increase innovation activity. They need to choose directions of development and develop appropriate strategies.

The number of non-oil energy companies is growing and their efficiency is improving. The oil business in oil importing countries, for example China, is focused on the development of innovations in the downstream.

It must be said that the overall low level of R&D expenditures in this industry can be explained by the fact that its subjects remain super-profitable. This industry is at the top in terms of wages and is viewed by many governments as the main source of tax revenue [35].

The development of innovative processes is also seriously hampered by the internal structure of Oil Companies, which is vertically integrated, in which the development of new technologies is limited to the activities of special units. Not all companied disclose this information. Nevertheless, for example, in British Petroleum and in Statoil, innovation processes are managed by a line manager, for whom this function is secondary [62].

The analysis shows that investments in R&D can be transformed into successfully implemented innovations that produce an economic effect only if there are special market strategies. The key perspectives directions for Oil Companies will be: drilling technologies, remote control of drilling, improving hydraulic fracturing technologies, ensuring environmental safety, new methods of studying the geology of wells at great depths.

At the end of the last century, the global average oil recovery rate did not exceed 15–20%. Since the beginning of this century, it has grown to 35% and in some countries, it has reached 50%. In modern oil production, the oil recovery coefficient ranges from 9–75% [63] but this coefficient differs in different countries: the highest is at one of the fields in Norway (66%), in Russia—up to 35%, in North America—35–37%, in Latin America and Southeast Asia—24–27% and in Iran 16–17% [64,65]. Meanwhile, an increase in oil recovery in all regions of the world by at least 1% would cover the needs of the world economy for 2–3 years ahead (according to Ernst & Young, by about 88 billion barrels [66]).

One of the most important directions in the development of the oil and gas industry is to increase the efficiency of operating the existing fields. The most challenging tasks are production optimization and maximum reduction of downtime costs. It can be solved through the introduction of automation systems that reduce labor costs for well maintenance, reservoir-to-surface modeling, pump control and the development and implementation of integrated technologies for Smart Field. It will be possible to maximize oil production from the reservoir through real-time control, flexible changes in the production schedule, adaptation to changing environmental conditions and reduction of energy and hydraulic costs. In this case, the life cycle of the field will be lengthened and environmental risks will decrease.

As the analysis of plans and projects of Oil Companies has shown, projects on the use of solar energy and wind energy are being implemented by Royal Dutch Shell, Chevron, Petrobras, Total, BP; projects related to geothermal energy are being developed by Royal Dutch Shell, Chevron, PetroChina; Chevron, Petrobras, PetroChina are interested in the implementation of projects for the production of biodiesel; BP is interested in the development of projects for obtaining biofuels and energy from inedible plant materials and R&D in the field of hydrogen energy belongs to ExxonMobil, Chevron, Total.

Due to the industry specifics, the development and implementation of innovations in the oil and gas industry initially implies a high level of R&D costs. In the near future, the demand for them may reach such volumes that will not be available even to supermajors. Realizing this, these companies are already looking for government support today.

4. Discussion

Despite the fact that the purpose of this article has been achieved and its hypothesis has been confirmed, the issues studied in this article can be further developed and become objects for scientific and practical discussions. All problems determine the prospects for the further development of the authors' ideas, both in terms of the further development of theory and in relation to the development of practical possibilities for applying the conclusions obtained in this article.

The object for further scientific research and for scientific and scientific-practical discussions becomes and another moment. It consists in the point that the scenario forecasting methodology can be developed in further research, which will study the problems associated with taking into account new trends, including those caused by the crisis caused by Covid-19 and the subsequent economic recovery. There are also important directions for the development of this approach through the use of quantitative methods for assessing the development of scenarios and it is the to expand of the research horizon.

Another discussion point, which requires further research, is the conclusion made by the authors and that the problem of insufficient investment in R&D in Oil Companies is a consequence but the reason consists of the existing and prevailing business model in the industry. Their disadvantages were identified in the article. They can lead to the collapse of the existing (modern) Oil Companies due to the impact on them of the fifth force of competition.

5. Conclusions

The hypothesis proposed in this study is confirmed. However, the confirmation does not cause optimism in connection with the results of forecasting. Indeed, most Oil Companies show low innovative activity, continuing to develop along an extensive trajectory. The transition to an intensive model is possible only through an active innovation policy, an increase in R&D costs and the effectiveness of the implementation of the results obtained, as in other industries.

Practical conclusions should be drawn from our analysis, which is that Oil Companies need not only come to understand that they are in a "pre-calaptic" state but also that it is necessary to develop new directions of their development that will allow them to achieve the best possible development scenario. As the main directions for the development of Oil Companies and this industry it should be noted that as the main directions of the industry development, several consolidated directions should be distinguished. The first of them consists of the implementation of innovative activities in the field of traditional processes associated with the production of hydrocarbons and the production of petroleum products, where R&D should be aimed at increasing the efficiency of production processes, increasing reservoir recovery. The second direction is to improve the environmental friendliness of the production processes of traditional hydrocarbon products. The third direction is associated with diversification, with the development of new sources of energy production. An important direction is also the improvement of the organizational structure of the management of R&D processes, the innovation policy of Oil Companies (fourth). As the fifth direction, one should also highlight the direction associated with the use of digital technologies in the production process—the use of big data technologies, Distributed ledger systems, AI, VR, and nanotechnology [67].

A source of accelerating the rate of innovative development can be the possibility of forming intersectoral scientific and technical alliances, developing scientific and technical cooperation with other companies through the creation of joint ventures and the organization of specialized portals.

The above aspects, on the one hand, define a wide field of discussion and on the other, they determine the further direction of research. In addition to the need to develop the above areas, to continue the process of monitoring R&D in the energy sector, we believe that it is important, firstly, to raise the issue of forming a concept for the development of the oil and gas business in the 4.0 format, as a symbol of a qualitative business transformation. So far, this sounds like a slogan [52] and therefore there is an urgent need for a transition to action—exploring opportunities, forming a methodology and forming special strategies, tactics and instruments.

Secondly, the development of an effective system of innovative marketing, which involves monitoring changing demand and opportunities for innovative development not only through portfolio but also through direct investment. The tactical component of new business models is to increase investment in R & D and more actively implement their results in production. The strategic component involves the development of innovative marketing principles and the transformation of Oil Companies into diversified energy producers.

Author Contributions: Conceptualization, Y.S.M. and E.V.; methodology, Y.S.M.; validation, Y.S.M., L.S. and Y.P.; formal analysis, Y.S.M.; investigation, Y.S.M.; resources, L.S.; data curation, E.V.; writing—original draft preparation, Y.S.M.; writing—review and editing, E.V.; visualization, E.V.; project administration, Y.P. All authors have read and agreed to the published version of the manuscript.

Funding: This research received no external funding.

Institutional Review Board Statement: Not applicable.

Informed Consent Statement: Not applicable.

Data Availability Statement: Not applicable.

Conflicts of Interest: The authors declare no conflict of interest.

References

1. Ardalan, F.; Almasi, N.A.; Atasheneh, M. Effects of contractor and employer's obligations in buy back contracts: Case study of oil exporting country. *Entrep. Sustain. Issues* **2017**, *5*, 345–356. [CrossRef]
2. Plenkina, V.; Osinovskaya, I. Improving the system of labor incentives and stimulation in oil companies. *Entrep. Sustain. Issues* **2018**, *6*, 912–926. [CrossRef]
3. Plenkina, V.; Andronova, I.; Deberdieva, E.; Lenkova, O.; Osinovskaya, I. Specifics of strategic managerial decisions-making in Russian oil companies. *Entrep. Sustain. Issues* **2018**, *5*, 858–874. [CrossRef]
4. Plėta, T.; Tvaronavičienė, M.; Della-Casa, S.; Agafonov, K. Cyber-attacks to critical energy infrastructure and management issues: Overview of selected cases. *Insights Reg. Dev.* **2020**, *2*, 703–715. [CrossRef]
5. Sansyzbayev, A. Influence of social partnership to the development of enterprise: On the example of oil industry. *Entrep. Sustain. Issues* **2019**, *7*, 1613–1627. [CrossRef]
6. Alkhathlan, K.A.; Alkhateeb, T.T.Y.; Mahmood, H.; Bindabel, W.A. Concentration of oil sector or diversification in Saudi economy: Consequences on growth sustainability. *Entrep. Sustain. Issues* **2020**, *7*, 3369–3384. [CrossRef]
7. Masood, O.; Tvaronavičienė, M.; Javaria, K. Impact of oil prices on stock return: Evidence from G7 countries. *Insights Reg. Dev.* **2019**, *1*, 129–137. [CrossRef]
8. Al-Mazrouei, M.A.; Khalid, K.; Davidson, R. Development and validation of a safety climate scale for United Arab Emirates oil and gas industries. *Entrep. Sustain. Issues* **2020**, *7*, 2863–2882. [CrossRef]
9. Humbatova, S.I.; Tanriverdiev, S.M.O.; Mammadov, I.N.; Hajiyev, N.G.-O. Impact of investment on GDP and non-oil GDP in Azerbaijan. *Entrep. Sustain. Issues* **2020**, *7*, 2645–2663. [CrossRef]
10. Hughes, J.E.; Lange, I. Who (else) benefits from electricity deregulation? Coal prices, natural gas, and price discrimination. *Econ. Inq.* **2020**, *58*, 1053–1075. [CrossRef]
11. World Production Gas, Natural Gas Liquids, Coal and Lignite, Electricity, Primary Energy. Oil & Energy Trends: Annual Sta-tistical Review. *Stat. Tables* **2019**, *40*, 1.
12. Porter, M.E. *Competitive Advantage, Creating and Sustaining Competitive Performance*, 1st ed.; The Free Press: New York, NY, USA, 1985.
13. Harvey, M.; Porter, M.E. Competition in Global Industries. *J. Mark.* **1988**, *52*, 150. [CrossRef]
14. Porter, M.E. Towards a dynamic theory of strategy. *Strat. Manag. J.* **1991**, *12*, 95–117. [CrossRef]
15. Obama, B. Oval Office Address to the Nation on BP Oil Spill Disaster. Available online: https://www.americanrhetoric.com/speeches/barackobama/barackobamabpoilspillovaloffice.htm (accessed on 20 November 2020).
16. Helman, C. Which Companies Pay the Most in Taxes? Available online: https://www.forbes.com/sites/christopherhelman/2012/04/16/which-megacorps-pay-megataxes/#31fd69d85586 (accessed on 30 September 2020).
17. Marshall, A. *Principles of Economics*, 8th ed.; Amherst Prometheus Books: New York, NY, USA, 1997; 356p.
18. Samuelson, P.A.; Nordhaus, W.D. *Economics: An Introductory Analysis*, 19th ed.; McGraw–Hill: New York, NY, USA, 2009.
19. Rapp, F.; Mensch, G. Das technologische Patt: Innovationen uberwinden die Depression. *Technol. Cult.* **1977**, *18*, 287. [CrossRef]
20. Hayek, F.A. Competition as a Discovery Procedure/F. A. Hayek, trans in English S. Marcellus. *Q. J. Austrian Econ.* **2002**, *5*, 9–23.
21. Kondratiev, N.D. *Bol'shie Cikly Ehkonomicheskoj Kon"Yunktury Problemy Ehkonomicheskoj Dinamiki [Big Cycles of an Economic Conjuncture. Problems of Economic Dynamics]*; EHkonomika: Moscow, Russia, 1989; pp. 172–226. (In Russian)
22. Schumpeter, J.A. *The Theory of Economic Development: An Inquiry into Profits, Capital, Credit, Interest and the Business Cycle; translated from the German by Redvers Opie*; Oxford University Press: Oxford, UK, 1961.
23. Schumpeter, J.A. *The Theory of Economic Development: An Inquiry into Profits, Capital, Credit, Interest and the Business Cycle, Translated from the German by Redvers Opie*; Transaction Publishers: New Brunswick, NJ, USA, 2008.
24. Nussbaum, F.L.; Schumpeter, J.A.; Schumpeter, E.B. History of Economic Analysis. *Am. Hist. Rev.* **1954**, *60*, 62. [CrossRef]
25. Nelson, R.R.; Winter, S.G. *An Evolutionary Theory of Economic Change*; Harvard University Press: Cambridge, MA, USA, 1982.
26. Bornstein, M.; Toffler, A. Future Shock. *Technol. Cult.* **1971**, *12*, 532. [CrossRef]
27. Bell, D. *The Coming of Post-Industrial Society: A Venture of Social Forecasting*; Basic Books: New York, NY, USA, 1973.
28. Galbraith, J.K. American Capitalism. *Am. Capital.* **2017**, *8*. [CrossRef]
29. Galbraith, J.K. Economics as a System of Belief. *Am. Econ. Rev.* **1970**, *60*, 469–478.
30. Santo, B. *Innovaciya kak Sredstvo Ehkonomicheskogo Razvitiya [Innovation as a Means of Economic Development]*; Progress: Moscow, Russia, 1990; p. 296. (In Russian)
31. Anchishkin, A.I. *Nauka—Tekhnika—Ehkonomika [Science—Technology—Economics]*, 2nd ed.; EHkonomika: Moscow, Russia, 1989; p. 386. (In Russian)
32. Glaziev, S.Y. *Teoriya Dolgosrochnogo Tekhniko-Ehkonomicheskogo Razvitiya [Theory of Long-Term Technical and Economic Devel-Opment.]*; VlaDar: Moscow, Russia, 1993; p. 310.
33. Glaziev, S.Y. *EHkonomicheskaya Teoriya Tekhnicheskogo Razvitiya [Economic Theory of Technical Development]*; Nauka: Moscow, Russia, 1990; p. 230.

34. Yakovec, Y.V. Innovacionnyj proryv: Social'nye i nauchnye aspekty [Innovative breakthrough: Social and scientific aspects]. *Filos. Hozyajstva Econ. Philos.* **2008**, *6*, 281–286. (In Russian)
35. Oligney, R.; Izquierdo, A.; Economides, M. *The Color of Oil: The History, the Money and the Politics of the World's Biggest Business*; Round Oak Publishing: Katy, TX, USA, 2000; p. 220.
36. Kamien, M.I.; Schwartz, N.L. Market structure and innovation: A survey. *J. Econ. Lit.* **1975**, *13*, 1–37.
37. Grabowski, H.G. The Determinants of Industrial Research and Development: A Study of the Chemical, Drug, and Petroleum Industries. *J. Politi Econ.* **1968**, *76*, 292–306. [CrossRef]
38. Grabowski, H.G.; Baxter, N.D. Rivalry in Industrial Research and Development: An Empirical Study. *J. Ind. Econ.* **1973**, *21*, 209. [CrossRef]
39. Mansfield, E.; Rapoport, J.; Schnee, J.; Wagner, S.; Hamburger, M. Research and Development: Characteristics, Organization, and Outcome. In *Research and Innovation in the Modern Corporation*; Springer Nature: New York, NY, USA, 1971; pp. 18–46.
40. Commanor, W.S.; Sherer, F.M. Patent Statistics as a Mesure of Technical Change. *J. Political Econ.* **1969**, *77*, 392–398. [CrossRef]
41. Phillips, A. Market Structure, Innovation and Investment. In *Patents and Progress: The Sources and Impact of Advancing Technology*; Alderson, W., Terpstra, V., Shapiro., J., Eds.; Irwin: Homewood, IL, USA, 1965; p. 251.
42. Shmookler, J. *Invention and Economic Growth*; Harvard University Press: Cambridge, MA, USA, 1966; p. 332.
43. Mastepanov, A. Perelom energeticheskoi filosofii [Fracture of energy philosophy]. *Neft' Ross. Oil Russ.* **2014**, *11*, 17–24. (In Russian)
44. Poleshchuk, M.S. Analiz prioritetnykh napravlenii innovatsionnoi deyatel'nosti rossiiskikh i inostrannykh neftegazovykh kompanii [Analysis of priority directions of innovative activity of Russian and foreign oil and gas companies]. *Probl. Ekon. Upr. Neftegazov. Kompleks. Probl. Econ. Manag. Oil Gas Complex* **2015**, *2*, 31–37. (In Russian)
45. Cherkasov, M. Kurs na povyshenie effektivnosti [Course on increasing efficiency]. *Neft' Ross. Oil Russ.* **2015**, *5*, 38–39. (In Russian)
46. Cherkasov, M. Intellektual'nye Tekhnologii Dlya Sovremennogo Mestorozhdeniya: «Umnoe Mestorozhdenie» [Intellectual Technologies for a Modern Deposit: "Smart Deposit"]. Delovoi Portal «Upravlenie Proizvodstvom» [Business Portal "Production Management"]. Available online: http://www.up-pro.ru/print/library/information_systems/production/umnoe-mestorojdenie.html (accessed on 5 September 2020). (In Russian).
47. Blaug, M. *The Methodology of Economics, or How Economists Explain*; Cambridge University Press: Cambridge, UK, 1992.
48. Abt, K.C.H.; Foster, R.N.; Ri, R.G. *Metodika Sostavleniya Scenariev [A Scenario Generating Methodology]. Rukovodstvo Po Nauchno-Tekhnicheskomu Prognozirovaniy. [Scientific and Technical Forecasting Guide]*; To Russian, Progress: Moscow, Russia, 1977; Volume 349, pp. 132–163. (In Russian)
49. Thomas Reuters. *State of Innovation 2015. The Future Is Open*; Thomson Reuters: Toronto, ON, Canada, 2015; Available online: https://www.thomsonreuters.com/en/press-releases/2015/october/thomson-reuters-reports-third-quarter-2015-results.html (accessed on 15 May 2020).
50. Thomson Reuters. *Disruptive, Game-Changing Innovation*; Thomson Reuters: Toronto, ON, Canada, 2016; Available online: https://www.thomsonreuters.com/en/press-releases/2016/november/thomson-reuters-reports-third-quarter-2016-results.html (accessed on 15 May 2020).
51. Clarivate Analytics. The State of Innovation Report. 2017. The Relentless Desire to Advance. Available online: https://clarivate.com/wp-content/uploads/2017/10/The_State_of_Innovation_Report_2017.pdf (accessed on 16 May 2020).
52. The Global Innovation Index 2020: Who Will Finance Innovation? Cornell University, INSEAD, and the World Intellectual Property Organization. 2020. Available online: https://www.wipo.int/edocs/pubdocs/en/wipo_pub_gii_2020.pdf (accessed on 15 October 2020).
53. Matkovskaya, Y. The influence of the pandemic to the energy development of the economy: Innovation should solve problems. *Druk. Vestnik* **2020**, 104–110. [CrossRef]
54. Hernández, G.H.; Grassano, N.; Tübke, A.; Potters, L.; Amoroso, S.; Dosso, M.; Gkotsis, P.; Vezzani, A. EU R&D Scoreboard: The 2017 EU Industrial R&D Investment Scoreboard. Available online: http://iri.jrc.ec.europa.eu/survey17.html (accessed on 15 October 2020).
55. Hernández, G.H.; Grassano, N.; Tübke, A.; Grassano, N.; Tübke, A.; Amoroso, S.; Csefalvay, Z.; Gkotsis, P. The 2019 EU Industrial R&D Investment Scoreboard. EU Science Hub. Available online: https://ec.europa.eu/jrc (accessed on 15 October 2020).
56. PWC. The 2017 Global Innovation 1000 Study. Investigating Trends at the World's 1000 Largest Corporate R&D Spenders. Available online: https://www.strategyand.pwc.com/innovation1000 (accessed on 15 October 2020).
57. PWC. The 2018 Global Innovation 1000 Study. What the Top Innovators Get Right. Available online: https://www.strategyand.pwc.com/gx/en/insights/innovation1000.html (accessed on 15 October 2020).
58. Porter, M. *Competitive Advantage: How to Achieve a High Result and Ensure its Stability/Michael Porter*, 2nd ed.; Alpina Business Books: Moscow, Russia, 2006; 715p.
59. International Energy Agency. Energy Technology Perspectives 2020. Available online: https://www.iea.org/reports/energy-technology-perspectives-2020 (accessed on 5 October 2019).
60. Energy Technology Perspectives. Special Report on Clean Energy Innovation. International Energy Agency. Available online: https://www.iea.org/reports/clean-energy-innovation (accessed on 5 October 2019).
61. International Energy Agency. World Energy Model Documentation 2020 Version. Available online: https://www.iea.org/reports/world-energy-model/documentation (accessed on 5 October 2019).

62. Pokhlebkin, D. Strategiya Innovatsii: Luchshaya Praktika [Innovation Strategy: Best Practice]. Available online: https://www.e-xecutive.ru/community/magazine/1585455-dmitrii-pohlebkin-strategiya-innovatsii-luchshaya-praktika (accessed on 19 May 2020). (In Russian).
63. Neftegaz.ru. Technologies. Oil Recovery (Oil Recovery Coefficient of CIN) and Methods of Increasing Oil Recovery (EOR). Neftegaz.ru. Available online: https://neftegaz.ru/tech_library/view/4420-Nefteotdacha-koeffitsient-izvlecheniya-nefti-KIN-i-metody-povysheniya-nefteotdachi-MUN (accessed on 6 September 2020).
64. Sergeev, L. The Growth of the Oil Recovery Factor is the Guarantee of the Economic Well-Being of the Country. Available online: https://forpost-sz.ru/a/2018-01-20/rost-koehfficienta-izvlecheniya-nefti-zalog-ehkonomicheskogo-blagopoluchiya-strany (accessed on 15 September 2020).
65. Lloyd's Register Energy. Oil and Gas Technology Radar 2014: An Assessment of the Sector's Innovation Trends and Drivers, London: Lloyd's Register. Available online: https://mediaserver.responsesource.com/fjd-profile/39737/Lloyds_Register_Energy_Oil_And_G.pdf. (accessed on 15 October 2019).
66. Primenenie Sovremennyh Metodov Uvelicheniya Nefteotdachi v Rossii: Vazhno ne Upustit' Vremya. Ernst end Yang (SNG) B.V. S. 1 [Application of Modern Methods of Enhanced Oil Recovery in Russia: It Is Important not to Waste Time. Ernst & Young (CIS) V.V. P. 1]. 2013. Available online: http://npf-its.com/wp-content/uploads/2015/01/.pdf (accessed on 15 May 2020).
67. Nagaraj, V.; Berente, N.; Lyytinen, K.; Gaskin, J. Team Design Thinking, Product Innovativeness, and the Moderating Role of Problem Unfamiliarity. *J. Prod. Innov. Manag.* **2020**, *37*, 297–323. [CrossRef]

energies

MDPI

Article

Measuring the Implementation of the *Agenda 2030* Vision in Its Comprehensive Sense: Methodology and Tool

Edvins Karnitis, Janis Bicevskis and Girts Karnitis *

Faculty of Computing, University of Latvia, Raina blvd. 19, LV1586 Riga, Latvia; edvins.karnitis@lza.lv (E.K.); janis.bicevskis@lu.lv (J.B.)
* Correspondence: girts.karnitis@lu.lv

Abstract: The contribution of the energy sector to human development (HD) is one of the aspects that requires measuring growth of the overall sustainability of HD. The UN program, *Agenda 2030*, has defined the vision of advanced development by introducing a sustainable HD paradigm, the balanced integrated development paradigm, in three dimensions: economic, social and environmental. The overall performance of countries is not currently assessed; there are several proposals for measuring sustainable HD level, but none have become widely accepted due to their weaknesses. The selected indirect measurement method reflects the interlinkage of development dimensions in real cases of low data availability. By combining the strengths of existing proposals and eliminating their weaknesses, the measuring methodology has been created and an appropriate tool—the Advanced Human Development Index (AHDI)—has been designed, which ensures unity and concordance of all included comprehensive dimension indices. The calculations confirm the accuracy and simplicity of the measurement. The proposed methodology and AHDI, as a simple, balanced index that is based on result-oriented headline indices, provide the *big picture*, which will be transparent, acceptable and usable for experts, politicians and the global community to assess the achieved development levels and to make strategic decisions for the coming period.

Keywords: *Agenda 2030*; sustainable human development; measuring development; energy security; energy sustainability; indirect measurement; composed indexes

Citation: Karnitis, E.; Bicevskis, J.; Karnitis, G. Measuring the Implementation of the *Agenda 2030* Vision in Its Comprehensive Sense: Methodology and Tool. *Energies* **2021**, *14*, 856. https://doi.org/10.3390/en14040856

Academic Editor: Miguel-Angel Tarancon
Received: 28 December 2020
Accepted: 1 February 2021
Published: 6 February 2021

1. Introduction

Academic and political understanding of human development, its scope and dimensions, interrelations and significance in the human development process is formed and changes during global socioeconomic and political development. Such a concept asks for the regular comprehensive measurement of progress in human development (let us remember Niels Bohr's expression, "Nothing exists until it is measured") to explicate the achieved level in statics and dynamics, as well as to make strategic decisions for the coming period.

A secure and stable supply of energy has been a key driver of development for both the world's leading economies and developing countries. From the beginning of human development, energy has made a direct contribution to economic growth [1]. "Energy and economy evolve in tandem" [2] is a well-established axiom. In addition to the economy, access to energy is a critical precondition for the social development of humanity; this is the conviction of politicians (e.g., "Energy is the lifeblood of our present-day civilization and culture" [3]) and experts (e.g., "The provision of adequate energy services is a precondition for socioeconomic development and human wellbeing" [4]).

Such comprehensive qualitative assessments can be confirmed and quantified by the indication of causation between energy consumption and human development. The last aspect is currently measured by the Human Development Index (HDI) as a parameter, which is globally accepted by experts and politicians as the indicator of national socioeconomic development level. Figure 1 shows that the correlation between both mentioned indices is strong. In general, access to exponentially increasing amounts of energy is necessary to achieve linear growth of the development level. There are no indications that this pattern is substantially weakened on a global scale. This causation should be taken into account in determining future energy (including energy efficiency and performance) policies to break growing energy consumption in many countries.

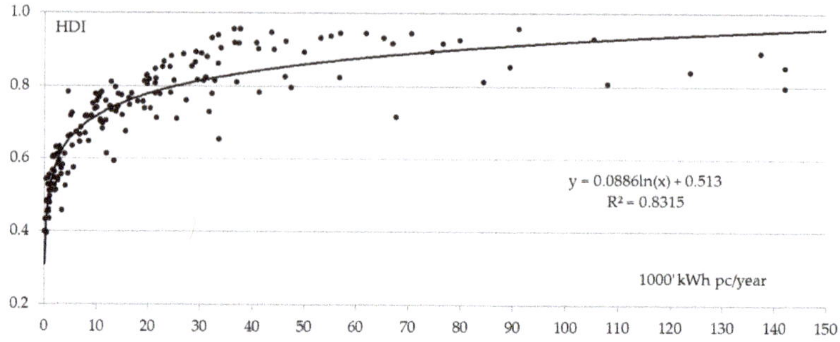

Figure 1. Human development index vs. final energy consumption per capita (2019 or the last available year); sources—Human Development Report and Our World in Data.

To shape human development as a global sustainable trend, which will secure welfare opportunities for future generations, the United Nations (UN) has adopted an advanced development concept, *Agenda 2030*, which includes an environmental dimension in addition to the economic and social dimensions. The concept forecasts energy as one of the key pillars for human wellbeing and further sustainable development in the 21st century: "Sustainable development of energy systems requires consideration of all three sustainability dimensions: environmental, economic and social" [5]. Issues such as long-term energy efficiency and energy security are becoming critical far beyond the context of energy: "It is clear that the energy sector must be at the heart of efforts to lead the world on a more sustainable pathway" [6].

Even in this case, which corresponds to the *Agenda 2030* vision, the real contribution of energy consumption to sustainable human development can best be indicated by measuring growth of the overall development level. However, the quantitative assessment in this case lacks a complex indicator of development level, which integrates all three dimensions and corresponds to the *Agenda 2030* vision.

The aim of this study is to develop a measurement methodology, which is consistent with the *Agenda 2030* vision on the advanced human sustainable development paradigm, as well as an appropriate tool—the Advanced Human Development Index (AHDI)—that reconciles the social and economic dimensions with the environmental dimension. Of course, the importance of creating such an index goes far beyond energy, which is certainly very important, but is only one aspect of human development.

In Section 2, an advancement of the human development paradigm and its measurement is analyzed to indicate academic and political context. Section 3 shows the selection of the optimum method for measuring sustainable human development. Section 4 is devoted to the creation of measurement methodology and calculation of the AHDI. Section 5 discusses the results obtained, while Section 6 outlines the introduction of the AHDI in the practice of experts and politicians.

2. An Advancing Paradigm of Human Development

The term development in general is described as "the steady growth of something so that it becomes more advanced, stronger, etc." [7]; in relation to humans, the term denotes a continuously advanced concept on a "comprehensive societal process that covers all aspects of life" [8].

For many centuries, the economic activities were considered as the single real driving force of human/national development and progress. Gross domestic product (GDP) (or its derivatives) has been considered as the core indicator in measuring the development level of a nation and its power (see, e.g., [9]).

From the 1960s onwards, experts and politicians have increasingly emphasized that human development is not just about economic growth. "It [GDP] measures neither our wit nor our courage, neither our wisdom nor our learning, neither our compassion nor our devotion to our country, it measures everything in short, except that which makes life worthwhile" [10]. Consequently, "National income ... [becomes] an inappropriate indicator of development" [11]; in terms of measuring development, this means that " ... such as the GDP are useful, development is now considered to be a much broader concept and must be more than a focus on economic growth [only]" [12].

Starting with a general thesis on social development, a consensus has gradually emerged on the most crucial dimensions for humans: health and longevity, good education and sufficiently high income. Instead of money, people became the priority of global development.

An integrated and, at the same time, simple indicator—the Human Development Index—was created and adopted within the UN framework. The HDI has convinced academics, politicians and the whole community for 30 years that they can and must value development not only with economic progress, but also with the increase of human well-being. In the first edition of subsequent annual Human Development Reports, it was declared that "It is about more than GNP growth, more than income and wealth and more than producing commodities and accumulating capital. It is about how development enlarges their [people] choices. The most critical of these wide-ranging choices are to live a long and healthy life, to be educated and to have access to resources needed for a decent standard of living" [13].

At the end of the 20th century, an extensive global discussion of experts and high-level politicians expanded on the next fundamental issue: how to increase the sustainability of human development.

In general, experts define sustainability as a long-term concept of various interrelated economic, social and environment aspects (see, e.g., [14]). Different priorities have been set; however, the broadest consensus has been achieved on the importance of protecting the natural environment. This resulted in the 21st Century Action Program Agenda 21 [15]; the UN Summit on Sustainable Development then expanded and modified it as the *Agenda 2030* [16]. The programs defined the extended vision, which is the kernel of the today's sustainable human development paradigm, as a holistic vector of global progress, incorporating balanced integrated solutions to the economic, social and environmental dimensions. Seventeen sustainable development goals (SDG) and 169 targets have been set for the period up to 2030, these are directed to the implementation of the multidimensional *Agenda 2030* vision.

Naturally, such a vision and sustainability concept asks for an advanced human development measuring procedure to characterize progress in statics and dynamics: " ... it is important to seriously start thinking about the performance measurement system associated with SDGs at the initial stages itself, to lay foundations for effective and evidence-based policy making" [17].

Although "...the concept of sustainable development is much broader than the protection of natural resources and the physical environment; it includes the protection of human lives in the future; after all, it is people, not trees, whose future options need to be protected" [13], it will take a long time to globally harmonize other aspects of sustainability (see, e.g., [18]). The current task is to measure sustainable human development exactly according to the current consensus, thus indicating the successes and problems in the implementation of *Agenda 2030*.

The COVID-19 pandemic and the subsequent difficult recovery period significantly increases the importance of sustainability in global and national long-term policy planning (see, e.g., [19]). The predicted inevitable decline in the level of globalization of processes and changes in value chains; the growing importance of self-sufficiency at the national level; and changes in the level and structure of consumption, the national isolation of countries and the growth of international competition (see, e.g., [20,21]), all increase the share of sustainability in reasoned policymaking. As a result, the importance of objective comprehensive measurements and monitoring of sustainable human development as a result of the implementation of post-COVID recovery policies becomes much higher.

There is a typical measuring problem for multidimensional programs: even many separate narrow profile indicators cannot show the overall progress of the program; therefore, composite indicators (indexes) are created, aggregating individual indicators. "Composite indicators can be used to summarize complex or multidimensional issues, in view of supporting decision-makers. Composite indicators provide the big picture . . . An index remains useful to make a point for action" [22].

Currently, the problem is a lack of a politically accepted methodology and tools to measure the implementation of the comprehensive *Agenda 2030* vision. At the time of elaboration of *Agenda 2030*, the UN Division for Sustainable Development recommended that "Indicators corresponding to the future SDGs are most important for monitoring future progress, but they will need to be complemented by composite indices of sustainable development progress" [23]. The task to improve the system of indicators for progress measurement already exists in the composition of SDGs: "By 2030, build on existing initiatives to develop measurements of progress on sustainable development" [24].

Experts have a similar position: "We must first decide where we are going—our overarching goal—to measure progress toward it. . . . We are certainly not recommending throwing out the [indicator] dashboard, but merely recognizing that the dashboard and an aggregated indicator of overall progress toward our shared goal are both necessary if we hope to achieve our goal" [25].

3. An Optimum Method for Measuring Sustainable Human Development

The selection of a measurement method is a very important step in the creation of any measurement procedure; unfortunately, its importance is not always fully appreciated. However, it is the method that is the kernel for further generation of the methodology and thus primarily determines the performance of the whole measurement procedure. We carried out a purposeful detailed investigation of the cohesion of the *Agenda 2030* vision and past experience in measuring human development to make an informed choice.

3.1. *Direct Measurement Method*

"In direct methods of measurement, the unknown quantity is directly compared against a standard and the result is expressed as a numerical number and a unit" (see, e.g., [26]. Direct (absolute) measurement of the achieved progress in sustainable human development and obtaining the measurement result as one quantity value is impossible due to lack of a standard (etalon).

That is why the SDG indicators' set, which was adopted by the UN General Assembly, consists of 232 narrow profile indicators only [27]. A huge advantage of the set is its relatively widespread use of hard statistical indicators; they provide a representative view of the relevant part of aspects of sustainable development. Nevertheless, not all SDG indicators can be quantified; the global data availability is low (about 50%, according to [28]), especially for reliable data. The similarly purposed Eurostat database provides the data on 100 indicators for EU countries only [29].

With such a huge number of indicators, uneven execution occurs in various aspects (targets); similar progress for all aspects is unthinkable. An analysis of the *Agenda 2030* implementation trend reflects that part of the individual indicators show progress, others show mining at the current level with some fluctuations while still others show regression in the relevant aspects.

The UN High-level Political Forum on Sustainable Development [30] was held to comprehensively review the achieved progress. In assessing progress, successes as well problems were mentioned only on individual aspects, e.g., access to safe drinking water and electricity, child and neonatal mortality, economic growth, quality education, poverty, hunger, gender equality and environment. Progress as a whole and the general dynamics had not been assessed; leaders and those falling behind were not identified. Due to the *Agenda 2030* key thesis on integrity of the sustainable human development process, the forum had to give a political conclusion on the overall progress, and it did: " ... yet it is clear that the world is not on track to meet the SDGs by 2030". However, it is difficult to consider it as a data-based conclusion.

Assessments at EU [28] and OECD [31] levels had an analogous fragmented structure. Analysts were also forced to draw conclusions about the overall trend based on individual indicators only and using an indefinite conclusion on " ... only limited progress" [32].

There are several additional reasons as to why obtaining a positive result by creation of a composite index aggregating many specific target-related indicators (the so-called bottom-up approach) in this case is questionable.

As the integrated essence of multidimensional sustainable human development, the overall progress of *Agenda 2030*'s implementation cannot be adequately characterized, even by a combination of many sectoral numerical indicators "because these [social, economic, and environmental] resources are interconnected, there are no simple solutions to the problems society causes" [33]; e.g., the economic growth affects investments in social and environment dimensions, knowledge drives the economy and creates awareness of the environmental issues and the polluted environment directly impacts health. Bennich, Weitz and Carlsen [34] provide an extensive analytic overview of these interlinkages.

In turn, Barbier and Burgess [35] show that many of the SDGs contribute to more than one dimension of sustainability; quantitative analysis of the interactions and complementarities convincingly confirms the existence of interconnections; 43% of the sustainable development indicators in the EU database are used to describe two or more SDGs. The actions taken and the relevant indicators do not always have a positive effect on all SDGs; a different and, in the medium term, even contradictory effect on overall progress is possible (e.g., the balance between economic and environmental activities is still being sought).

Nevertheless, there have been several attempts to create the index that directly integrates SDG-related indicators.

The Sustainable Development Goals Index (SDGI) methodology is directly based on the structure of 17 SDGs that is an advantage of the approach [21]. The SDGI is created by summarizing part of the target indicators, which reflects the SDG settings; averaging the normalized indicators takes place first within the SDGs and then across SDGs.

Several weaknesses in the index methodology have been mentioned [36,37]. Only 60 (issue 2016) and 85 (issue 2020) of the accepted 232 SDG indicators are aggregated depending on the data availability to date. It is not known whether these are the determining indicators; this creates uncertainty in the calculated result. The methodology does not account for the interlinkages and the integrity of the SDGs.

The Systemic Indicator of Sustainable Development (SISD) is an academic attempt to remedy the last shortcoming of the SDGI and to measure the sustainable human development process in a more holistic manner [38]. Applying the systems thinking theory and approach, the dataset of 47 indicators (the same data availability problem), which relate to 17 SDGs, is transformed into another set of 14 factors taking into account the complexity of sustainable development and interlinkages of the SDGs. These factors are aggregated as secondary indicators to create the SISD. At the current stage of elaboration of the index, the result has become difficult to interpret; the question of interlinkage at the factor level also remains.

There are also proposals for the creation of composite indexes at the level of individual SDGs [39]. However, the bids are for some goals only and their aggregation to obtain a common index has not been elaborated.

3.2. Indirect Measurement Method

The essence of the indirect measurement method is synthesis of the unknown quantity from measurements made by direct methods of measurement of some other quantity linked to the unknown one by a defined relationship (see, e.g., [26]). In the practice of measuring human development, several quantities (indicators), which are related to achievements of human development as a whole or its dimensions, are usually used. This (so-called top-down) approach makes the problems of the direct measurement method irrelevant: a lack of a part of individual indicators is insignificant, interdimensional links and impacts are observed and the impact of indicators on several targets and SDGs is considered.

Dimension indicators typically are aggregated as components in the composed index. Experts, politicians as well as the global community, already have long-term experience and habits for creating and using an exactly indirect measurement method of the achieved human development level.

"Gross domestic product is an aggregate measure of production" [40], which is formed by the value added by all households, companies, public bodies and nonprofit institutions across the economy. It is a comprehensive indicator, which is constructed using datasets of the National Accounts and is based on the economic output, which is a principal advantage for data-driven high-level political decision-making. In fact, GDP not only summarizes the economy in one number, but "it drives government policy and sets priorities in a variety of vital social fields—from schooling to healthcare" [41]. Therefore, "the success of GDP is based on the fact that, with them, politicians were from the outset able to pursue a whole array of goals beyond just documenting economic processes" [42].

HDI is a politically accepted index, it has been used successfully for many years as "the best available alternative to GDP per capita" [43], although problems have already been identified (see, e.g., [44]). HDI is calculated as the geometric mean of three normalized dimension indexes (see Figure 2), thus ensuring a balance between the dimensions [45].

$$HDI = (I_{Life})^{1/3} \times (I_{Education})^{1/3} \times (I_{Income})^{1/3} \quad (1)$$

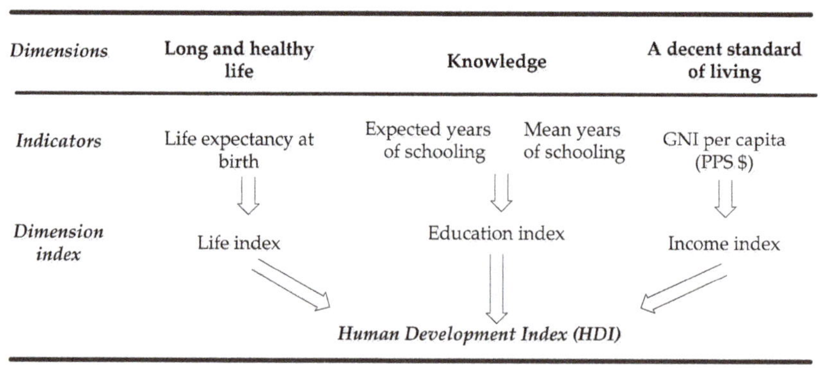

Figure 2. Design of the HDI; source—Human Development Report.

There are more suggestions for indirect measurement of sustainable human development compared with the direct one. Two of them propose the use of one or more narrow profile environmental indicators, others use integrated environmental indices. It should be noted that there are also offers that include other aspects such as peace, happiness and so on (see, e.g., [46]).

The Human Sustainable Development Index (HSDI) [47] was proposed by Togtokh and Gaffney "as a small step ahead to promote sustainable development" [48]. It is created by the addition of an environmental dimension to the HDI. Although environmental health and the services, which ecosystems provide for humans, are mentioned as factors that increase quality of life, CO_2 emissions per capita is proposed as the only indicator representing the environmental dimension. HDI methodology is preserved; there is an equal weighting of all four dimensions (standard of living, longevity, knowledge and environment):

$$HSDI = (I_{Life})^{1/4} \times (I_{Education})^{1/4} \times (I_{Income})^{1/4} \times (I_{CO_2})^{1/4}. \quad (2)$$

There are not many worldwide complex environmental indices that cover various environmental aspects as a whole.

The Ecological Footprint (EF) created by the World Wildlife Fund [49] reflects use of the natural capital (soil, air, water, minerals and living creatures) as well as the amount of generated carbon emissions. This should be evaluated as an insufficient composition for its use as an environmental indicator. The authors of the report themselves acknowledge that "National Footprint Accounts are not capable to quantify human environmental damage or pollution". Nevertheless, several authors have used EF to create sustainable human development indices.

The Ecological Well-being Performance indicator (EWP) [50] is composed as the ratio of the HDI and the logarithmically normalized value (EFn) of the EF to measure the achieved level of sustainable human development in relation to ecological assets consumed:

$$EWP = HDI/EFn. \quad (3)$$

Analysis shows that there is quite a strong positive correlation between the HDI and EF (see Figure 3): a more developed economy and society is generally more resource intensive. The impact of the environmental dimension (EFn) on the calculated composite indicator is extremely strong, resulting in a moderate negative correlation between HDI and EWP. Such an introduction of the environmental factor crosses the economic, longevity and educational achievements of society.

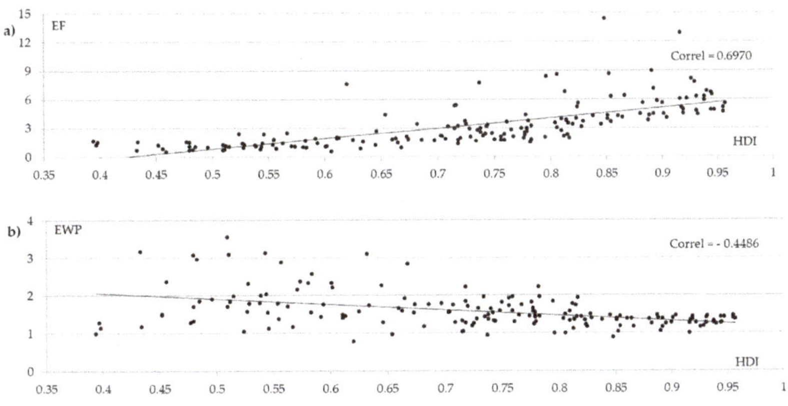

Figure 3. Ecological Footprint (**a**) and The Ecological Well-being Performance indicator (**b**) vs. HDI; sources—Human Development Report and World Wildlife Fund.

The Sustainable Development Index (SDI) philosophy [51], in general, is similar to that of EWP. HDI is slightly modified (HDIm), e.g., Gross National Income per capita (GNI pc) levels have been limited at $20,000 PPS (Purchasing Parity Standard). Unlike EWP, the environmental dimension is formed by a complex function of material footprint (MF) and CO_2 emissions:

$$SDI = HDIm/f\ \{MF, CO_2\}. \quad (4)$$

Correlations between the HDI and both MF and CO_2 are strongly positive. The result in this offer is very similar to that of the EWP.

There is a clear need for a more integrated environmental indicator that covers as many aspects as possible and thus assesses the inevitable pressure of human development on the environment.

The Environmental Performance Index (EPI), which is compiled by experts from several prominent universities, has much wider coverage of environmental issues [52]. It covers 11 high-priority environmental aspects (air and water quality and pollution, biodiversity degradation, climate change and emissions, waste management, etc.), for which 32 statistical indicators have been compiled for characterization. The index provides a balanced assessment of environmental health, evolving with increasing levels of knowledge and well-being, and the viability of ecosystems, which suffer from industrialization and urbanization.

Experts of the EC Joint Research Centre acknowledge the EPI "as a reliable composite indicator to measure environmental performance worldwide" [53].

The Environmental Human Development Index (EHDI) takes a holistic approach [54] to the synthesis of human development (which is characterized by the HDI) and environmental factor (EPI, as the most comprehensive environmental index is chosen). Based on the assumption of equal importance (resp. weights) of both mentioned components on sustainable human development, the index is calculated as the geometric mean of HDI and EPI:

$$\mathrm{EHDI} = (\mathrm{HDI})^{1/2} \times (\mathrm{EPI})^{1/2}. \quad (5)$$

Accounting for formula (1):

$$\mathrm{EHDI} = (\mathrm{I_{Life}})^{1/6} \times (\mathrm{I_{Education}})^{1/6} \times (\mathrm{I_{Income}})^{1/6} \times (\mathrm{I_{Environment}})^{1/2}. \quad (6)$$

The environmental dimension has the same weight as the life, education and income dimensions in total. For example, a 5% growth in one of the HDI dimensions results in a 0.82% increase of the EHDI, while a 5% growth in the environment dimension provides a 2.47% increase of the EHDI. This is not as strong an impact of the environmental dimension as we have seen in the cases of EWP and SDI; nevertheless, it does not correspond to the *Agenda 2030* setting on interdimensional balance. In fact, these examples reflect the regular efforts of radical environmental activists to limit the concept of sustainability to the natural environment only.

3.3. *Selection of the Optimum Measurement Method*

Several postulates arise from the investigation of the key principles of *Agenda 2030* and past experience in measuring human development (see [55]). Currently, when experts and politicians are looking for a successor of the HDI to measure sustainable human development, it is important to take them into account to select the most appropriate measurement method:

- The *Agenda 2030* general vision—"achieving sustainable development in its three dimensions—economic, social and environmental"—describes not only the development process but, firstly, the result achieved. Thus, respect is given to the principle already defined in the 1st Human Development Report [13]: "Human development has two sides: the formation of human capabilities—such as improved health, knowledge and skills—and the use people make of their acquired capabilities". The indirect measurement method makes it possible to create the AHDI by aggregating the indicators characterizing the results achieved in various dimensions of sustainable human development; this opportunity should be appreciated.
- Assuming that any of the 17 SDGs, 169 targets and 232 indicators is, to some extent, the driving force of the *Agenda 2030* vision, it is necessary to consider all indicators in the sustainable human development measurement process, if the bottom-up method is used. Many of these indicators could be insignificant and could be dropped, but which ones? It would be worthwhile to determine them using data mining methods, but before that, data on all indicators are necessary. While data for many indicators

are not available, several indicators are not quantifiable at all, this is a fundamental obstacle to obtaining an adequate measurement. Applying the top-down method does not result in such a problem if appropriate result-oriented indices are used to describe dimensions of sustainable human development.

- The interlinkages and integrated nature of the SDGs are of crucial importance when implementing the *Agenda 2030* vision; an indivisible, integrated approach is one of the basic principles of *Agenda 2030*: "As already pointed out, many of the goals included in the SDGs are multidimensional, covering more than one dimension of sustainability. Many synergies and complementarities can exist among the different SDGs. But trade-offs are also possible where improvements in one dimension could trigger negative results in another" [36]. When aggregating many individual indicators in a composite index, these features may be lost. When working with the resulting data, they will be considered.

The postulates show that the use of the direct measurement method is associated with great difficulties in creating the composite index for measuring sustainable human development, which will "be helpful in setting policy priorities and in benchmarking or monitoring performance" [56], and will be understandable, transparent, acceptable and practicable for experts, politicians and the global community. The existing offers confirm that there is a long way to go in solving various problems, and there is no guarantee of a successful result.

Accounting for the SDGs' and indicators' crosslinks is of crucial importance. The final documents of the SDG Summit [30] reiterate the necessity of a systemic and all-inclusive approach: "Focusing on one goal or target at a time and working in silos is therefore unlikely to result in overall success. Transforming to a socially, economically and ecologically sustainable society requires strengthening the focus toward all the SDGs and ensuring cooperation among the different measures or levers of action". Furthermore: "The widely accepted *ceteris paribus* condition, which analyses the behavior of each system independently, cannot be considered helpful in the *Agenda 2030* context. More holistic approaches are required as the contextual evaluation of several systems together is fundamental" [36].

Experience shows that the indirect measurement method (used by both GDP and HDI) has been popular and practically applicable. "Top-down approach is less time and resource intensive and has been generally applied at larger geographical scale; ... it is driven by the opinions of experts/researchers, through which the framework is designed based on broad common issues" [57]. This is exactly what is needed to carry out sustainable human development planning, monitoring and evaluation tasks on a global and national scale.

The indirect measurement method also allows minimizing potential discussion of one's or another country's national priorities and interests within the scope of the SDGs, targets and indicators. The core dimensions of *Agenda 2030*—economic, social (i.e., life and education) and environmental—are important for any country and society.

Overall, it shows that the indirect measurement of sustainable human development offers a much wider perspective. We selected the indirect measurement method for creation of the methodology and measurement tool, the AHDI.

4. Methodology and the AHDI Calculation

Similarly, observing the principles of *Agenda 2030* and the experience gained in measuring human development, we put forward several postulates to generate the framework for the measuring methodology and selection of the environmental dimension indicator:

- Heredity, evolution rather than revolution; in creating the HDI as a substitute of GDP, the very popular GDP was not discarded. It was included in the set of HDI dimension indexes, supplemented by the longevity and knowledge indexes. There is neither reason nor need to stop using this principle in the future. Usage of the HDI, which is globally accepted at the expert and political level, is expedient in the creation of the AHDI to describe social and economic dimensions. Significantly, all described indirect measurement proposals are designed exactly by supplementing the HDI with some environmental indicators.

- Simplicity of the indicator and the use of few dimension indices provides understanding to non-economists, politicians and the community, which is important for political decision-making. The main advantage of GDP, which has enabled it to gain a strong political acceptance and become "the most powerful statistical figure in human history" [41], is its simplicity of use. Likewise, "the HDI's simplicity, coupled with the transparency assured by the utilization of data published by international organizations, has been one of the main drivers behind the success of the HDI in the past twenty years" [58]. The aspect of simplicity in creating the AHDI should not be underestimated; the design of AHDI, using only a few indicators, will contribute to its popularity. Only one comprehensive headline indicator should be chosen to characterize the environmental dimension.
- An important feature of HDI's design is its unity and concordance of all three dimension indices; this is confirmed by strong positive correlations between the indices (Table 1). In turn, the HDI methodology guarantees a balanced effect of the dimension indices on the HDI value. An equal weight of united and consistent dimension indicators in the AHDI is the best decision following *Agenda 2030*'s politically accepted settings on development in a "balanced and integrated manner".
- Any dimension indicator should provide the fullest possible coverage of the aspects of the *Agenda 2030* dimension concerned. The chosen environmental indicator should also cover various aspects of environment pollution, degradation and preservation; given the diversity of environmental aspects, an integrated dimension index will have to be used in practice.
- There is no rational reason for developing a new comprehensive environmental index; instead, it is desirable to use a stable existing index that is elaborated by high-level experts and that is already politically accepted.

Table 1. Interdimensional correlations of HDI 2020.

	Income Index	Life Index	Education Index
Income index	xxx	0.8411	0.8675
Life index		xxx	0.8193
Education index			xxx

Table 2 summarizes the compliance of the proposed indices for sustainable human development measurement with the proposed postulates.

All proposed indices are based on the HDI expanding, nevertheless there is no such consensus on other postulates. However, new constructions of ecological indicators do not fully cover the problems of environmental pollution and degradation, while unduly complicating the index; EF is, by definition, contradictory to economic development.

The HSDI and EHDI methodologies are most appropriate for the defined set of postulates. The chosen narrow-profile environmental indicator (CO_2 per capita) is a shortcoming of the HSDI, while the weak point of the EHDI is a dimensional imbalance, as the impact of the environmental index on the EHDI significantly exceeds that of other dimension indices. Possible further advancement is achievable by an integration of the strengths of both methodologies: the HSDI methodology that provides a balance of all dimensions in the index by partial opening HDI calculation as well as by the EHDI methodology that uses the EPI as an environmental indicator, providing incomparably wider coverage of the environmental aspects in comparison with that of the other options.

To bring the HDI in line with today's concept of sustainability (balanced long-term integration of economic, social and environmental dimensions), its existing dimension base had been complemented by an environmental dimension (Figure 4). The AHDI methodology was elaborated by combining the advantages of EHDI and HSDI, thus achieving a significantly higher compliance of the AHDI with the principles of the *Agenda 2030*. In doing so, the HDI calculation methodology [45] was used precisely to maintain the multidimensional balance already achieved. We rely on the expertise and authority of the UN, so we accept the calculation methodology of the HDI as time-tested and sufficiently stable.

Table 2. Compliance of the proposed indices with the defined postulates.

Postulates	HSDI	EWP	SDI	EHDI
Heredity, usage of the globally accepted HDI	2	2	1	2
Simplicity, a small set of "headline indicators"	2	2	0	2
Balanced and integrated, united and consistent set of dimension indicators	2	0	0	1
Wide coverage of environment pollution, degradation and preservation	0	1	1	2
Existing and politically accepted environmental index/indicator	2	1	0	2

HSDI—The Human Sustainable Development Index; EWP—The Ecological Well-being Performance indicator; SDI—The Sustainable Development Index; EHDI—The Environmental Human Development Index; (2—good, 1—moderate, 0—does not comply).

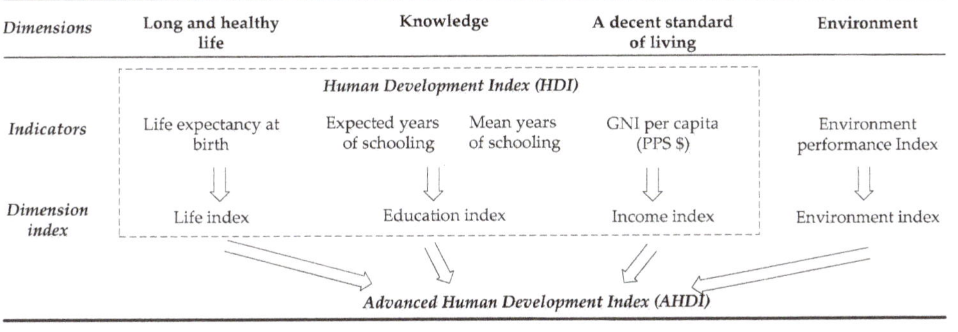

Figure 4. Design of the AHDI. Created by the authors.

By selecting the EPI as an environmental index (wide coverage, accepted index) for expanding HDI (simplicity, result-orientated indices) and using the HSDI's methodological principle for composing the index (balance), we obtain the formula of AHDI:

$$\text{AHDI} = (\text{I}_{\text{Life}})^{1/4} \times (\text{I}_{\text{Education}})^{1/4} \times (\text{I}_{Income})^{1/4} \times (\text{I}_{\text{Environment}})^{1/4}. \quad (7)$$

The EPI values must also be normalized analogously to other dimension indices on a single scale between 0 and 1; the min–max normalization is used as in HDI. Additionally, as in the case of the HDI, to keep the EPI index constant for new entrants and for several years, minimum and maximum threshold values (goalposts) have been adopted with some reserve (Table 3).

Table 3. Threshold values for normalization.

Indicators	Minimum		Maximum	
	Actual National Values	Threshold Values	Actual National Values	Threshold Values
Life expectancy at birth (years)	53.3	20	83.6	85
Expected years of schooling (years)	5	0	22	18
Mean years of schooling (years)	1.6	0	14.2	15
Gross national income per capita ($ PPS 2017)	754	100	88,155	75,000
Environment Performance Index	22.6	20	82.5	90

5. Results and Discussion

The aggregated AHDI values were calculated applying formula (7) for 178 countries, the data of which are included in both the HDI 2020 and EPI 2020 reports (Figure 5).

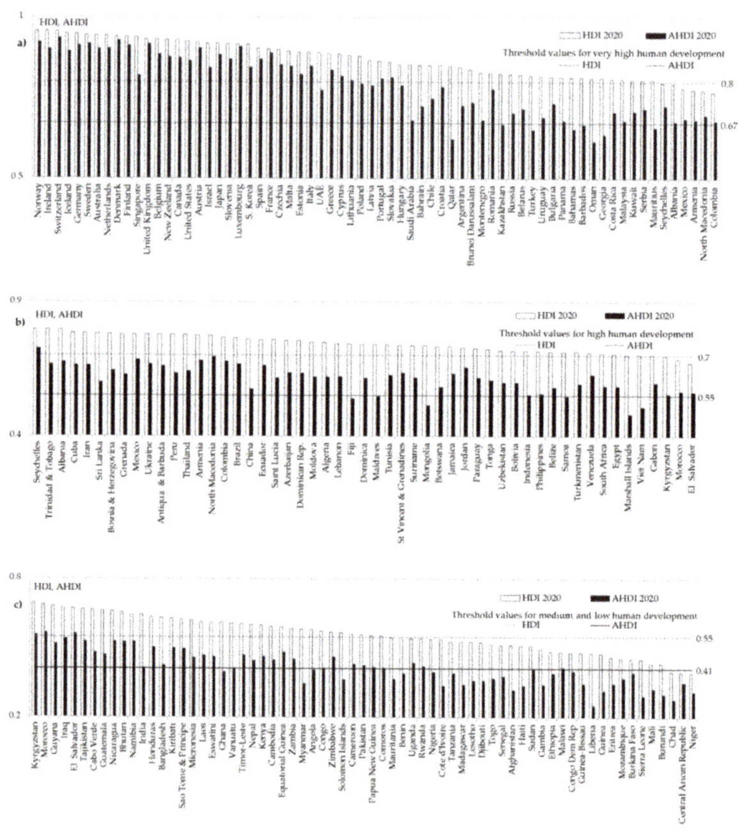

Figure 5. AHDI ranking. (**a**) very high development; (**b**) high development level; (**c**) medium and low development level achieved. Authors' calculations.

There are strong positive Pearson correlation coefficients between AHDI and dimension indices (Table 4); all values show a remarkable impact of all dimensions on the AHDI. At the same time, the similarity of correlations to those in the HDI show that potential data redundancy would not create problems.

Table 4. Correlation between dimension indices and composite indexes.

Composite Indexes	Dimension Indices			
	Income	Life	Education	Environment
HDI	0.9589	0.9113	0.9607	xxx
AHDI	0.9281	0.8841	0.9359	0.9614

Table 5 indicates the strong correlations between dimension indices; it shows that there are no contradictions between the dimension indices. This is a very substantial positive feature of the AHDI as the functions of all dimension indices are integrated; there is no need to compromise with others when improving one of them.

Table 5. AHDI interdimensional correlations and VIFs.

Dimension Indexes	Income Index	Life Index	Education Index	Environment Index
	Pearson Correlation Coefficients			
Income index	xxx	0.8411	0.8675	0.8345
Life index	3.42	xxx	0.8193	0.7981
Education index	4.04	2.91	xxx	0.8462
Environment index	3.29	2.75	3.52	xxx
	Variance Inflation Factors (VIF)			

At the same time, the cross-independence level of the dimension indices is not so high as to cause unreliability by usage of interdependent indicators (so-called multicollinearity problem). To examine the potential problem, we used the typical tool, variance inflation factors (VIFs), which characterizes the mutual correlation between any pair of dimension indices (see, e.g., [59]). The reliability criterion is

$$\text{VIF} = 1/(1 - R^2) < \delta.$$

There are various evaluations of the critical value δ; however, in general, VIF values greater than 5 are considered moderately (undesirably) high. In our case, the maximum VIF value (4.04) satisfies even conservative recommendations.

Figure 5 shows a relatively different impact of the EPI on assessment of human development. For 55 countries, the difference of position in the AHDI ranking compared to that in the HDI ranking is 3 positions or less, which can be considered as statistically insignificant. In these countries, environmental policy keeps pace with the country's socioeconomic development.

Inclusion of the environmental dimension in the AHDI has shifted another 55 countries down in the ranking. In these countries, the environmental policy and activities do not meet the level of human development that is shown by the three-dimensional HDI.

We carried out the grouping of countries into categories according to the level of sustainable human development achieved (very high, high, medium, low); the selection of cutoff points was performed using the HDI methodological principles (Table 6). As shown, 18 countries moved to a lower development category when compared by HDI ranking.

Table 6. Cutoff points and grouping of countries by HDI and AHDI rankings.

Categories of Development	Cut Off Points		Number of Countries			
	HDI	AHDI	HDI	Shift Down	Shift Up	AHDI
Very high			62			60
	0.8	0.67		8↓	6↑	
High			50			49
	0.7	0.55		6↓	3↑	
Medium			36			38
	0.55	0.41		4↓	3↑	
Low			31			32

A more detailed analysis shows that for the eight countries that were excluded from the very high development category (Table 7), many of the positions in the global rankings per several key environment indicators (components of the EPI) are only in the second hundred. This confirms that the downward shift is reasonable; such pollution and ecosystem degradation levels cannot be considered as a very high sustainable human development feature.

Table 7. Ranks of the countries that have lost positions in the very high development category.

Country	Waste Management		Air Quality		Pollution							Ecosystem	
					Growth Rates								
	Solid Waste	Waste-Water	Ozone	PM2.5	CO_2	SO_2	NO_x	CH_4	N_2O	Black Carbon	CHG pc	Nitrogen	Biodiversity
Bahamas	133	100			137	168	162	107	125	172	131	168	
Barbados		104				159				164		170	177
Georgia	133		113	119	109	176	174		130	148		163	126
Kazakhstan	124		167	138				177	105	127	170		128
Mauritius						123	101	131	131			154	170
Oman	133		135	171	134		157	133	172	172	168		159
Qatar			150	169	160		134	177	174	172	172	134	129
Turkey	113		160	117	110				174				175

6. Conclusions

Although the existing proposed indices for measuring sustainable human development seem insufficient to accurately measure the implementation of the *Agenda 2030* vision in its comprehensive sense, their strengths and weaknesses strongly helped to improve the measuring methodology and tool. The performed integrated investigation of the key principles of *Agenda 2030* and experience gained in measuring human development provided an opportunity to combine the strengths of existing proposals for measuring sustainable human development and to eliminate their weaknesses.

The selected indirect measurement method allows, in real cases of insufficient data availability, to reflect the results achieved in the implementation of the *Agenda 2030* vision, including accounting for the mutual relationships and the inter-impact of the development dimensions. The set of postulates formulated in the study is the basis for the elaborated optimum measurement procedure that is consistent with the *Agenda 2030* vision on sustainable human development.

The created measurement methodology ensures the balance and integrity of economic, social and environmental dimensions of the *Agenda 2030* vision. An appropriate tool, the Advanced Human Development Index, was designed. This provides unity and concordance of all included, well-known and accepted comprehensive dimension indices.

Figure 6 shows that the causality between final energy consumption and the Advanced Human Development Index is strong; the AHDI is usable for measuring the linkage between supplied energy and sustainable development level in countries worldwide. Further, to reiterate, sustainable development also asks for energy.

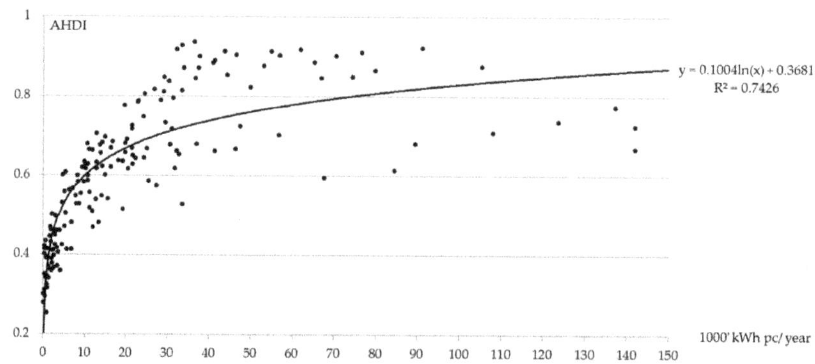

Figure 6. Advanced human development index vs. final energy consumption per capita (2019 or the last available year); sources—Our World in Data and authors' calculation.

Two groups of countries are visible. For some countries, energy is an efficient driver of the sustainable development; an increase in energy consumption above 20,000 kWh pc/year has enabled further growth in sustainability of development, reaching very high levels of AHDI of around 0.9. These countries are realizing sustainable energy policies and effective actions to improve energy security and efficiency, with an emphasis on environment friendly practice. However, growth in energy consumption above 40,000 kWh pc/year is not directly related to an increase in SD level.

For other countries, energy is a less efficient driver of sustainable development. At energy consumption above 20,000 kWh pc/year, the growth of SD slows down; it is hampered by growing environment pollution problems (see Table 7); even at consumption above 100 kWh pc/year, these countries do not reach an AHDI level of 0.8.

AHDI calculations confirm the accuracy and simplicity of the measurement, as well as its usability for experts, politicians and the community. They show objective and reasonable shifts in conventional human development assessments due to the inclusion of the environment component. The countries with unilateral development policies and practices cannot claim sustainability.

For five years now, the *Agenda 2030* has been pursued without a sound understanding of the overall trend at both national and global levels. The proposed AHDI, as a simple balanced index that is based on result-oriented headline indices, provides the big picture, which will be understandable, transparent, acceptable and usable for experts, politicians and the global community. The COVID-19 pandemic has sharply exacerbated the need for an efficient general assessment of the performance of national recovery policies. The developed measurement procedure makes it possible to meet these needs immediately.

Author Contributions: Conceptualization and methodology, E.K.; data analysis, E.K., G.K., J.B.; writing—original draft preparation, E.K., J.B., G.K.; writing—review and editing, E.K., G.K., J.B.; visualization, E.K., G.K.; resources, supervision and project administration, G.K.; funding acquisition, E.K., G.K., J.B. All authors have read and agreed to the published version of the manuscript.

Funding: This research was funded by Latvian Council of Science, grant Nr. lzp-2020/2-0073, University of Latvia Nr. LZP2020/69.

Institutional Review Board Statement: Not applicable.

Informed Consent Statement: Not applicable.

Data Availability Statement: Data are publicly available in indicated sources.

Conflicts of Interest: The authors declare no conflict of interest.

References

1. Tvaronaviciene, M. Towards sustainable and secure development: Energy efficiency peculiarities in transport sector. *J. Secur. Sustain. Issues* **2018**, *7*, 719–725. [CrossRef]
2. Kaya, Y.; Yokobori, K. (Eds.) Energy and Economy. In *Global Energy Assessment (Toward a Sustainable Future)*; Cambridge University Press: Cambridge, UK; New York, NY, USA, 2012; pp. 385–423.
3. Opinion of the European Economic and Social Committee on the 'Proposal for a Decision of the Council on a Supplementary Research Programme for the ITER Project (2014–2018). Available online: https://eur-lex.europa.eu/LexUriServ/LexUriServ.do?uri=OJ:C:2012:229:0060:0063:EN:PDF (accessed on 20 December 2020).
4. International Institute for Applied Systems Analysis. Available online: https://iiasa.ac.at/web/home/research/researchPrograms/Energy/About-Energy-Program.en.html (accessed on 20 December 2020).
5. Santoyo-Castelazo, E.; Azapagic, A. Sustainability assessment of energy systems: Integrating environmental, economic and social aspects. *J. Clean. Prod.* **2014**, *80*, 119–138. [CrossRef]
6. Birol, F.; International Energy Agency. Energy Is at the Hearth of the Sustainable Development Agenda to 2030. 2018. Available online: https://www.iea.org/commentaries/energy-is-at-the-heart-of-the-sustainable-development-agenda-to-2030 (accessed on 20 December 2020).
7. Oxford Learner's Dictionaries. Available online: https://www.oxfordlearnersdictionaries.com (accessed on 20 December 2020).
8. Rabie, M. *A Theory of Sustainable Sociocultural and Economic Development*; Palgrave Macmillan: London, UK, 2016; p. 243.
9. Coyle, D. *GDP; a Brief but Affectionate History*; Princeton University Press: Princeton, NJ, USA; Oxford, MS, USA, 2015; p. 184.
10. Kennedy, J.F. Remarks at the University of Kansas. 1968. Available online: https://www.jfklibrary.org/learn/about-jfk/the-kennedy-family/robert-f-kennedy/robert-f-kennedy-speeches/remarks-at-the-university-of-kansas-march-18-1968 (accessed on 20 December 2020).

11. Seers, D. What are We Trying to Measure? In *Measuring Development; the Role and Adequacy of Development Indicator*; Baster, N., Ed.; Routledge: Abingdon, UK, 1972; pp. 21–36.
12. Morse, S. *Indices and Indicators in Development: An Unhealthy Obsession with Numbers*; Earthscan Publications: London, UK, 2013; p. 224.
13. UN Development Programme. Human Development Report 1990: Concept and Measurement of Human Development. Available online: http://hdr.undp.org/en/reports/global/hdr1990 (accessed on 20 December 2020).
14. de Boer, B.; Hueting, R. Sustainable National Income and Multiple Indicators for Sustainable Development. In *Measuring Sustainable Development; Integrated Economic, Environmental and Social Frameworks*; OECD Publishing: Paris, France, 2004; pp. 39–52.
15. United Nations Sustainable Development; Agenda 21. UN General Assembly. Available online: https://sustainabledevelopment.un.org/content/documents/Agenda21.pdf (accessed on 20 December 2020).
16. Resolution Adopted by the General Assembly on 25 September 2015. Transforming Our World: The 2030 Agenda for Sustainable Development. UN General Assembly. Available online: https://www.un.org/ga/search/view_doc.asp?symbol=A/RES/70/1&Lang=E (accessed on 20 December 2020).
17. Jacob, A. Mind the Gap: Analysing the Impact of Data Gap in Millennium Development Goals' (MDGs) Indicators on the Progress toward MDGs. *World Dev.* **2017**, *93*, 260–278. [CrossRef]
18. Balkyte, A.; Tvaronaviciene, M. Perception of competitiveness in the context of sustainable development: Facets of "Sustainable competitiveness". *J. Bus. Econ. Manag.* **2010**, *11*, 341–365. [CrossRef]
19. Cizo, E.; Lavrinenko, O.; Ignatjeva, S. Analysis of the relationship between financial development and economic growth in the EU countries. *Insights Reg. Dev.* **2020**, *2*, 645–660. [CrossRef]
20. Worldwide Cost of Living 2020: How is Covid-19 Affecting the Prices of Consumer Goods? The Economist Intelligence Unit. Available online: https://www.eiu.com/ (accessed on 20 December 2020).
21. Sachs, J.; Schmidt-Traub, G.; Kroll, C.; Lafortune, G.; Fuller, G.; Woelm, F. *Sustainable Development Report 2020; The Sustainable Development Goals and COVID-19*; Cambridge University Press: Cambridge, UK, 2020; pp. 1–99.
22. Saltelli, A. Composite Indicators between Analysis and Advocacy. *Soc. Indic. Res.* **2007**, *81*, 5–77. [CrossRef]
23. Measuring Progress. In *Prototype Global Sustainable Development Report*; UN Department of Economic and Social Affairs, Division for Sustainable Development: New York, NY, USA, 2014; pp. 73–92. Available online: http://sustainabledevelopment.un.org/globalsdreport/2014 (accessed on 20 December 2020).
24. UN Department of Economic and Social Affairs. Sustainable Development Goal 17. Available online: https://sdgs.un.org/goals/goal17 (accessed on 20 December 2020).
25. Costanza, R.; Daly, L.; Fioramonti, L.; Giovannini, E.; Kubiszewski, I.; Mortensen, L.F.; Pickett, K.E.; Ragnarsdottir, K.V.; De Vogli, R.; Wilkinson, R. Modelling and Measuring Sustainable Wellbeing in Connection with the UN Sustainable Development Goals. *Ecol. Econ.* **2016**, *130*, 350–355. [CrossRef]
26. Bewoor, A.K.; Kulkarni, V.A. *Metrology and Measurement*; Tata McGraw-Hill Education: New Delhi, India, 2009; pp. 1–21.
27. Resolution Adopted by the General Assembly on 6 July 2017: Work of the Statistical Commission Pertaining to the 2030 Agenda for Sustainable Development. UN General Assembly. 2017. Available online: http://ggim.un.org/documents/A_RES_71_313.pdf (accessed on 20 December 2020).
28. *Sustainable Development in the European Union; Monitoring Report on Progress towards the SDGs in an EU Context*, 2020 ed.; Publications Office of the European Union: Luxembourg, 2020; pp. 1–34.
29. Eurostat. Sustainable Development Indicators. Available online: https://ec.europa.eu/eurostat/data/database (accessed on 20 December 2020).
30. SDG Summit 2019; Summary of the President of the General Assembly. UN General Assembly. 2019. Available online: https://sustainabledevelopment.un.org/content/documents/25200SDG_Summary.pdf (accessed on 20 December 2020).
31. OECD. Measuring Distance to the SDG Targets; an Assessment of where OECD Countries Stand. 2017. Available online: http://www.oecd.org/sdd/OECD-Measuring-Distance-to-SDG-Targets.pdf (accessed on 20 December 2020).
32. Moyer, J.D.; Hedden, S. Are we on the right path to achieve the sustainable development goals? *World Dev.* **2020**, *127*, 104749. [CrossRef]
33. Flint, R.W. Practice of Sustainable Community Development. In *Basics of Sustainable Development*; Springer: New York, NY, USA; Heidelberg, Germany, 2013; pp. 25–54.
34. Bennich, T.; Weitz, N.; Carlsen, H. Deciphering the scientific literature on SDG interactions: A review and reading guide. *Sci. Total Environ.* **2020**, *728*, 138405. [CrossRef] [PubMed]
35. Barbier, E.B.; Burgess, J.C. Sustainable development goal indicators: Analyzing trade-offs and complementarities. *World Dev.* **2019**, *122*, 295–305. [CrossRef]
36. Miola, A.; Schiltz, F. Measuring sustainable development goals performance: How to monitor policy action in the 2030 Agenda implementation? *Ecol. Econ.* **2019**, *164*, 106373. [CrossRef] [PubMed]
37. Papadimitriou, E.; Neves, A.R.; Becker, W. *JRC Statistical Audit of the Sustainable Development Goals Index and Dashboards [Technical Report]*; Joint Research Centre. Publication Office of the EU: Luxembourg, 2019; pp. 1–30.
38. Eustachio, J.H.; Caldana, A.C.; Liboni, L.B.; Martinelli, D.P. Systemic Indicator of Sustainable Development: Proposal and Application of a Framework. *J. Clean. Prod.* **2019**, *241*, 118383. [CrossRef]

39. Kynclova, P.; Upadhyaya, S.; Nice, T. Composite Index as a Measure on Achieving Sustainable Development Goal 9 (SDG-9) Industry-related Targets: The SDG-9 Index. *Appl. Energy* **2020**, *265*, 114755. [CrossRef]
40. Callen, T.; International Monetary Fund. Gross Domestic Product: An Economy's All. Available online: https://www.imf.org/external/pubs/ft/fandd/basics/gdp.htm (accessed on 20 December 2020).
41. Fioramonti, L. *Gross Domestic Problem: The Politics Behind the World's Most Powerful Number*; Zed Books: London, UK; New York, NY, USA, 2013; p. 208.
42. Lepenies, P. *The Power of a Single Number, a Political History of GDP*; Columbia University Press: New York, NY, USA, 2016; p. 208.
43. Klasen, S.; UN Human Development Report Office. Human Development Indices and Indicators: A Critical Evaluation [Background Paper]. 2018. Available online: http://hdr.undp.org/sites/default/files/klasen_final.pdf (accessed on 20 December 2020).
44. Mazzanti, M.; Mazzarano, M.; Pronti, A.; Quatrosi, M. Fiscal policies, public investments and wellbeing: Mapping the evolution of the EU. *Insights Reg. Dev.* **2020**, *2*, 725–749. [CrossRef]
45. UN Development Programme. Human Development Report 2020; Statistical Annex. Available online: http://hdr.undp.org/sites/default/files/hdr2020.pdf (accessed on 20 December 2020).
46. Prakash, R.; Garg, P. Comparative assessment of HDI with Composite Development Index (CDI). *Insights Reg. Dev.* **2019**, *1*, 58–76. [CrossRef]
47. Togtokh, C.; Gaffney, O.; UN University. Human Sustainable Development Index. 2010. Available online: https://ourworld.unu.edu/en/the-2010-human-sustainable-development-index (accessed on 20 December 2020).
48. Bravo, G. The Human Sustainable Development Index: New Calculations and a First Critical Analysis. *Ecol. Indic.* **2014**, *37*, 145–150. [CrossRef]
49. World Wildlife Fund. Ecological Footprint. 2020. Available online: https://wwf.panda.org/knowledge_hub/all_publications/ecological_footprint2/ (accessed on 20 December 2020).
50. Zhu, D.; Zhanga, S.; Sutton, D.B. Linking Daly's Proposition to Policymaking for Sustainable Development: Indicators and Pathways. *J. Clean. Prod.* **2015**, *102*, 333–341. [CrossRef]
51. Hickel, J. The Sustainable Development Index: Measuring the Ecological Efficiency of Human Development in the Anthropocene. *Ecol. Econ.* **2020**, *167*, 106331. [CrossRef]
52. Environmental Performance Index. 2020. Available online: https://epi.yale.edu/downloads (accessed on 20 December 2020).
53. Papadimitriou, E.; Neves, A.R.; Saisana, M. *JRC Statistical Audit of the 2020 Environmental Performance Index [Technical Report]*; Joint Research Centre, Publication Office of the EU: Luxembourg, 2020; p. 24.
54. Maccari, N. Environmental Sustainability and Human Development: A Greening of Human Development Index. 2014. Available online: https://ssrn.com/abstract=2426073 (accessed on 20 December 2020).
55. Pinter, L.; Hardi, P.; Bartelmus, P. Sustainable Development Indicators; Proposals for the Way Forward (Discussion Paper). International Institute for Sustainable Development. 2005. Available online: https://www.iisd.org/pdf/2005/measure_indicators_sd_way_forward.pdf (accessed on 20 December 2020).
56. *Handbook on Constructing Composite Indicators*; OECD Publications: Paris, France, 2008; Available online: https://www.oecd.org/sdd/42495745.pdf (accessed on 20 December 2020).
57. Kwatra, S.; Kumar, A.; Sharma, P. A Critical Review of Studies Related to Construction and Computation of Sustainable Development Indices. *Ecol. Indic.* **2020**, *112*, 106061. [CrossRef]
58. Klugman, J.; Rodríguez, F.; Choi, H. The HDI 2010: New Controversies, Old Critiques. *J. Econ. Inequal.* **2011**, *9*, 249–288. [CrossRef]
59. Hair, J.F.; Babin, J.B.; Anderson, R.E.; Black, W.C. Multivariate Data Analysis. In *Multiple Regression*, 8th ed.; Cengage Learning EMEA: London, UK, 2018; pp. 246–314.

energies

MDPI

Article

Modeling and Management of Power Supply Enterprises' Cash Flows

Darya Pyatkina [1,2], Tamara Shcherbina [3], Vadim Samusenkov [4], Irina Razinkina [5] and Mariusz Sroka [6,*]

1 Department of Applied Informatics and Probability Theory, Peoples' Friendship University of Russia (RUDN University), 117198 Moscow, Russia; pyatkina-da@rudn.ru or dapyatkina@fa.ru
2 Department of Data Analysis and Machine Learning, Financial University under Government of the Russian Federation, 115583 Moscow, Russia
3 Department of Corporate Finance and Corporate Governance, Financial University under the Government of the Russian Federation, 115583 Moscow, Russia; gero-tamara@yandex.ru
4 Department of Prosthetic Dentistry, The State Education Institution of Higher Professional Training, I.M Sechenov First Moscow State Medical University, 119991 Moscow, Russia; croc@bk.ru
5 Department of Management and Innovation, Financial University under the Government of the Russian Federation, 115583 Moscow, Russia; Irina-razinkina@yandex.ru
6 Faculty of Management, Technical University of Czestochowa, 42-201 Czestochowa, Poland
* Correspondence: mariusz.sroka@pcz.pl

Citation: Pyatkina, D.; Shcherbina, T.; Samusenkov, V.; Razinkina, I.; Sroka, M. Modeling and Management of Power Supply Enterprises' Cash Flows. *Energies* **2021**, *14*, 1181. https://doi.org/10.3390/en14041181

Academic Editor: Manuela Tvaronavičienė

Received: 17 January 2021
Accepted: 14 February 2021
Published: 23 February 2021

Abstract: The purpose of the study is to assess the efficiency of cash flow management at power supply companies of the CIS (Commonwealth of Independent States) countries. A methodological approach to cash flow forecasting with the use of linear and polynomial regression has been developed. The study is based on the data provided by 12 power supply companies operating in CIS member countries. Forecasting based on the generated polynomial models of multiple regression of cash flow for the power supply companies under study confirms the strong possibility of extrapolating the studied trends to future periods. Compared to the linear model, the polynomial one confirms higher values of the determination coefficients for the majority of power supply companies. The projected volumes of cash inflow, cash outflow, and net cash flows of power supply companies with the application of the described polynomial multiple regression models have a fairly high degree of approximation. The correlations between operating cash flows and outflows, between total cash inflow and outflow of the majority of power supply companies are high. The low level of synchronization between cash inflows and outflows of the companies under study is associated with the specifics of their financial and investment activities and the cash flow management policy. It has been proven that energy enterprises' financial stability significantly depends on the synchronization and uniformity of cash flows. The proposed methodological approach allows identifying enterprises by the criterion of riskiness from the standpoint of the synchronization and homogeneity of their cash flows.

Keywords: forecasting; linear regression; planning; polynomial regression; power supply

1. Introduction

Currently, power supply companies are an integral component of the energy system. Economic stability as well as national and international energy security depend on the reliability of their operation. The transition to new systems of incentive tariff setting, the implementation of which affects the size and structure of cash flows of power supply companies in different countries and regions, requires a proper theoretical and methodological justification [1–6].

The efficiency of enterprise cash flow management determines their ability to achieve current financial goals as well as ensure solvency, financial stability, and balance in the long term [7]. In the context of the highly dynamic internal and external environment of enterprises, a decrease in their solvency and financial stability is an urgent problem of imbalance and unevenness of cash flows. The economic and legal consequence of this is the insolvency and bankruptcy of a significant number of power supply enterprises [8]. These problems in a large number of enterprises indicate that they do not have an effective cash flow planning system. The management of enterprises gives little attention to cash flow management both at the operational and strategic levels. Cash flow planning is a practical tool that allows companies to ensure the synchronization of cash flows [9].

According to the modern scientific paradigm of corporate finance, cash flows are one of the key factors in the formation of enterprise value. The greater volume of cash flows generated by the assets of the enterprise contributes to its greater market value. Maximization of the market value of enterprises and the welfare of shareholders is a strategic goal of the enterprise financial management [10]. In this regard, the study is aimed at the improvement of practical tools for planning and managing cash flows to ensure a high level of solvency, achieve financial balance, and increase the market value of power supply companies in the short and long term.

2. Literature Review

The issue of determining the economic nature of cash flows is extremely important in the theoretical and practical aspect, as the existing approaches to the interpretation of the essence of cash flows will further determine the foundations of their planning and control. In addition, the issue of determining the place and role of cash flows in corporate finance management in order to improve the current system of financial management of enterprises is still relevant [11]. As evidenced by financial theory and practice, there is no unified approach for determining the economic essence of the definition of enterprise cash flow, as there is a controversy among scientists and practitioners over this issue [12].

Any financial transaction can be associated with a certain cash flow, that is, a set of payments (outflows) and receipts (inflows) of funds distributed over time and arising from managerial decisions within operating, investing, and financial activities [13]. Semantically, cash flow is a quantitative expression of the money that an enterprise has at a particular point of time. From the perspective of an investor, this is the expected return on investment. From the point of view of business leaders, cash flow is a projection of cash flow, etc. In either case, the dynamics of the company cash flow is a continuous process [14]. However, this approach to understanding the essence of cash flows is too fragmentary and does not allow reflecting all aspects of this phenomenon.

The amount of money that an enterprise needs depends on the degree of predictability of both revenue receipts from economic activities and payments (demand) to suppliers and employees. Cash flow is the receipt of funds that allows the company to meet the demand for them [15]. An alternative to this is the availability of external investors who are willing to finance any cash deficit. However, in order to attract external investment, an enterprise has to confirm its ability to achieve a positive cash flow that will allow it to pay interest and ultimately pay the investors back [16]. There are three approaches to the essence of enterprise cash flows: static, dynamic, and aggregated. Each of these approaches discloses the essence of cash flows taking into account their specific characteristics. The advocates of the static approach define cash flow as the net cash flow in a given period of time, while at the dynamic level, cash flow is the movement of funds [17].

Cash flow planning, on the one hand, is the process of developing a system of projects and targets for the development of various types of these flows for the operating, investing, and financial activities of an enterprise in the next period. On the other hand, it is a set of measures and tools for forecasting and managing the cash flow cycle, which consists of two parts: projected cash inflows and outflows [18]. Cash flow planning can be viewed as the process of developing cash flow plans for various activities with the help of specific means and tools. It involves the creation of a system of projected financial performance indicators, on the basis of which cash flow plans are drawn up and deviations from the desired values are monitored (estimated, identified, and corrected) [19]. The cash flow planning system of enterprises consists of three levels: strategic (long-term) planning; current (tactical) planning; and operational planning. The system of strategic cash flow planning is designed to develop and forecast the most important targets for the development of cash flows. Tactical cash flow planning aims to ensure that negative and positive cash flows are balanced and synchronized throughout the year. Tactical planning of enterprise cash flows involves two major documents related to the income and expenditure budget, and the cash flow budget. Based on the cash flow budget, it is possible to determine the projected cash inflow and outflow of the company; it allows the assessment of the structure of settlements with suppliers and buyers within a given period [20].

The initial aspect of the formation of a cash flow management model for a power supply enterprise is the definition of its strategic goal [21]. Based on the principles of building a cash flow management system, this goal should be subordinated, on the one hand, to the implementation of the main goal of financial management (meeting stakeholders' interests) [22,23], and on the other hand, to ensure the implementation of an enterprise's general development strategy [24]. When forming a strategy for managing cash flows of a power supply enterprise, conceptual aspects are determined, the implementation of which implies the achievement and implementation of the strategic goal of managing its cash flows [25]. The power supply enterprise's cash flow management strategy defines the main priorities that must be implemented in the process of managing its cash flows [26,27]. Key criteria for the effectiveness of cash flow management are identified. They are divided into components and are communicated to the employees of an enterprise who are responsible for them [28].

For the effective functioning of the cash flow management system of a power supply enterprise, making optimal management decisions, it is necessary to establish stably functioning information channels that will provide company's management with timely, reliable, and necessary information [29,30]. It should be noted that these should be both internal and external information channels, which are formed as a result of monitoring the environment of an enterprise [31–33].

Compliance with the time frames and volumes of payments on loans is an important component of control of outgoing cash flows of a power supply company [34,35]. If borrowed capital is attracted by issuing bonds, then the cash flow generated by this financial transaction is clearly determined in the bond issue prospectus. At the same time, it defines the time of coupon payments and the terms of bond redemption. The payment schedule on bonds is the basis for planning this type of initial cash flow [36]. At the same time, it also acts as a tool for controlling the initial cash flow from the financial activities of a power supply company. Compliance with it is an important component of ensuring the solvency of a power supply company [37,38]. An important factor that will influence the cash flows of power supply companies is the introduction of incentive tariff regulation [39,40].

To effectively control the intended use of investment funds, a power supply company can use current accounts with a special mode of use [41]. The accumulated investment resources have a designated purpose and can be used to finance the activities of the required investment program [42,43]. Power supply companies can use financial ratios based on cash flows as planning and cash flow control tools [44]. In particular, the planned values of the coefficients of cash flows can act as target indicators in the planning of cash flows [45]. At the same time, they can also be tools for controlling cash flows in the process of comparing actual values with planned ones [46]. To balance the cash flows of a power supply company in the long run, it is important to predict the occurrence of a cash gap, which manifests itself in the inability to fulfill the company's obligations at a certain point in time due to the lack of funds in the account [47]. The most common way to cover cash gaps is a short-term loan [48,49].

To ensure the proper and timely synchronization of cash flows, it is necessary to create an effective system of financial planning and forecasting of cash flows that involves cash balance assessment at the beginning of the period, maintenance of the balance of cash inflows and outflows over the current period, calculation of the deficit (or excess) cash flow and the possibility of its balancing, and support for the creation of a cash reserve for making payments at the end of the period [50]. The alignment of cash flows is aimed at smoothing their volumes in the context of individual intervals of the time period under consideration. The use of this assessment method makes it possible to eliminate the impact of seasonal and cyclical fluctuations on the cash flow creation, optimize the average cash balance, and increase liquidity indicators [51,52].

However, from the perspective of the applied challenges of our time, the issues of theoretical and methodological substantiation of enterprise cash flow planning and management are still found in the scientific and practical literature. The present paper aims to eliminate this gap in science by improving methodological tools for modeling and managing the cash flows of power supply companies. The purpose of the research is to assess the effectiveness of cash flow management of power supply companies in the CIS (Commonwealth of Independent States) countries in the context of planning and control by projecting their long-term values. To achieve this goal, the following research tasks have been set:

- To deepen the conceptual foundations of planning and control of the cash flow of enterprises;
- To reveal the methodological basis for cash flow planning and control;
- To analyze cash flows from operating, investing, and financial activities of power supply companies;
- To identify the analytical possibilities of using financial ratios to assess the cash flows of enterprises; and
- To develop an econometric model of balancing the cash flows of power supply companies and to forecast them.

At the same time, the study formulated the following mutually exclusive hypotheses:

Hypotheses 0 (H0). *The level of synchronization of cash flows does not depend on the level of financial stability and balance of power supply companies;*

Hypotheses 1 (H1). *The level of financial stability and equilibrium of power supply enterprises largely depends on the level of synchronization and uniformity of cash flows.*

3. Materials and Methods

In order to effectively analyze the cash flow from the operating activities of power supply companies, the study takes into account the fact that cash flow planning and forecasting processes should also be considered. Therefore, in the conceptual model of cash flow planning, it is assumed that it is feasible to start the process of planning cash flows from operating activities along with the development of a long-term forecast of operating cash flow for the period of 5 years: up to 2025. In order to make a long-term forecast on operating cash flow, the capabilities of Microsoft Excel were used. Operating cash flow forecasting differs from planning—it is a guideline of probabilistic nature. The Microsoft Excel "Forecast" function made it possible to predict the values of operating cash inflows and outflows of power supply companies for the next 5 years with the help of a linear approximation based on the data on their actual volumes for the period of 2015–2019. The projected values of the volume of cash flow from operating activities of power supply companies calculated in the study can be used as long-term indicators in the process of cash flow planning.

The planned volume of net cash flow from operating activities ($OCF_{net(p)}$) is determined by the formula:

$$OCF_{net(p)} = OCF_{in(p)} - OCF_{out(p)}, \tag{1}$$

where $OCF_{in(p)}$ is the planned volume of operating cash inflow; $OCF_{out(p)}$ is the planned volume of operating cash outflow.

The planned volume of cash inflow from operating activities is determined by the formula:

$$OCF_{in(p)} = EE_p + AE_p + I_p + D_p + TF_p + TE_p + OE_p, \tag{2}$$

where EE_p is projected revenue from electricity supplies; AE_p is projected revenue from advance payments; I_p is the planned volume of cash inflow from bank interest; D_p is the planned volume of cash inflow from debtors; TF_p is the planned amount of cash inflow from target financing; TE_p is the projected tax and obligatory payment refund; OE_p is the planned volume of other cash inflows from operating activities.

The planned volume of cash outflow from operating activities is determined by the formula:

$$OCF_{out(p)} = EP_p + AP_p + S_p + TP_p + OP_p, \quad (3)$$

where EP_p is projected payments for the purchase of electricity; AP_p is the planned volume of advance payments by the power supply company; S_p is the planned volume of employee payments; TP_p is the planned amount of tax, fee, and other mandatory payments; OP_p is the planned amount of other payments for operating activities.

The planned cash inflow from electricity supplies is determined as follows:

$$EE_p = \sum_{i=1}^{n} Q_i \times t_{i(plan)} \times K_{i(plan)} \quad (4)$$

where Q_i is the planned volume of supply to the ith consumer of electricity; $t_{i(plan)}$ is the average electricity rate for the ith consumer in the planning period; and $K_{i(plan)}$ is the planned payment ratio of the ith consumer. The planned payments for the purchase of electricity are determined as follows:

$$EE_p = Q_p \times t_{p(plan)} \times K_{p(plan)} \quad (5)$$

where Q_p is the total volume of electricity purchase in the planning period; $t_{p(plan)}$ is the planned weighted average electricity price in the planning period; and $K_{p(plan)}$ is the planned payment coefficient.

The proposed methodology for determining the planned volume of net cash flow from operating activities of power supply companies ($OCF_{net(p)}$) takes into account the impact of key factors that determine operating cash flows. The planned volumes of advance payments determine the cash flows that will come from consumers in the planning period to pay for future electricity supplies. They are determined on the basis of contractual conditions between the supplier and the consumer of electricity. To plan cash flows from operating activities of power supply companies, it is necessary to determine the volumes of advance payments in the planning period. The planned volume of cash inflows from debtors is determined on the basis of the analysis of the claims work with debtors to make them pay off penalties for the late electricity bill payment. The planned volume of cash inflow from the current account interest rate is determined based on the balance of available funds and the interest rate. The planned volume of income from target financing is determined on the basis of the analysis of state programs providing for the provision of preferences to certain groups of electricity consumers.

In addition to the volume of supplies (purchase) of electricity and the weighted average electricity supply (purchase) price, the methodology for planning cash flows from operating activities includes payment ratios for each individual electricity consumer (K_i) and the power supply company itself. On the one hand, they reflect the level of settlement discipline of electricity consumers (and the power supply company itself), and on the other hand, they determine the real volumes of cash inflows and outflows in the process of operating activities of power supply companies. We assume that the coefficients of consumer and power supply company payments can be effective tools for the efficient management of cash flows from the operating activities of power supply companies. We propose to calculate the coefficient of electricity consumer payments (K_i) as the ratio of the amount of money that has been paid (in case of the actual payment coefficient) or will be paid (in case of the planned payment coefficient) for the electricity consumed to the cost of the electricity supply. It is calculated as follows:

$$K_i = \frac{OCF_{in(i)}}{E_i}, \tag{6}$$

where $OCF_{in(i)}$ is the amount of money that has been received (or will be received) from the ith consumer; and E_i is the cost of electricity supplied to the ith consumer. From the perspective of cash inflow, the most favorable value of this coefficient is >1. It describes a situation when the consumer has no debt for the consumed electricity, but there are advance payments for future electricity supplies. In the context of the confidentiality of information on the settlements of individual consumers on the basis of the data found in the financial statements of power supply companies, a consolidated modification of the consumer payment coefficient is proposed. It can be defined as the ratio of cash inflows from the sales of products (goods, services), from buyers and customers, to the cash inflow from the sales of products over the period under study.

The starting point of the operating cash inflow management $\left(OCF_{in(i)}\right)$ of power supply companies is the calculation of the actual value of the coefficient $K_{i(fact)}$. Monitoring of the actual value of the payment coefficient of the ith consumer is based on the value of the coefficient of payments $K_{i(fact)} \geq 1$, as well as the actual value of the coefficient of payments of the ith consumer along with the planned one $K_{i(fact)} \geq K_{i(plan)}$. Furthermore, the identification of the type of consumer considering the value of the payment ratio has been performed. The value of the coefficient shows whether the consumer is a debtor (<1) or a customer that pays in advance (>1). At the same time, the identification of the type of consumer is an important task of the management of operational cash flows of power supply companies, as the use of specific methods and tools for planning and managing cash flows is required to manage cash inflows from each group of electricity consumers. The next stage in the management of the operational cash inflow of power supply companies is the comparison of the actual value of the payment ratio with its projected value. When the actual value of the payment ratio exceeds the planned one, the projected volume of the cash flow of the power supply companies is adjusted upward. When the actual value of the payment ratio is lower than the planned one, the projected volume of cash inflow from operating activities decreases.

The Microsoft Excel Forecast function is used to forecast cash flows; it projects the value of cash inflows and outflows based on their current values. Thus, based on the financial statements of power supply companies for the period of 2015–2019 and the use of Microsoft Excel Linest and Forecast functions, linear regression equations for the net cash flow, cash inflow, and outflow of power supply companies have been constructed. As a result, the following model has been obtained:

$$TCF_{in(i)} = a_{TCFin} + b_{TCFin} \times t_i \tag{7}$$

$$TCF_{out(i)} = a_{TCFout} + b_{TCFout} \times t_i \tag{8}$$

$$TCF_{net(i)} = a_{TCFnet} + b_{TCFnet} \times t_i \tag{9}$$

where $TCF_{in(i)}$, $TCF_{out(i)}$, $TCF_{net(i)}$ represent the value of the total cash inflow, cash outflow, and net cash flow in the i-th period; a_{TCFin}, a_{TCFout}, a_{TCFnet} represent the constant parameters of the regression model of the total cash inflow, cash outflow, and net cash flow; b_{TCFin}, b_{TCFout}, b_{TCFnet} represent the regression coefficients of the total cash inflow, cash outflow, and net cash flow; and t_i represents the numerical order of the period from 1 (2015) to 5 (2019), which are actual values, while if t_i = 6 (2020) or 7, 8, etc., then they are planned values. Based on the use of the Microsoft Excel "Linest" function and the data on the cash inflows, cash outflows, and net cash flows of power supply companies for the period of 2015–2019, the following values of the regression coefficients and constant parameters of the linear regression model of cash flows were obtained. The constructed linear regression models can be used to calculate the projected value of the cash inflow, cash outflow, and net cash flow of power supply companies for subsequent periods. Based on the proposed models, the values of the cash inflow, cash outflow, and net cash flow of power supply companies for 2025 (t_i = 11) were calculated [53,54].

The study is based on the data provided by 12 power supply companies operating in the CIS member countries. The main criterion for the selection of companies was the CIS membership of the country in which they are located, the period of operation over 5 years, and the availability of the required financial information in the context of cash flows. The research sample included Azerenerji JSC (Azerbaijan), Electric Networks of Armenia CJSC (Armenia), Telasi JSC (Georgia), ADEC JSC (Kazakhstan), NES Kyrgyzstan JSC (Kyrgyzstan), Belenergo SPA (Belarus), ERES SUE (Moldova), Rosseti PJSC (Russian Federation), Pamirenergy PJSC (Tajikistan), Turkmenenergo (Turkmenistan), Kyivenergo PJSC (Ukraine), and Karakalpak enterprise of territorial power grids (Uzbekistan).

4. Results

The level of settlement discipline of the consumers of each enterprise under consideration is different. This is evidenced by the indicators of the consolidated ratio of consumer payments shown in Figure 1.

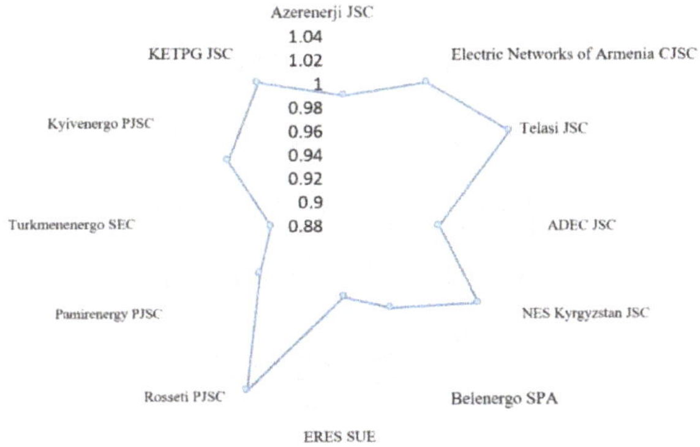

Figure 1. The consolidated ratio of consumer payments in 2019. Source: own development.

The coefficient is <1 for seven out of 12 enterprises and it is >1 for the remaining five enterprises. Thus, in 2019, seven out of 12 surveyed enterprises had problems related to settlement discipline.

In order to reflect the generalized trends that are inherent in the structural ratios of cash flows, Figure 2 shows the average coefficients of the share of operating, financial, and investing cash flows in the total cash inflow and outflow of power supply companies in 2015–2019.

As for the average structural financial ratios, which reflect the specifics of the formation of the total cash outflow of power supply companies in 2015–2019, it should be noted that the structural proportions inherent in the cash flow are also preserved here. Thus, for the majority of power supply companies over the period of 2015–2019, the basis of all payments was the cash outflow from operating activities. The highest average share of the operating cash outflow in the total cash flow can be observed in the following power supply companies: Azerenerji JSC, ADEC JSC, ERES State Unitary Enterprise, Turkmenenergo SEC, and Kyivenergo PJSC. At the same time, the average share of investing and financing cash outflows in the total cash flow is the greatest in Belenergo SPA, NES Kyrgyzstan JSC, and Telasi JSC.

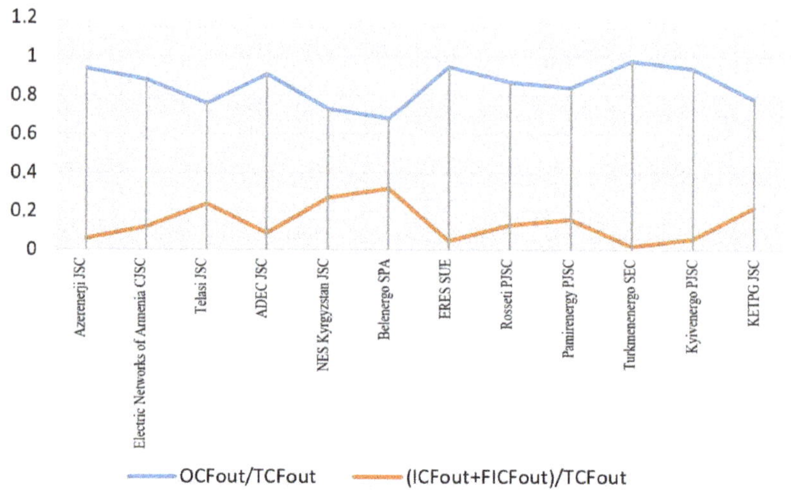

Figure 2. Average values of the coefficients of the share of cash outflows of power supply companies. Source: own development.

In order to determine the uniformity of cash inflows (cash flow) of power supply companies for the period of 2015–2019, the average value of the cash flow (*CFin (av)*), its standard deviation (*σCFin*), and the coefficient of variation (*CVCFin*) were assessed (Figure 3). The average deviation of the cash inflow reflects the range of fluctuations in the cash flow of power supply companies in 2015–2019.

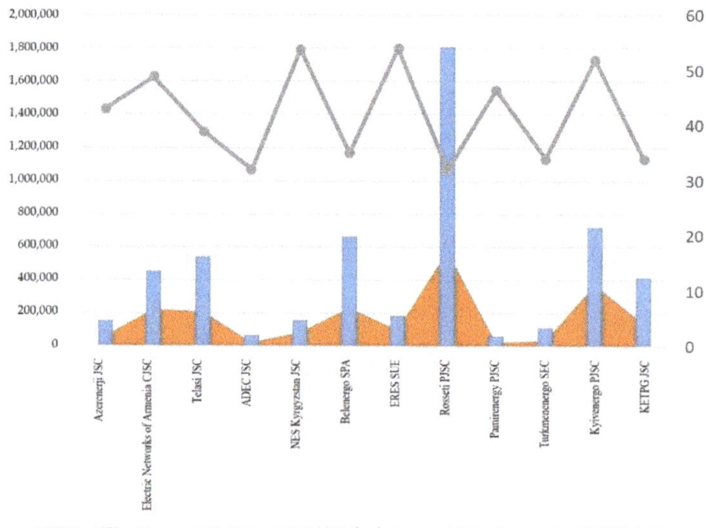

Figure 3. Indicators of the level of uniformity of the cash flow of the power supply companies under consideration. Source: own development.

On the one hand, the absolute value of the root-mean-square deviation depends on the uniformity of cash inflows over the studied period, and on the other hand, it depends on the size of the power supply company. Taking into account the above, it is rather difficult to assess the cash flow variance on the basis of the absolute value of the standard deviation. Therefore, the coefficient of variation is a fairly informative relative indicator that allows comparing the variation in the cash flow of various power supply companies. Figure 3 shows that NES Kyrgyzstan JSC, ERES SUE (54%) and Kyivenergo PJSC (52%) are typically characterized by high cash flow variance. These enterprises have the most volatile level of cash flow. At the same time, there are power supply companies with a stable cash flow level. These are ADEC JSC (32%), Rosseti PJSC (32%), Turkmenenergo SEC (34%), and KETPG JSC (34%). The rest of the power supply companies have an average level of cash flow uniformity, and the coefficient of cash flow variance ranges from 35 to 45%.

Another important component of the study of the uniformity of cash flows is the assessment of the uniformity of cash outflows of power supply companies over the period of 2015–2019, which is shown in Figure 4.

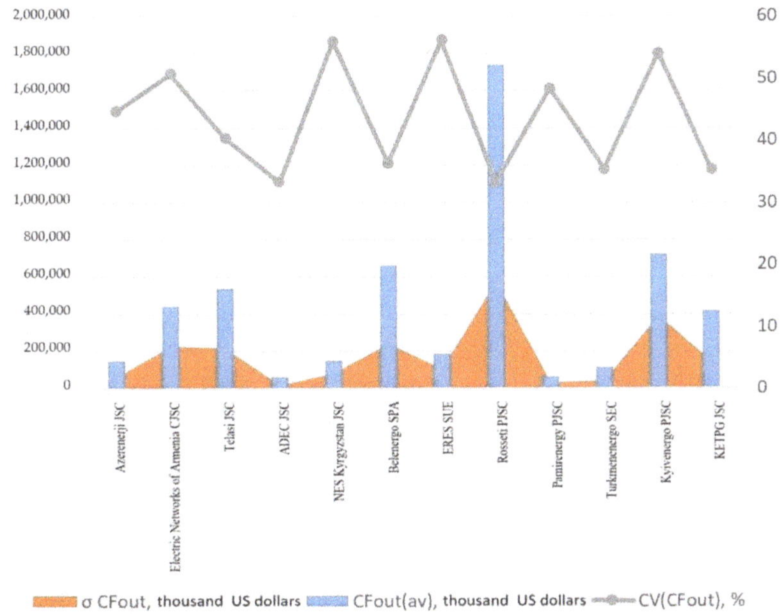

Figure 4. Indicators of the level of uniformity of the cash outflow of the power supply companies under consideration. Source: own development.

The obtained indicators (Figure 4) demonstrate that over the period under study, the volume of the cash outflow of power supply companies varies significantly. This is evidenced by the value of the standard deviation of the cash outflow and the coefficient of variation, which preserve the change in the variation of the cash inflow uniformity. In general, the values of the coefficients of variation of the cash outflow of power supply companies reflect the same trends as the values of the coefficients of variation of the cash flow of power supply companies. From the perspective of solvency, it is quite important to assess the uniformity of the net cash flow of the company, which is the basis for increasing cleared funds in the company account.

In order to project the volumes of cash inflow, cash outflow, and net cash flows of power supply companies under consideration for subsequent periods, namely until 2025, linear regression models were proposed. In particular, subsequent values can be projected on their basis. Thus, $t_i = 5$ for 2019, $t_i = 6$ for 2020, etc., while the values of the regression coefficients (b_{TCFin}, b_{TCFout}, b_{TCFnet}) and the values of the constant parameters of the linear regression model (a_{TCFin}, a_{TCFout}, a_{TCFnet}) remain fixed for all subsequent periods (Table 1).

Table 1. Projected volumes of the cash inflow, cash outflow, and net cash flows for the studied power supply companies for 2025 (based on linear regression models).

Company	R^2 (TCFin)	R^2 (TCFout)	TCFin, Thousand US Dollars	TCFout, Thousand US Dollars	TCFnet, Thousand US Dollars
Azerenerji JSC	0.93	0.94	164,590	161,298	3292
Electric Networks of Armenia CJSC	0.91	0.91	478,021	477,065	956
Telasi JSC	0.22	0.22	531,135	531,967	−832
ADEC JSC	0.91	0.92	69,577	68,463	1113
NES Kyrgyzstan JSC	0.85	0.86	149,440	149,231	209
Belenergo SPA	0.92	0.92	651,525	650,645	880
ERES SUE	0.46	0.45	180,369	180,380	−11
Rosseti PJSC	0.93	0.94	1,864,319	1,827,032	37286
Pamirenergy PJSC	0.65	0.66	66,338	65,011	1327
Turkmenenergo SEC	0.56	0.57	116,777	115,598	1179
Kyivenergo PJSC	0.86	0.86	708,179	708,675	−496
KETPG JSC	0.91	0.92	431,700	426,908	4792

Source: own development.

The obtained values of the coefficients of determination (R^2) obviously show that the linear regression quite accurately describes the trajectory of the dynamics of the cash inflow and outflow of such companies as Azerenerji JSC, Electric Networks of Armenia CJSC, ADEC JSC, Belenergo SPA, Rosseti PJSC, and KETPG JSC, whose coefficients of determination exceed 0.9. The coefficient of determination for the other companies has low or close to zero values, which indicates the low predictive power of linear regression models for these companies. Linear regression models are quite simple and versatile tools that allow projecting the values of cash inflow, cash outflow, and net cash flows of enterprises. However, in modern economic realities, linear relationships are not common.

Therefore, for the purpose of a deeper study, the functional possibilities of explaining the trends that are observed in the cash flows of enterprises were determined, and, accordingly, polynomial models were used to build their forecasting models as compared to linear and exponential models, which can more accurately reproduce the cash flow trends that took place in the previous periods. The level of detail of the models should be above 0.9 to indicate their high quality. As a result of the simulation of cash flows of power supply companies for the period of 2015–2019, polynomial multiple regression models were obtained (Table 2).

Table 2. Polynomial multiple regression models of the cash flow of the power supply companies under consideration.

Company	R^2 (TCFin)	Polynomial Multiple Regression Models of Cash Inflow, (TCFin)	R^2 (TCFout)	Polynomial Multiple Regression Models of Cash Outflow, (TCFout)
Azerenerji JSC	0.96	$y = 1531x^2 - 1987x + 2395$	0.96	$y = 1552x^2 - 2762x + 2498$
Electric Networks of Armenia CJSC	0.77	$y = 4136x^2 - 3843x + 22383$	0.78	$y = 4116x^2 - 3219x + 18391$
Telasi JSC	0.79	$y = 3639x^2 - 6591x + 18349$	0.81	$y = 3713x^2 - 5892x + 16396$
ADEC JSC	0.94	$y = 388x^2 + 1701x + 6720$	0.95	$y = 314x^2 + 2474x + 5729$
NES Kyrgyzstan JSC	0.96	$y = 1392x^2 - 2122x + 4768$	0.97	$y = 1352x^2 - 1648x + 4287$
Belenergo SPA	0.97	$y = 6051x^2 - 8739x + 14933$	0.97	$y = 5858x^2 - 6832x + 16498$
ERES SUE	0.97	$y = 1898x^2 - 5295x + 9548$	0.96	$y = 1858x^2 - 472\,1x + 8340$
Rosseti PJSC	0.97	$y = 12859x^2 + 24837x + 36786$	0.98	$y = 13275x^2 + 17314x + 32894$
Pamirenergy PJSC	0.93	$y = 794x^2 - 3472x + 8734$	0.93	$y = 775x^2 - 3105x + 6323$
Turkmenenergo SEC	0.18	$y = 1105x^2 - 1874x + 4591$	0.19	$y = 1048x^2 - 1549x + 6812$
Kyivenergo PJSC	0.91	$y = 6951x^2 - 13573x + 21847$	0.91	$y = 6634x^2 - 9857x + 18649$
KETPG JSC	0.91	$y = 2643x^2 + 8934x + 11782$	0.92	$y = 2895x^2 + 6205x + 7892$

Source: own development based on the financial statements of power supply companies for the period of 2015–2019.

This type of regression has only one explanatory variable, which is included in the regression equations with various degrees (1, 2, 3, 4); based on the maximum magnitude of the obtained model, regression of the second, third, and fourth order is distinguished. In the context of this approach, the study assumes that the values of the cash inflow and outflow of power supply companies are formed under the influence of a large number of factors for which there is no information. In this case, the future value of the cash flow does not depend on its actual value. The values of the coefficients of determination of most obtained polynomial multiple regression models of the cash inflow and outflow of power supply companies exceed 0.9, which indicates the high possibility of extrapolating the trends of 2015–2019 to future periods. Table 3 shows that compared to the linear model, the polynomial one has higher values of the coefficients of determination; in particular, in most power supply companies, they are >0.95. Another advantage of polynomial models is the absence of multicollinearity, as different degrees of "x" are linear functions. At the same time, the regression coefficients included in the model are linear, which makes it possible to apply the ordinary least squares method. The above indicates that the projected values of the cash inflow and outflow calculated based on polynomial models fully take into account the trends that took place in the cash flows of the power supply companies under consideration over the period of 2015–2019. It is impossible to build a qualitative polynomial model for Turkmenenergo SEC on the basis of the information for the period under study. There are also companies whose polynomial models have a lower coefficient of determination that ranges between 0.77 and 0.9. The projected values of the cash flows obtained with the help of such models cannot fully reflect the trends that have been formed in the cash inflow and outflow.

Table 3. Projected values of cash inflow, cash outflow, and net cash flow of the power supply companies under study for 2025 (based on polynomial models).

No.	Company	TCFin, Thousand US Dollars	TCFout, Thousand US Dollars	TCFnet, Thousand US Dollars
1	Azerenerji JSC	165,789	159,908	5881
2	Electric Networks of Armenia CJSC	480,566	481,018	−452
3	Telasi JSC	531,169	530,481	688
4	ADEC JSC	72,379	70,937	1442
5	NES Kyrgyzstan JSC	149,858	149,751	107
6	Belenergo SPA	650,975	650,164	811
7	ERES SUE	180,961	181,227	−266
8	Rosseti PJSC	1,865,932	1,829,623	36,309
9	Pamirenergy PJSC	66,616	65,943	673
10	Turkmenenergo SEC	117,682	116,581	1101
11	Kyivenergo PJSC	713,615	712,936	679
12	KETPG JSC	429,859	426,442	3417

Source: own development.

Table 3 shows the projected values of the cash inflow, cash outflow, and net cash flows of power supply companies calculated based on the proposed polynomial multiple regression models. The results obtained can be used as projections that have a rather high degree of approximation. An important practical result, which is achieved in the process of projecting the values of cash flows of power supply companies, is the projected value of the net cash flow. If the projected value of the net cash flow is positive, the surplus cash flow is expected. When the value of the projected cash outflow is greater than the projected value of the cash flow, there will be cash flow deficit.

It is expected that most power supply companies will have a small net cash flow deficit, which in most cases can be covered by cleared funds. However, there are also companies that will have to look for opportunities to balance cash flows as high cash flow deficit and surplus are projected. The indicators for calculating the main criteria that allow us to assess the relationship between the cash inflow and outflow of power supply companies are shown in Table 4, which contains the values of the correlation coefficients between the cash inflow and outflow and operating cash inflow and outflow.

Table 4. The correlation between the cash inflow and outflow of the power supply companies under consideration.

No.	Company	p(OCFin/OCFout)	p(TCFin/TCFout)
1	Azerenerji JSC	0.9955	0.9932
2	Electric Networks of Armenia CJSC	0.9772	1.0000
3	Telasi JSC	0.9980	0.9973
4	ADEC JSC	0.9992	0.9999
5	NES Kyrgyzstan JSC	0.9997	0.9999
6	Belenergo SPA	0.9995	0.9983
7	ERES SUE	0.9995	0.9996
8	Rosseti PJSC	0.9999	1.0000
9	Pamirenergy PJSC	0.9987	0.9998
10	Turkmenenergo SEC	0.9992	0.9969
11	Kyivenergo PJSC	0.9983	0.9997
12	KETPG JSC	0.9955	0.9932

Source: own development.

For most companies, the correlation coefficients between operating cash inflows and outflows, as well as between the total cash inflow and outflow for the period of 2015–2019 range between 0.99 and 1.00. This indicates the synchronicity between cash inflows and outflows over the period being considered. Slightly lower correlation coefficients between operating cash inflows and outflows are typically found in Electric Networks of Armenia CJSC. The projected cash flows based on the polynomial models have obvious discrepancies in the results. At the same time, two models show that in ERES SUE, there will be deficit cash flow. Companies such as Electric Networks of Armenia CJSC (according to polynomial models), Telasi JSC, and Kyivenergo PJSC (according to linear models) are also at risk. Based on all of the above, it is possible to exclude hypothesis H0 and accept hypothesis H1, since a significant degree of dependence of the level of financial stability and equilibrium of power supply enterprises on the level of synchronization and uniformity of their cash flows has been proven. One should consider that some enterprises are characterized by a high variation in incoming and outgoing cash flows, while for others, it is relatively insignificant. This feature makes it possible to identify enterprises according to the criterion of riskiness of cash flows as more or less risky from the standpoint of synchronization and uniformity of receipt (or spending) of funds. The weaker synchronicity between the cash inflow and outflow of these corporations is primarily due to the specifics of their financial and investment activities and cash flow management policies.

5. Discussion

Operating cash flow management based on the proposed methodological approach to monitoring, which is based on the calculation of planned and actual indicators of cash flows and their comparison, should be primarily implemented in the context of large consumers of electricity, which may include industrial consumers, budgetary and municipal institutions, and others. This is due to the fact that the operating cash inflow of power supply companies is most sensitive to the level of payment by the largest electricity consumers [55,56]. Considering the above, it is proposed to continuously monitor the payment ratios of the largest electricity consumers in order to control the cash flow from operating activities. Power supply companies have their own specifics in the use of funds received from their electricity consumers. It should also be taken into account that the cash flows from electricity consumers are credited to a special distribution account. In the absence of overdue debt, all funds from the distribution account are transferred to the current account of the power supply company. In the presence of overdue debt, a portion of funds can be credited to the current account, and another portion can be used to repay the debt [57,58]. The specified feature of the state regulation of cash flows of power supply companies should be taken into account when planning and managing cash flows.

Based on the analysis of the operating cash flows of power supply companies over the period of 2015–2019, it can be asserted that the funds received from electricity consumers in the form of advance payments have the greatest share in the structure of operating cash flows. In this regard, the proposed methodology for the cash flow management and planning is exclusively focused on planning those operating cash flow components, which together make up more than 90% of its volume [13]. At the same time, the operating cash outflow of power supply companies depends on the volume of electricity purchased, average electricity rates, the amount of taxes and duties paid by power supply companies, and employee benefits [59]. In recent years, the introduction of new techniques has had the greatest impact on the cost of electricity, which leads to an increase in the cost of electricity produced by thermal power plants and combined heat and power (CHP) power stations [60]. Planned cash flow contribution ratios can reflect the planned (target) structure of cash inflow and outflow of power supply companies [61]. At the same time, the coefficients of the contribution of cash flow can be tools for controlling cash flows in the process of comparing their actual values with the planned and industry average values [24,62]. These properties of financial ratios make them versatile tools for planning and controlling the cash flows of enterprises.

The approach to long-term forecasting of cash flows developed in the study makes it possible to determine long-term benchmarks of operating cash flows that can be used in the process of planning cash flows from operating activities. The methodology for planning cash flows from the operating activities proposed in this section allows us to determine the volume of operating cash flow for the planning period. When planning cash flows from operating activities, the focus should be placed on the factors that most significantly affect the volume of operating cash flow. These factors include the cost and volume of the electricity supplied to consumers, the level of energy settlement, and rates. Consideration of the expected trends in the specified factors is the basis for planning and managing operating cash flows [56,63]. The proposed methodological approach to the management of cash flows from the operating activities of power supply companies should be based on monitoring the payment ratios of both the consumer and the company itself. The ratios described in the study are effective tools for planning and managing the operating cash flows of power supply companies [63,64].

A significant advantage of the proposed methodological approach is the fact that the coefficients of the contribution of operating, investing, and financial cash flows to the total cash inflow (outflow) can indicate target planned proportions reflecting cash inflow (outflow) for certain types of activities of power supply companies [60]. In this regard, on the one hand, they can be used as target structural indicators of the formation of planned cash inflow and outflow, and on the other hand, they can be effective tools for the management of planned cash flows.

The limitation of this study is the impossibility of its practical application in other economic sectors and spheres. This is due to the fact that the ratios of cash flows should be adjusted in order to fully take into account the specifics of a particular industry or type of activity. At the same time, the positive aspect is that the contribution coefficients of cash flows are universal tools for planning and managing cash flows, and their standard values are individual and should take into account the specifics of the enterprise.

In the future, the study can be expanded in terms of the number of companies, industries, and geographical boundaries studied. At the same time, industry averages can be formed to be used as control indicators in the cash flow management and planning.

6. Conclusions

Based on the indicators of the consolidated consumer payment ratio of the power supply companies under study, it can be argued that the level of settlement discipline of their consumers varies. In most companies, there are settlement discipline problems.

The analysis of the average structural financial ratios, which reflect the specifics of the formation of the total cash inflow and outflow of power supply companies in 2015–2019, confirmed the preservation of the structural proportions inherent in the cash flow. At the same time, for the majority of power supply companies, the basis of all payments is the cash outflow from operating activities.

The uniformity of cash inflows of the power supply companies based on the assessment of the average value of the cash flow, its standard deviation, and the coefficient of variation made it possible to determine the highest level of cash flow variance in NES Kyrgyzstan JSC, ERES SUE, and Kyivenergo PJSC. These companies are characterized by the most volatile level of cash flow. At the same time, a group of companies with stable cash inflows has been identified: ADEC JSC, Rosseti PJSC, Turkmenenergo SEC, and KETPG JSC. The uniformity of cash flows based on the assessment of the distribution of payments in the power supply companies being studied indicates a significant variation in their cash outflow. In this case, the values of the standard deviation of the cash outflow and the coefficient of variation preserve the trend of changes in the cash inflow uniformity. Generally, the values of the coefficients of variation of the cash outflow of power supply enterprises reflect the same patterns as their cash flow variation coefficients.

Forecasting the volumes of cash inflow, cash outflow, and net cash flows for the power supply companies under consideration for 2025 based on the obtained coefficients of determination made it possible to determine the companies (Azerenerji JSC, Electric Networks of Armenia CJSC, ADEC JSC, Belenergo SPA, Rosseti PJSC, and KETPG JSC) for which linear regression can fairly accurately describe the trajectory of the cash inflow and outflow. However, for the other companies, which make up 50% of the studied enterprises, the coefficient of determination has low or close to zero values, which indicates the low predictive power of the linear regression models for these companies.

Based on the polynomial cash flow multiple regression models designed for the power supply companies under study, the values of the determination coefficients were determined, which exceed 0.9 for most enterprises. This confirms the strong possibility to extrapolate the studied trends to future periods. Thus, compared to the linear model, the polynomial one confirms higher values of the determination coefficients (>0.95) for the majority of power supply companies.

The projected volumes of cash inflow, cash outflow, and net cash flows of power supply companies with the application of the described polynomial multiple regression models have a fairly high degree of approximation. At the same time, most power supply companies are expected to have a small net cash deficit that can be covered by the available cash assets. However, there are also companies that will be forced to look for opportunities to ensure balanced cash flow, as the projected cash deficit and excess cash flow are quite high. The correlations between operating cash flows and outflows, between total cash inflow and outflow of the majority of power supply companies, are high and range between 0.99 and 1. The low level of synchronization between cash inflows and outflows of the companies under study is associated, first of all, with the specifics of their financial and investment activities and the cash flow management policy.

Hypothesis H1 was adopted taking into account the proven significant dependence of power enterprises' financial stability on the synchronization and uniformity of cash flows. Since the hypotheses put forward are mutually exclusive, hypothesis H0 was rejected. The proposed methodological approach allows identifying enterprises according to the criterion of cash flow riskiness from the standpoint of cash flow synchronization and homogeneity.

Author Contributions: Conceptualization, T.S. and V.S.; methodology, D.P. and I.R.; software, I.R.; validation, D.P., T.S., V.S., and I.R.; formal analysis, D.P. and M.S.; investigation, T.S. and V.S.; resources, I.R.; writing—original draft preparation, D.P.; writing—review and editing, T.S. and M.S.; visualization, V.S.; supervision, T.S.; project administration, D.P.; funding acquisition, I.R. and M.S. All authors have read and agreed to the published version of the manuscript.

Funding: This research received no external funding.

Institutional Review Board Statement: Not applicable.

Informed Consent Statement: Not applicable.

Data Availability Statement: Data will be available on request.

Acknowledgments: This paper has been supported by the RUDN University Strategic Academic Leadership Program.

Conflicts of Interest: The authors declare no conflict of interest.

References

1. Neves, M.E.; Henriques, C.; Vilas, J. Financial performance assessment of electricity companies: Evidence from Portugal. *Oper. Res.* **2019**, *1*. [CrossRef]
2. Strielkowski, W.; Lisin, E.; Astachova, E. Economic sustainability of energy systems and prices in the EU. *Entrep. Sustain. Issues* **2017**, *4*, 591–600. [CrossRef]
3. Rogalev, A.; Komarov, I.; Kindra, V.; Zlyvk, O. Entrepreneurial assessment of sustainable development technologies for power energy sector. *Enterp. Sustain. Issues* **2018**, *6*, 429–445. [CrossRef]
4. Vlasov, A.I.; Shakhnov, V.A.; Filin, S.S.; Krivoshein, A.I. Sustainable energy systems in the digital economy: Concept of smart machines. *Entrep. Sustain. Issues* **2019**, *6*, 1975–1986. [CrossRef]
5. Ospanov, A.; Timurbekova, A. New hypothesis of energy of crushing. *J. Hyg. Eng. Des.* **2019**, *27*, 87–89.
6. Nasr, A.K.; Kashan, M.K.; Maleki, A.; Jafari, N.; Hashemi, H. Assessment of barriers to renewable energy development using stakeholders approach. *Entrep. Sustain. Issues* **2020**, *7*, 2526–2541. [CrossRef]
7. Persson, M.E. Implications of cash flow accounting for management. In *Harold Cecil Edey: A Collection of Unpublished Material from a 20th Century Accounting Reformer*; Emerald Publishing Limited: Bingley, UK, 2019.
8. Abzalbekova, M.; Arenova, L.; Zbinskaya, Y.; Sultanbekova, G.; Yessentaeva, N. Validity and motivation of decisions in criminal proceedings. *J. Adv. Res. Law Econ.* **2018**, *9*, 2209–2217.
9. Bian, Y.; Lemoine, D.; Yeung, T.G.; Bostel, N.; Hovelaque, V.; Viviani, J.L.; Gayraud, F. A dynamic lot-sizing-based profit maximization discounted cash flow model considering working capital requirement financing cost with infinite production capacity. *Int. J. Prod. Econ.* **2018**, *196*, 319–332. [CrossRef]
10. Gleißner, W. Cost of capital and probability of default in value-based risk management. *Manag. Res. Rev.* **2019**, *42*, 1243–1258. [CrossRef]
11. van der Burg, J.; Song, X.; Tippett, M. A hyperbolic model of optimal cash balances. *Eur. J. Financ.* **2019**, *25*, 101–115. [CrossRef]
12. Figueroa, N.; Leukhina, O. Cash flows and credit cycles. *J. Bank. Financ.* **2018**, *87*, 318–332. [CrossRef]
13. Dufour, D.; Luu, P.; Teller, P. The influence of cash flow on the speed of adjustment to the optimal capital structure. *Res. Int. Bus. Financ.* **2018**, *45*, 62–71. [CrossRef]
14. Lewin, P.; Cachanosky, N. Substance and semantics: The question of capital. *J. Econ. Behav. Organ.* **2018**, *150*, 423–431. [CrossRef]
15. Kryzia, D.; Kopacz, M.; Kryzia, K. The valuation of the operational flexibility of the energy investment project based on a gas-fired power plant. *Energies* **2020**, *13*, 1567. [CrossRef]
16. Kucuksari, S.; Erdogan, N.; Cali, U. Impact of electrical topology, capacity factor and line length on economic performance of offshore wind investments. *Energies* **2019**, *12*, 3191. [CrossRef]
17. Vaidya, R.; Patel, P. Determinants of profitability of capital-intensive firms in the Indian capital market: A static and dynamic panel approach. *IUP J. Account. Res. Audit Pract.* **2019**, *18*, 33–51.
18. Zameer, H.; Wang, Y. Energy production system optimization: Evidence from Pakistan. *Renew. Sustain. Energy Rev.* **2018**, *82*, 886–893. [CrossRef]
19. Qin, X.; Huang, G.; Shen, H.; Fu, M. COVID-19 pandemic and firm-level cash holding—Moderating effect of goodwill and goodwill impairment. *Emerg. Mark. Financ. Trade* **2020**, *56*, 2243–2258. [CrossRef]
20. Then, D.; Spalthoff, C.; Bauer, J.; Kneiske, T.M.; Braun, M. Impact of natural gas distribution network structure and operator strategies on grid economy in face of decreasing demand. *Energies* **2020**, *13*, 664. [CrossRef]
21. Wellmer, F.W.; Buchholz, P.; Gutzmer, J.; Hagelüken, C.; Herzig, P.; Littke, R.; Thauer, R.K. *Raw Materials for Future Energy Supply*; Springer: Berlin/Heidelberg, Germany, 2019.
22. Khuong, N.V.; Ha, N.T.T.; Thu, P.A. The relationship between real earnings management and firm performance: The case of energy firms in Vietnam. *Int. J. Energy Econ. Policy* **2019**, *9*, 307–314.
23. Kasperowicz, R.; Bilan, Y.; Štreimikienė, D. The renewable energy and economic growth nexus in European countries. *Sustain. Dev.* **2020**, *28*, 1086–1093. [CrossRef]
24. Giones, F.; Brem, A.; Berger, A. Strategic decisions in turbulent times: Lessons from the energy industry. *Bus. Horiz.* **2019**, *62*, 215–225. [CrossRef]
25. Zhu, N.; Bu, Y.; Jin, M.; Mbroh, N. Green financial behavior and green development strategy of Chinese power companies in the context of carbon tax. *J. Clean. Prod.* **2020**, *245*, 118908. [CrossRef]

26. Klimek, D.; Jędrych, E. A model for the sustainable management of enterprise capital. *Sustainability* **2021**, *13*, 183. [CrossRef]
27. Hussain, H.I.; Slusarczyk, B.; Kamarudin, F.; Thaker, H.M.T.; Szczepańska-Woszczyna, K. An investigation of an adaptive neuro-fuzzy inference system to predict the relationship among energy intensity, globalization, and financial development in major ASEAN economies. *Energies* **2020**, *13*, 850. [CrossRef]
28. Afy-Shararah, M.; Rich, N. Operations flow effectiveness: A systems approach to measuring flow performance. *Int. J. Oper. Prod. Manag.* **2018**, *38*, 2096–2123. [CrossRef]
29. Sarkar, B.; Guchhait, R.; Sarkar, M.; Cárdenas-Barrón, L.E. How does an industry manage the optimum cash flow within a smart production system with the carbon footprint and carbon emission under logistics framework? *Int. J. Prod. Econ.* **2019**, *213*, 243–257. [CrossRef]
30. Peshkova, G.Y.; Samarina, A.Y. Digital economy and recruitment potential: Strategical interconnection and prospects. *Educ. Sci. J.* **2018**, *20*, 50–75. (In Russian) [CrossRef]
31. Busu, M. Applications of TQM processes to increase the management performance of enterprises in the romanian renewable energy sector. *Processes* **2019**, *7*, 685. [CrossRef]
32. Dudin, M.N.; Frolova, E.E.; Protopopova, O.V.; Mamedov, A.A.; Odintsov, S.V. Study of innovative technologies in the energy industry: Nontraditional and renewable energy sources. *Entrep. Sustain.* **2019**, *6*, 1704–1713. [CrossRef]
33. Gavurova, B.; Kocisova, K.; Behun, M.; Tarhanicova, M. Environmental performance in OECD countries: A non-radial DEA approach. *Acta Montan. Slovaca* **2018**, *23*, 206–215.
34. Geshniani, Y.V.; Kamranrad, R.; Emami, I. A multi-objective scheduling payment pattern for project cash flow by considering resource constraints (A case study in power transfer system). *Int. J. Ind. Syst. Eng.* **2020**, *35*, 299–313.
35. Laborda, J.; Salas-Fumás, V.; Suárez, C. An endogenous approach to the cyclicality of R&D investment under credit constraints: Firms' cash flow matters! *J. Open Innov. Technol. Mark. Complex.* **2020**, *6*, 33.
36. Tang, D.Y.; Zhang, Y. Do shareholders benefit from green bonds? *J. Corp. Financ.* **2020**, *61*, 101427. [CrossRef]
37. Li, P.; Zhou, R.; Xiong, Y. Can ESG performance affect bond default rate? Evidence from China. *Sustainability* **2020**, *12*, 2954. [CrossRef]
38. Hinz, F.; Schmidt, M.; Möst, D. Regional distribution effects of different electricity network tariff designs with a distributed generation structure: The case of Germany. *Energy Policy* **2018**, *113*, 97–111. [CrossRef]
39. Mikita, M.; Kolcun, M.; Špes, M.; Vojtek, M.; Ivančák, M. Impact of electrical power load time management at sizing and cost of hybrid renewable power system. *Pol. J. Manag. Stud.* **2017**, *15*, 154–162. [CrossRef]
40. Jishkariani, M. Electricity tariffs in Georgia. *World Sci.* **2019**, *1*, 20–22. [CrossRef]
41. Carbonara, N.; Pellegrino, R. Public-private partnerships for energy efficiency projects: A win-win model to choose the energy performance contracting structure. *J. Clean. Prod.* **2018**, *170*, 1064–1075. [CrossRef]
42. Steffen, B. The importance of project finance for renewable energy projects. *Energy Econ.* **2018**, *69*, 280–294. [CrossRef]
43. El Iysaouy, L.; El Idrissi, N.E.; Tvaronavičienė, M.; Lahbabi, M.; Oumnad, A. Towards energy efficiency: Case of Morocco. *Insights Reg. Dev.* **2019**, *1*, 259–271. [CrossRef]
44. Andoni, M.; Robu, V.; Flynn, D.; Abram, S.; Geach, D.; Jenkins, D.; Peacock, A. Blockchain technology in the energy sector: A systematic review of challenges and opportunities. *Renew. Sustain. Energy Rev.* **2019**, *100*, 143–174. [CrossRef]
45. Saługa, P.W.; Szczepańska-Woszczyna, K.; Miśkiewicz, R.; Chłąd, M. Cost of equity of coal-fired power generation projects in Poland: Its importance for the management of decision-making process. *Energies* **2020**, *13*, 4833. [CrossRef]
46. Vayas-Ortega, G.; Soguero-Ruiz, C.; Rojo-Álvarez, J.L.; Gimeno-Blanes, F.J. On the differential analysis of enterprise valuation methods as a guideline for unlisted companies assessment (I): Empowering discounted cash flow valuation. *Appl. Sci.* **2020**, *10*, 5875. [CrossRef]
47. Awwad, B.S.A.; Hamdan, A.M.M. Gulf cooperation council energy sectors governance and dividend policy. *Int. J. Econ. Financ. Issues* **2018**, *8*, 163.
48. Alonso, A.C. The Role of the Commercial Banks in the Financing of the Renewable Energy Industry. In *Renewable Energy Finance: Funding The Future Of Energy*; World Scientific Publishing: Singapore, 2020; p. 217.
49. Nangia, R. Securing Asia's energy future with regional integration. *Energy Policy* **2019**, *132*, 1262–1273. [CrossRef]
50. Zhu, Z.; Zhu, Z.; Xu, P.; Xue, D. Exploring the impact of government subsidy and R&D investment on financial competitiveness of China's new energy listed companies: An empirical study. *Energy Rep.* **2019**, *5*, 919–925.
51. Boichenko, K.S.; Tepliuk, M.A.; Rekova, N.Y.; Stashkevych, I.I.; Morkunas, M. Management of fluctuation of financial and economic integrated development of innovative enterprise. *Financ. Credit Act. Probl. Theory Pract.* **2019**, *3*, 62–69. [CrossRef]
52. Salas-Molina, F. Risk-sensitive control of cash management systems. *Oper. Res.* **2020**, *20*, 1159–1176. [CrossRef]
53. Potočnik, P.; Šilc, J.; Papa, G. A comparison of models for forecasting the residential natural gas demand of an urban area. *Energy* **2019**, *167*, 511–522.
54. Luo, S.; Weng, Y. A two-stage supervised learning approach for electricity price forecasting by leveraging different data sources. *Appl. Energy* **2019**, *242*, 1497–1512. [CrossRef]
55. Mu, Q.; Gao, Y.; Yang, Y.; Liang, H. Design of power supply package for electricity sales companies considering user side energy storage configuration. *Energies* **2019**, *12*, 3219. [CrossRef]
56. Shi, M. Overinvestment and corporate governance in energy listed companies: Evidence from China. *Financ. Res. Lett.* **2019**, *30*, 436–445. [CrossRef]

57. He, Y.X.; Jiao, J.; Chen, R.J.; Shu, H. The optimization of Chinese power grid investment based on transmission and distribution tariff policy: A system dynamics approach. *Energy Policy* **2018**, *113*, 112–122. [CrossRef]
58. Connor, R.; Heffron, R.J.; Khan, A.A.; Perkins, E. Legal strategies for the mitigation of risk for energy infrastructure projects. *J. Energy Nat. Resour. Law* **2019**, *37*, 47–66. [CrossRef]
59. Andalib, R.; Hoseini, A.; Gatmiri, B. A stochastic model of cash flow forecasting considering delays in owners' payments. *Construct. Manag. Econ.* **2018**, *36*, 545–564. [CrossRef]
60. Vasylieva, T.; Lyulyov, O.; Bilan, Y.; Streimikiene, D. Sustainable economic development and greenhouse gas emissions: The dynamic impact of renewable energy consumption, GDP, and corruption. *Energies* **2019**, *12*, 3289. [CrossRef]
61. Gordon, E.A.; Hsu, H.T. Tangible long-lived asset impairments and future operating cash flows under US GAAP and IFRS. *Account. Rev.* **2018**, *93*, 187–211. [CrossRef]
62. Sabishchenko, O.; Rębilas, R.; Sczygiol, N.; Urbański, M. Ukraine energy sector management using hybrid renewable energy systems. *Energies* **2020**, *13*, 1776. [CrossRef]
63. Kwon, G.J. Comparative value relevance of accounting information among Asian countries. *Manag. Financ.* **2018**, *44*, 110–126. [CrossRef]
64. Postigo Marcos, F.; Mateo Domingo, C.; Gómez San Román, T.; Cossent Arín, R. Location and sizing of micro-grids to improve continuity of supply in radial distribution networks. *Energies* **2020**, *13*, 3495. [CrossRef]

energies

MDPI

Article

Innovation Management in Polish Real Estate Developers in the Renewable Energy Sources Context

Marcin Sitek [1,*] and Manuela Tvaronavičienė [2,3,*]

[1] The Management Faculty, Czestochowa University of Technology, 42-201 Czestochowa, Poland
[2] Department of Business Technologies and Entrepreneurship, Vilnius Gediminas Technical University (Vilnius Tech), Sauletekio 11, 10223 Vilnius, Lithuania
[3] General Jonas Zemaitis Military Academy of Lithuania, Silo 5A, 10322 Vilnius, Lithuania
* Correspondence: marcin.sitek@pcz.pl (M.S.); manuela.tvaronaviciene@vilniustech.lt (M.T.)

Abstract: This paper analyses innovative activities, including renewable energy sources (RES) in the housing market, the motivations for their introduction, effectiveness, benefits, limitations and management—which are open and current problems of Polish and international sustainable construction. This problem is part of a research gap concerning, among others, the role of developers and entities responsible for introducing energy innovations into housing construction. The aim of the paper is to analyse innovations, with particular emphasis on RES, introduced by residential developers in Poland in the context of global trends. The work is based on the results of surveys conducted among developers of the primary housing market. The research of 130 questionnaires received from entities such as multi-storey buildings and multi-family houses in Poland, was carried out on a nationwide sample using the CATI Computer Assisted Telephone Interview method. The results of the survey research were summarized by setting research hypotheses, which were verified using the significance test based on the Laplace normal distribution. The research conducted has shown that residential developers in Poland point at, among others, competitive pressures, concern about the quality of products and services offered, increasing market share or satisfaction and increasing customer requirements. This clearly shows that the implementation of innovation is perceived through the prism of increasing competitiveness. Furthermore, developers operating in the more competitive markets—nationwide and on the market of the five largest Polish cities, usually larger companies, showed a greater willingness and even the need to introduce innovation in their activities. In the study, developers as one of the barriers of introducing innovations, apart from the lack of adequate support, mainly from public entities in the field of, inter alia, appropriate law favouring the implementation of innovations and financial support, indicate difficulties of the client's market manifested by the lack of knowledge and identification of needs in the field of innovation, and the lack of willingness to pay a higher price in regard to the product with higher innovation level. In the paper, the authors recommend developers, among others, to implement in the management process the identification of possible to implement innovations and the customers' needs in innovations with education in the field of possible innovations and its benefits. The article indicates the need to intensify the implementation of innovations in housing projects in order to increase competitiveness and to meet the European Union's requirement regarding the use of renewable energy sources.

Keywords: green building; housing projects management; innovations; sustainable construction; renewable energy sources; real estate developers

Citation: Sitek, M.; Tvaronavičienė, M. Innovation Management in Polish Real Estate Developers in the Renewable Energy Sources Context. *Energies* **2021**, *14*, 1702. https://doi.org/10.3390/en14061702

Academic Editors: Álvaro Gutiérrez and Chi-Ming Lai

Received: 2 February 2021
Accepted: 15 March 2021
Published: 18 March 2021

1. Introduction

Continuous processes of intense globalization and accelerated economic growth are stimulated by increased competitiveness. In the case of the European Union, the main problem for economic success at a time when global energy markets are changing as a result of the transition to clean energy is to maintain competitiveness. This is why the

European Commission presented the Clean Energy for all Europeans package in Brussels on 30 November 2016 stressing that the transition to clean energy will be the main growth sector (smart money) in the future [1]. The Clean Energy for all Europeans package is a set of ambitious requirements whose authors have set themselves the goal of building a better, more efficient and sustainable energy system for future generations. As tools to achieve this objective, solutions that are as free-market as possible, thus increasing competition for, and ensuring efficiency in, the microeconomic sense have been adopted [2–5].

The proposed package of measures has three main objectives: energy efficiency as a priority, to make the European Union a leader in renewable energy, and to ensure fair treatment of consumers (European Commission 20192). In 2015, clean energy sources worldwide attracted investments worth €300 billion. The European Union is well prepared to use its research, technological development and innovation policies to turn the transition to clean energy into concrete industrial opportunities. By mobilizing public and private investments of €177 billion per year from 2021, this package can generate close to 1% of GDP growth over the next decade and create 900,000 new jobs [6].

The Clean Energy for all Europeans package of legislative proposals addresses issues of energy efficiency, renewable energy, electricity market structure, the security of energy supply and governance for the Energy Union. It also addresses actions to accelerate clean energy innovations and those that will help mitigate the social impact of the transition to clean energy.

In October 2014, the European Council agreed on a climate and energy policy framework for the EU by 2030, setting an ambitious objective of reducing greenhouse gas emissions from all sectors of the economy by at least 40% by 2030 (especially CO_2). The Paris Agreement on Climate Change confirms the EU approach. The implementation of the climate and energy policy framework agreed by the European Council until 2030 is a priority issue in relation to climate the Paris Agreement and the Energy Union strategy, one of the priorities of the Juncker's Commission [7].

In 2015, the Commission proposed to reform the EU Emissions Trading System (EU ETS), whereas, in summer 2016, it prepared proposals to accelerate the transition to an European low-emission economy. The package currently presented addresses the remaining key elements necessary to fully implement the EU's climate and energy policy framework, in particular concerning renewable energy and energy efficiency.

Therefore, it can be concluded that renewable energy represents a cost-effective source of energy that isolates energy markets and consumers from instability, promotes economic, and stimulates sustainable growth.

As the main problem of economic success for the EU in times of changing the global energy market is maintaining competitiveness, besides the abovementioned EU programs, the HORIZON EUROPE 2021–2027 framework program for research and innovation is becoming the most important (the Next EU Research and Innovation Program 2021–2027, as a continuation of HORIZON 2020 program) [8]. It is the largest project implemented by the EU, with a budget of EUR 95.5 billion.

All legislative proposals for the Energy Union presented in 2015 and 2016 are treated as a priority by the European Parliament and the Council.

Since the construction sector generates as much as half of the waste generated in the world and buildings consume 40% of the energy produced in EU countries, it is very important to implement and manage innovation in real estate. In property and asset management (PAM) innovations plays, nowadays, one of the key roles. PAM can be divided into three pillars: asset management, property management and facilities management [9].

Activity of housing developers refers to asset management because developers—investors usually sell flats after the investment is completed, sporadically keep them in their portfolio for rent. Therefore, their objectives are maximizing value and increasing returns from property. Innovation will increase the value of property but not necessarily the rates of return as the innovations will also increase costs—that is why proper management of the investment project is very important in this case. The framework of the management process of the construction process implementing the innovations indicates the need for identification of external environment variables like location, procurement form, innovation acceptance of the client, regulation degree and critical variables of the internal environment like service offer, knowledge strength, cooperative behaviour, financial strength and time needs [10]. Innovation's management in construction life cycle starts from opportunity exploration, where market trends, client insights, technology trends, data analytics, regulatory and competitor information plays an important role to manage the pro-jects or the contractors to investigate new innovations. At the front end of innovation level the project needs are studied, if a new tool, method or technology is identified and whether the new innovation is suitable, usable, and scalable. Then user experience tests are carried and whether the project team will be able to sustain using this new innovation is determined. During the backend design level aspects such as usability, serviceability, robustness and assembly are considered. At the commercialization level, change and leadership that are required for continued operational excellence are applied. At the daily management and project engagement level, training and project team engagement are important, and reporting and innovation implementation progress monitoring is applied. The final phase is improvement, where the effects from applying this new innovation is evaluated and innovation evolvement is investigated [11–15].

In the field of property management, and in particular facility management, innovations implemented in the construction phase allow for achieving higher efficiency by reducing the operating costs of buildings, improving the functionality, the environment in-side the building, or the impact on the natural environment through the emission of greenhouse gases or materials used in construction. These elements, especially reducing the operating costs of buildings, such as the consumption of energy needed to central heating, water heating, electricity, water consumption or generated waste, increase the value of the property. The innovations implemented in a building are often, nowadays confirmed by certificate granted by specially entities, which are an indisputable factor in increasing the value of real estate. For the most prestigious certifications we can include LEED (Leadership in Energy and Environmental Design) and BREEAM (Building Research Establishment's Environmental Assessment Method).

The aim of the paper is to analyse innovations, with particular emphasis on RES, introduced by residential developers in Poland in the context of global trends.

The research topics, covered in this paper, include:

1. The assessment of the position regarding the introduction of energy innovation in housing in the light of benefits and barriers.
2. Analysis of the introduction of innovations in Poland and worldwide in terms of RES.

2. Literature Review

In the case of the construction sector, the implementation of sustainable urban development assumptions, especially those concerning housing, is the responsibility of engineers, managers and research and development entities, focused on the implementation of innovations already in the processes of preparing materials for use in construction. In the case of the real estate market, especially residential real estate, the implementation of the assumptions of environmental sustainability is the responsibility of real estate developers and property buyers [16–18].

Green housing development requires a complex approach based on using green or technologies and on the behaviour of housing market actors. Many governments across the world introduce requirements or incentives to introduce innovations according to green housing and especially with energy efficiency. The Chinese government promotes energy efficiency in the new urban building sector by requiring 50% of new buildings to be green buildings [19,20]. Moreover, the European Union energy objective 2020 assumed having 20% of the energy consumption acquired from renewable energy sources. That leads to an increase in sustainability through the introduction of renewable energy innovations [21]. In UK nearly one third of total consumption of energy devoted to the domestic household sector, therefore sustainable housing developments have an important part to play in reducing greenhouse gas emissions and implementation of renewable energy sources in order to combat climate change [22]. Furthermore, in the perspective of the future 20–30 years the consumption of energy will be continuously increasing because of upward trends of using primary electric heating or electric cars. That is why the role of governments in implementation of RES is so important [23–25].

According to Uidhir et al. [26] in Ireland, improving the energy efficiency of residential flats will play a key role in achieving energy efficiency and GHG emissions reduction targets in 2020 and then to 2030. More than 80% of residential buildings have an energy efficiency rating (BER) of C. Ireland's Climate Action Plan 2019 aims to achieve 500,000 retrofits to a B2 standard by 2030, including the installation of 400,000 heat pumps. As exposed, the implementation of RES in residential buildings is becoming a standard, not the innovation.

Introducing the RES could also be an alternative for progressive tariffs. For example, in Italy and California, they seem to function well; however, they are not easily introduced. A free social minimum energy supply paid by progressive prices for high-end users can seem to be unfair. As exposed, the implementation of RES in residential buildings is becoming a standard, not the innovation [27].

Zhang et al. [28] research is interesting, and it is found that one of the engines of innovation in the Chinese real estate market is uncertainty. He shows that uncertainty has a positive impact on firms' innovation performance, including R&D investment, innovation level and other innovation outcomes.

Increasing of using RES is also important in the context of the existing trend in housing exactly growing floor space per capita. Forecasts for the EU increase from 20 m^2 in 1960 to 45 m^2 until now per person in the UK or 15 m^2 in post-war Germany to 45 m^2 in 2016 to over 50 m^2 per person until 2030. Increasing of using RES is also important in the context of the existing trend in housing, exactly growing floor space per capita [29].

Furthermore, trends referring to other than EU countries are the same, e.g., in the USA, an average is 74 m^2/cap. Despite the increasing trend, the EU figures compared with an average of 18 m^2/cap in Shanghai, 20 m^2/cap in Japan and 30 m^2/cap in Singapore are relatively high [30]. More floor space needs more energy for heating, cooling, ventilation and lighting. In this context, a good example is the GAG Immobilien AG (Köln, Germany), an incorporated housing company owned by the city of Cologne, that decided to withdraw from the stock market after 60 years of being listed, to do not have to maximize profit but instead be able to build smaller flats, thus, less energy-consuming ones [31].

Killip et al. ([32]) report that to investigate the process of implementing in investments, the analysis should be carried out of two markets: the first market of repair, maintenance and improvement (RMI) of homes, in which energy efficiency is not the most important; and the second market for deep retrofit, where energy performance is the main objective. The first RMI market is dominated by small enterprises, especially micro-businesses, working in local markets. These companies are not profiting maximisers or focused on building energy performance. In the second market, companies are focused on energy efficiency and are usually small and medium or large enterprises cooperating at the regional and national market. The same situation is in the primary housing market. According to Brown et al. ([33]), several elements can be distinguished for a successful business/management model for retrofit, including:

- A value-based on comfort, well-being, health and aesthetics
- Guaranteed energy savings
- Integrated supply chains able to provide a whole-home approach
- A single point of contact for customers
- An integrated low-cost financing model
- Coordination of all these elements

As we can see, energy efficiency is one of those basic elements. Introducing the innovations on the housing market refers not only to the use of homes but, also, nevertheless important, to resource efficiency in the industrialization of housing production. Industrialized Housing Construction (IHC) has been a strategy to deploy emerging innovations for resource-efficient housing construction [34,35]. It is important because the building construction industry alone is estimated to comprise 25% of virgin wood and 40% of raw stone, gravel and sand globally each year. It is also responsible for 40–50% of the global output of greenhouse gas [36]. In the process of management and value creation within IHC, the authors have distinguished subthemes presented in Figure 1, which include, among others, innovative and industrial materials and operational energy efficiency. It shows that innovations, including RES, are factors creating value in the real estate market.

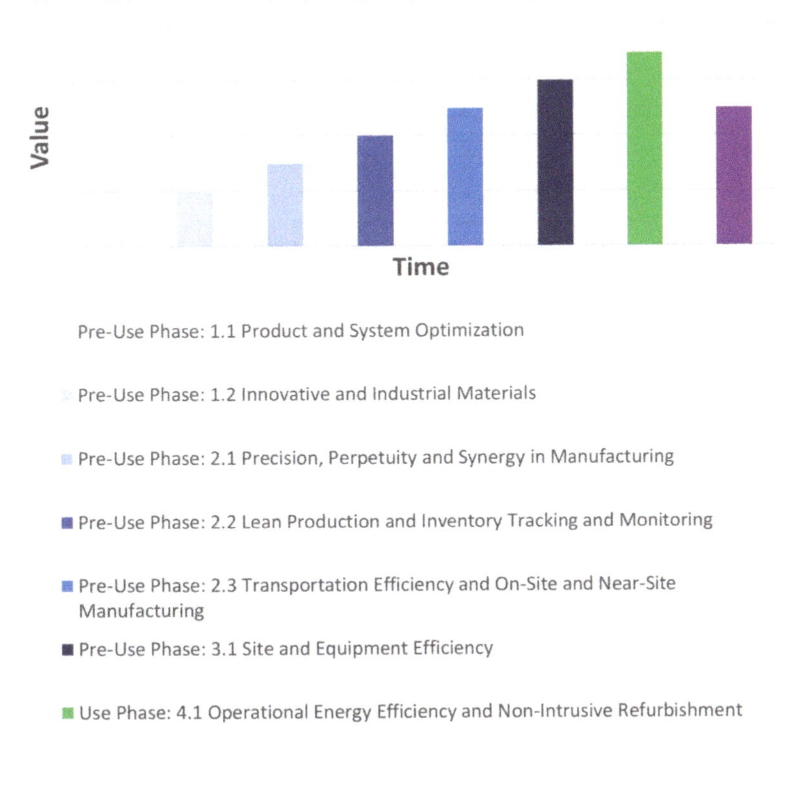

Figure 1. Subthemes in the process of management and value creation within IHC. Based on source: [34].

In relation to management, it should be stated that the core of modern society in the 21st century [37] (pp. 9–39), is not technology, information, or productivity: it is an investment managed as an instrument of society, responsible for producing certain results. In the case of RES, these results include, on the one hand, environmental protection by reducing gas emissions. On the other hand, they mean a reduction in operating costs of buildings and total or partial independence from external energy suppliers. A thorough analysis of management, according to the new quality paradigm, clearly shows that achieving and maintaining a high level of competitiveness requires continuous restructuring, including the improvement and introduction of new products or services, the development, search, and implementation of new technologies, the improvement of business management systems, production processes, sales, and marketing [38] (p. 13).

Therefore, it can be concluded that the basic determinant of global competitiveness in the 21st century will be the ability to create, share, and use knowledge, including, above all, generating and implementing innovation [39–41].

In the case of the construction sector, the implementation of sustainable urban development assumptions, especially those concerning housing, is the responsibility of engineers, managers and research and development entities focused on the implementation of innovations already in the processes of preparing materials for use in construction. In the case of the real estate market, especially residential real estate, the implementation of the assumptions of environmental sustainability is the responsibility of real estate developers and property buyers.

As presented above, innovations, especially renewable energy sources, play a significant role. The RES in the global energy mix is increasing from year to year. According to the International Renewable Energy Agency (IRENA), there were 2537 GW of renewable energy installations worldwide at the end of 2019. In 2019 alone, 176 GW of new power installed in RES was added.

The world's renewable energy sources are dominated by hydropower with an installed capacity of 1310.9 GW and wind power with an installed capacity of 622.7 GW.

The latest reports of the International Renewable Energy Agency [42,43] showed that the structure of RES in the world after 2019 was:

- 2537 GW of renewable energy installations (cumulative power),
- 176 GW was the renewable power added only in 2019,
- 54% accounted for new power installed in Asia in 2019,
- 72% of new generation capacities installed worldwide in 2019 were RES installations,
- the percentage of new wind and solar power in 2019 was 90%.

In Poland, the Institute of Renewable Energy (IEO) has already published the eighth edition of the report "Photovoltaic Market in Poland 2020" [39]. The report is a complete summary of the state and trends in the photovoltaic market in Poland.

The photovoltaics market is developing the fastest of all RES sectors in Poland. The total installed capacity of photovoltaic sources at the end of 2019 was almost 1500 MW, and, already in May 2020, it exceeded 1950 MW. The largest increase in new power is currently observed in the micro-installation segment, which illustrates the high activity of individual and business prosumers. In 2019, Poland achieved an increase in the new power about 0.9 GW and ranked among the top five in the European Union. New PV power installed in Poland accounted for 5.5% of the EU power growth.

According to the forecast, Poland will maintain its installed capacity growth rate this year and will be ranked 5th in the European Union. The Institute of Renewable Energy (IEO) estimates that at the end of 2020, the capacity installed in PV in Poland may reach 2.5 GW. IEO forecasts also indicate that the turnover in the photovoltaic market will increase this year compared to the previous one by as much as 25% and will exceed 5 billion PLN.

The most recent report by the Agency, containing a detailed overview that shows where and how much the percentage of renewable energy has increased, reveals that global RES increased by 7.6% last year. The leader is Asia with the growth accounting for as much as 54% of total added power. Although the expansion of RES slightly slowed down last year, the total growth of renewable energy still outpaced the growth of the use of fossil fuels (2.6 times), continuing the dominance of RES in the energy expansion achieved for the first time in 2012. Solar and wind energy accounted for 90% of the total renewable capacity added in 2019 [44,45] (pp. 58–80, 81–131).

Solar energy added 98 GW in 2019, with 60% being new power added in Asia. Nearly 60 GW increased wind power due to growth in China (26 GW) and the USA (9 GW). Both technologies generated 160 GW of global renewable power in 2019. It should be noted that wind and solar power, with 623 GW and 586 GW of installed capacity, respectively, cover almost half of global renewable capacity [44].

Hydropower, bioenergy, geothermal and marine energy showed a slight year-on-year increase of 12 GW, 6 GW, 700 MW and 500 MW, respectively. In the case of hydropower, an exceptionally low growth rate was recorded. The experts explain this situation by the fact that some large projects did not meet the expected completion dates. Most of the expansion occurred in China and Brazil, with each country increasing capacity by over 4 GW.

Continuous processes of intense globalization and accelerated economic growth are stimulated by increased competitiveness. In the case of the European Union, the main problem for economic success at a time when global energy markets are changing as a result of the transition to clean energy is to maintain competitiveness. Analysing the above literature review, it can be concluded that the basic determinant of global competitiveness in the 21st century will be the ability to create, share, and use knowledge, including, above all, generating and implementing innovation.

The literature review revealed gaps in the presented research, in particular regarding the introduction of innovation, including RES, the management of this process, the causes and barriers to innovation, and the effects of not undertaking innovation in the realities of housing developers' activity on the real estate market in Poland. In addition, the confrontation of the presented international scientific achievements regarding innovation on the real estate market in the context of renewable energy sources with Polish literature in this area clearly indicates competitiveness as the main element of the EU economic development strategy, while there are no reports on the impact of the market on the demand for this type of construction in the context of Poland.

Problems related to renewable energy sources in the context of real estate and innovation management, as well as the evaluation of the statement of developers regarding RES in the context of sustainable development, allow formulating hypotheses which, after verification and confirmation, will summarize the role of innovation and management in the context of renewable energy sources in Poland [17,45–49].

As the effect of literature review the following hypothesis were proposed:

Hypothesis H1. *The innovations introduced by developers into the housing market depend on the scale/range of their activities;*

Hypothesis H2. *Lack of innovation by housing developers is caused by the lack of demand for such construction;*

and

Hypothesis H3. *Barriers to innovation by housing developers limit their competitiveness.*

3. Materials and Methods

Research on innovation among developers of the primary housing market was conducted from 24 to 28 August 2020. The survey was conducted in the form of a survey on a sample of 130 entities: developers of the primary housing market implementing multi-family housing investments.

The survey was carried out using the CATI method (telephone survey using an online questionnaire) on a nationwide sample. The questionnaires were filled and returned from 130 entities, with the number of the entities that built blocks of flats, i.e., multi-storey, multi-family buildings, estimated as on 13 July 2020 at 324 entities.

The questionnaire consisted of 6 questions of the record, such as company name, its headquarter, geographic scope of the company's activity, origin of its capital, internship in the real estate market and employment and 25 substantive questions.

The survey dealt with innovations among developers of the primary housing market, where RES were only a small part of this survey. Since the most important characteristics of RES power are measured as the maximum net capacity of power plants and installations using RES for electricity generation while the obtained and generally available data reflect the installed and connected capacity at the end of the calendar year. The data were obtained from various sources, including a survey and secondary data such as IRENA questionnaires,

Renewables Global Status Reports, IEO, ERO (Polish Energy Regulatory Office), Statistics Poland reports, industry association reports and information articles.

Descriptive statistics were used to present the research results. It was found that the main elements resulting from the implementation of innovations in the primary housing market are the concept of intelligent sustainable construction, having an impact on energy management. The basic research method used in the study was a survey. The research tool used was the original CATI questionnaire a method of collecting information in the quantitative market and public opinion research). The significance test based on normal Laplace distribution was chosen to verify these hypotheses [46,47] (pp. 411–470). The data for the analysis were prepared based on the author's questionnaire.

The applied method of statistical inference allows in a relatively simple way, based on the experimental data, to verify the substantive hypothesis on the basis of the probability calculus. Hypothesis verification methods or significance tests are often used to test the relationship between variables, e.g., the causality test or the cointegration test or significance tests as hypothesis verification [50].

4. Results

The research part of the study concerns the analysis of the survey conducted on innovation among the developers of the primary housing market. The analysis of innovative activity on the housing market is most often carried out in the nearest market environment using specialized market research and reports and results of analyses prepared and published by public and private institutions such as IRENA, IEO reports, URE and Renewable Global Status Reports. The survey conducted in the paper enabled conclusions to be drawn on the motivations for introducing innovations in the housing market, types of innovations, benefits, barriers, support and management of innovations.

The survey concerned the research on innovation in development companies operating on the primary real estate market. Innovation research among residential construction companies on the primary market was carried out in August 2020. The survey among enterprises was conducted using the CATI method. Research using this technique is often used to develop a company's strategy or marketing activities. The study was conducted on a nationwide sample. The research sample consisted of 130 entities, with the general population constituting (± 5%) 324 entities building multi-storey, multi-family houses in Poland.

Developers—respondents (presented in Figure 2), representing companies operating in various geographical areas of the country, i.e., companies operating on the local market—61 companies; on the voivodeship market—52 companies; on the Polish national market—14 companies; and in the five largest Polish cities—three companies. Moreover, they represented companies with different origins of capital: Polish capital—122 companies; foreign capital—four companies; and mixed capital—four companies.

The presented companies were characterized by different years of operation from 5–9 years, for 10–15 years, to 15–24 years. Among the analysed companies, 65 companies employ up to 10 employees, 37 companies have employees from 10 to 49, five companies have employees from 50 to 249, two companies have more than 249 employees while the remaining companies did not want to disclose the number of employees.

The substantive questions are divided into four blocks—the motives for introducing innovations to the housing market and their nature, the reasons for cooperation and support for innovative activities, the effects of introducing innovations and their functional features, and managing the innovation portfolio.

The surveyed residential developers answered the question of whether they introduced innovations in their operations, giving 31.5% positive answers. Figure 3 graphically shows the answer to the question.

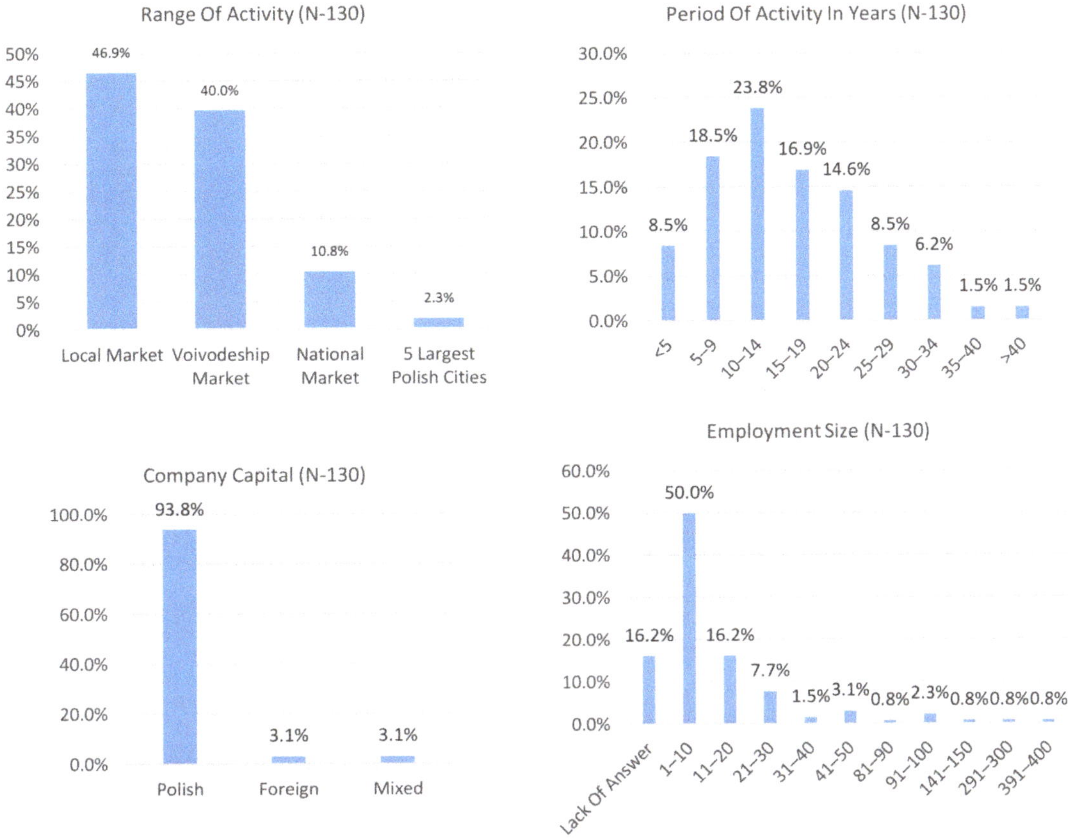

Figure 2. Characteristics of the research sample—entities building multi-storey, multi-family houses in Poland.

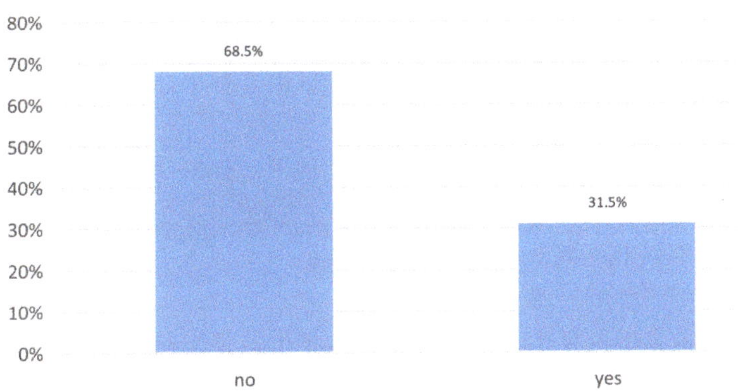

Figure 3. Do you introduce innovations in your activities in the housing market? (N = 130).

Other developers (68.5%) responded negatively. As can be seen, most of the enterprises studied do not introduce innovations in their operations, which may negatively affect their operations in the future. Furthermore, 53.9% of the respondents indicated that their customers chose conventional construction to implement innovative solutions. As arguments for not innovating, they mentioned the specificity of the enterprise (50%), attachment to conventional construction (37.5%), lack of demand (33.3%), or preferences for best practices (33.3%).

It is important to analyse the motivations and conditions for introducing innovations in the housing construction sector.

As regards the motivations for introducing innovations in the activity of a real estate developer on the housing market, the respondents mentioned fast sales of apartments, cost reduction, and growing customer requirements. Companies with a voivodeship range preferred especially customer satisfaction, while for companies with a nationwide range, the most important motivation was to improve the quality of products (apartments) and services. Reducing the environmental impact and increasing the company's market share are the most important motivations for developers operating in the five largest Polish cities.

While answering the question concerning the types of innovations, the developers listed technical and technological innovations as the most important (mean $\overline{36.25}$%), followed by procedural and organizational ($\overline{27}$ %), marketing ($\overline{23.75}$%) and financial ($\overline{13.5}$%) innovations. The details of the answers to this question are presented in Table 1.

Table 1. Types of innovations introduced by developers in the housing market.

	What Innovations Do You Use?			
	Local	Within the Voivodeship	Nationwide	The five largest Polish Cities
Process and organizational innovations	29%	27%	27%	25%
Technical and technological innovations	43%	37%	40%	25%
Marketing innovations	23%	27%	20%	25%
Financial innovations	6%	10%	13%	25%

It is technical and technological innovations that determine the directions of development for innovations in housing construction, especially related to RES. Their percentage is on average 36.25% of all technical and technological innovations. However, it is the RES innovations, regardless of the scope of the company's activity, that qualify for technical and technological innovations, which may favour the implementation of RES in the activity of real estate developers carrying out investment projects.

In the developers' opinion, the calculation of benefits and barriers determines cooperation with other entities in introducing innovations and also determines the direction of support for these activities.

The processes associated with innovative activities in the housing market are accompanied by a balance of benefits and barriers. When asked about the benefits of introducing innovations to the housing market, developers ranked them in the following order higher selling price of apartments, their faster sale, higher quality of apartments in a finishing stage, and a significant benefit, i.e., maintaining employment, which helps stabilize the company and guarantees its development.

According to housing developers, the most significant barriers include complicated legal regulations, lack of sufficient qualifications of employees, lack of financial resources, time pressure and lack of recognition of market needs for innovation.

When asked about the support needed for investment activity by residential developers, 50% of the companies answered that there was no need for support. Developers paid attention to technical support (25%) appreciated by companies implementing investment projects. They also emphasized financial, equipment and personnel support (25%).

Quite an interesting and varied answer was given by the developers to the question from the block on the effects of introducing innovations and their functional features.

As a result of introducing innovations in the field of housing construction, developers mentioned a higher selling price of apartments (33%), their faster sales (16.7%), higher quality of apartments in the developer standard (16.7%) and an increase in customer confidence through the company's image and reducing damage to the environment.

Innovations as factors influencing the interest in housing construction (functional features) are, according to developers, its high functionality (28.5%), ease of use of the installed innovative systems (24.5%), reduction of building operating costs (24%) and the possibility of expansion and remote access (23%).

Local, provincial and national enterprises most often indicate high functionality, while only enterprises operating in the largest five Polish cities indicate all of the above-mentioned features.

In response to the question of whether the problem of innovation management results from the company's strategy, 31.7% of the developers answered positively (due to gaining a leading market position among the competitors). Furthermore, 22% of respondents also confirmed the role of management in the adopted strategy due to the need to adapt products to the level offered by the competitors, while 46.2% did not give a clear answer.

For companies, it is important to guarantee their future business activity and continuous development.

While answering the topic of management and corporate strategy, 56.1% of the housing developers stated that introducing innovations to housing projects leads to difficulties in terms of managing them. Furthermore, 43.9% of them answered negatively. According to 100% of the respondents, the company's development strategy and investment project management depend on funds, of which 14.3% confirmed the dependence of management on the type of investment. Of the respondents, 31.7% stated that the introduction of innovations is an element of the company's development strategy due to gaining a leading position on the market among competitors. A total of 14.6% denied this statement, and 31.7% had no opinion. The answers to this question show that management assessed from the perspective of innovation must change with the company's development, technology development, knowledge and competence.

The developers also commented on the environmental benefits of bringing innovations to the housing market (the answer to this question was given by developers on a scale from 1 to 6, 6 being the maximum score, and 1 being the minimum score). A table of environmental benefits and their importance resulting from the introduction of innovations into the housing market is presented below.

Regardless of the scope of the activities of real estate developers, the environmental benefits were ranked by the developers by determining the arithmetic mean of each indication. They ranked the items as follows: water management ($\overline{5.325}$), reduced CO_2 emissions ($\overline{5.25}$), green technologies and recycled materials ($\overline{5.2}$), rational use of RES ($\overline{5.175}$), waste management ($\overline{4.95}$) and improving the environment in sustainable buildings ($\overline{4.85}$). The ranking determined based on developers' statements is surprising, because, in recent years (2018–2020), Poland has seen a rapid and very dynamic development of renewable energy sources (RES), with the benefits of using this innovation being disproportionate to the costs of these installations, although they are spread over time. The answer to this question also points to climate-friendly innovations.

The analysis of innovation effectiveness is most often performed from the perspective of the nearest market environment. There is a substantial variation in the assessment of the effectiveness of the innovations introduced. The most conclusive answers are those concerning the regional and national market. On the national market, the introduction of innovations results in a higher selling price of apartments, faster sales and higher quality of the final product, i.e., apartments in a finishing stage. A quite important factor, in this case, is the increase in customer confidence and improved corporate image.

The environmental benefits resulting from the introduction of innovations into the housing market are particularly based on energy efficiency through the rational use of energy as presented in Table 2. The most frequently chosen assessment of future activity by local developers is the attitude of maintaining an average level of innovation characteristic of the market in which developers operate. Real estate developers with a voivodeship and nationwide scope operating in the largest five Polish cities most often choose the attitude of conducting as many innovations as possible in their activities, which allows for minimization of the risk of bankruptcy and gaining competitive advantage. In terms of assessing the effectiveness of innovation, the resources involved in innovative activities are producing the expected results.

Table 2. Environmental benefits connected with bringing innovation to the housing market.

Please Prioritize the Environmental Benefits of Bringing Innovations to the Housing Market?				
	Local	Within the Voivodeship	Nationwide	Five Largest Polish Cities
Energy efficiency through the rational use of energy (use of renewable energy sources)	5.0	5.4	4.3	6.0
Improvement of the environment in sustainable buildings	4.5	5.1	3.8	6.0
Reducing harmful CO_2 emissions	4.8	5.2	5.0	6.0
The use of green technologies and recycled materials	4.6	5.2	5.0	6.0
Effective water economy	4.9	5.1	5.3	6.0
Rational waste management	4.4	4.6	4.8	6.0

Generally, thanks to the effects of introducing innovations in the field of housing construction, it was possible to achieve a higher selling price of flats, faster sales of flats, higher quality of the final product, i.e., flats in a finishing stage, and, importantly, to maintain employment. Furthermore, by introducing innovations in terms of housing construction, it has been possible to increase customer confidence in the company, improve the company's image, and reduce environmental impacts.

Some of the survey questions related to renewable energy are very important for ecological and economic reasons. Therefore, in order to confirm and summarize the results of the survey on the implementation of innovations, substantive hypotheses were proposed for verification.

The significance test based on normal Laplace distribution was chosen to verify these hypotheses [46,47] (pp. 411–470).

The level of statistical significance in the analysis was set at $\alpha = 0.05$. If the distribution of a trait in the population studied is normal, with expected value x_0 and standard deviation 6, and the parameter 6 is known, the significance test for $x = x_0$ will be met by first calculating the sample mean $\overline{x}$ and the value of the random variable u:

$$u = \frac{\overline{x} - x_0}{6}\sqrt{n}, \quad (1)$$

where n is the sample size;

x_0 is the expected value; and

6 is the standard deviation.

It seems that it would be more reasonable to take the weighted arithmetic average instead of the expected value x_0 [50] (pp. 589–595).

Since the weighted average is successfully used to calculate the average value, and all uncertainties, X_{ij}, are independent. Moreover, the weighted average gives the correct result only if the weights are independent, i.e., they are not correlated with each other.

For a pre-set level of significance α, the value of u_α is read from the table of cumulative distribution function, i.e., from the table of Laplace function, such that $|u| \geq u_\alpha$ is called the critical area for this test. If the value:

$$u = \frac{\overline{x} - x_0}{6}\sqrt{n}, \quad (2)$$

is such that $|u| \geq u_\alpha$, the hypothesis that $x = x_0$ is rejected. Otherwise, there is no reason to reject the hypothesis.

The significance test adopted in the paper verifies the hypothesis that certain features of set A and set B are statistically independent. Using the research tool such as the statistical method, the author considers the assumed hypotheses, which are calculated using the formulae reflecting the examination of the normality of statistical distribution.

It is essential to assess the statement of developers regarding renewable energy sources in the context of sustainable development. Therefore, data for the analysis were prepared based on the author's questionnaire. The data and calculations used to confirm the validity of the hypotheses are presented in Table 3.

Table 3. Empirical data used to verify the hypotheses.

n	H1	H2	H3
Local market	15	3	13
Voivodeship market	19	6	13
Polish market	6	13	9
Market of the five largest cities in Poland	1	1	6
	$\overline{x} = 10.25$ $x_0 = 12$	$\overline{x} = 6$ $x_0 = 7$	$\overline{x} = 10.25$ $x_0 = 12$

The sample mean was first calculated by $\overline{x}$, and a random alternate u value for a given sample size was calculated secondly. For a predetermined level of significance α, the table of cumulative distribution function, i.e., the Laplace function table is used to read the value of $u_\propto$ such that: $P\{\lceil u \rceil \geq u_\propto\} = \alpha$. The $u_\propto$ value is called the critical value for this test and is 1.96.

Since the questions about the specifics of the expected investments, planned implementations in the future, their positive effects and ecological benefits show that especially renewable energy innovations such as photovoltaics, heat/energy, solar collectors and Smart Home technologies (home automation such as heating, lighting, alarms), gates, roller shutters, ventilation, ingress protection IP cameras (protection level)—a combination of a camera and a computer-based on energy sources, controlling installations and devices remotely) are expected by customers; therefore, the hypotheses were related mostly to RES.

Hypothesis H1:

Innovations introduced by developers to the residential market depend on the scale/scope of their operations. Hypothesis H1 concerns the location of housing markets as local, provincial, national, and the market of the five largest cities in Poland.

$6^2 = -50.68$; $6 = 7.12$; $u = -0.50$; $u_\alpha = 1.96$

There is a relationship: $|u| \geq u_\propto$ i.e., $|-50| \geq 1.96$, which is untrue, i.e., there are no grounds to reject the H1 hypothesis.

Hypothesis H2:

The lack of innovation by housing developers is caused by the lack of demand for such construction. Hypothesis H2 is presented by questions about the strong increase in housing prices, better quality and higher and faster sales.

$6^2 = 17.75$; $6 = 4.21$; $u = 0.48$; $u_\alpha = 1.96$

There is a relationship: $|u| \geq u_\propto$ i.e., $|0.48| \geq 1.96$, which is untrue, i.e., there are no grounds to reject the H2 hypothesis.

Hypothesis H3:

Barriers to innovation by residential developers limit their competitiveness.

Hypothesis H3 was verified based on the answers of the survey respondents concerning maintaining employment, reducing environmental impact and improving the corporate image on the global market as well as taking advantage of the opportunities offered by the common European market. Therefore, it can be concluded that the introduction of innovations results from the corporate strategy, which is ultimately expected to guarantee the economic growth of the country.

$6^2 = 8.69$; $6 = 2.95$; $u = 1.19$; $u_\alpha = 1.96$

There is a relationship: $|u| \geq u_\alpha$ i.e., $1.19 \geq 1.96$, which is untrue, i.e., there are no grounds to reject the H3 hypothesis.

It can be seen that the verification of the hypotheses summarizing the results of the survey analysis confirmed their truthfulness for this market. It should be noted that, despite this fact, there is a good knowledge of the benefits and barriers to innovation in renewable energy on the real estate market, Poland was only a moderate innovator in the ranking of innovation systems of the EU Member States in 2020 [51].

5. Discussion

The percentage of energy generated from renewable energy sources remains small in Poland. The Polish power industry is still dominated by coal. The structure of electricity production in Poland in 2019 is presented in Table 4.

Table 4. The structure of electricity production in Poland in 2019.

Type	Percentage (%)
Utility hard coal-fired power plants	49.25
Utility brown coal-fired power plants	26.14
Wind power plants and other renewable energy sources	9.03
Utility gas-fired power plants	7.62
Industrial power plants	6.41
Utility hydropower plants	1.55

Source: [52].

Table 4 shows that the percentage of utility power plants in electricity generation in 2019 was 75.39% and renewable energy sources only 9.03%, while the percentage of utility hydropower plants was 1.55%. The capacity for renewable energy installations in Poland is almost 9.5 GW. The capacity for renewable energy installations in Poland is almost 9.5 GW. The largest source of electricity from RES is wind. Photovoltaic sources increased their percentage from 5% to 7%, according to the ERO publication on the capacity of individual types of installations based on renewable energy sources, as on 30 June 2020. During this period, the share of wind energy production in renewable energy is almost 64%, biomass—16%, water—10%, sun—7.5% and biogas—2.5%. From this, it follows that wind installations in Poland are the most numerous among the renewable energy sources, and their number is 1207. The second place is occupied by photovoltaics, with 1104 installations and water—771, biogas—371 and biomass—62. However, the ERO's comparison does not show the rapidly growing number of photovoltaic micro-installations, the owners of which settle accounts in the discount system or in the system of selling surpluses electricity at the average price from the wholesale market from the previous quarter. In 2020, the sector continued to develop very intensively. Home micro installations are currently experiencing a market boom due to the possibility of obtaining subsidies for the already relatively cheap PV panels.

RES in relation to multi-family and multi-storey houses can be used in two ways. First by introducing innovations-RES in heat plants supplying entire districts and even cities. This way of central heating and water heating is the most popular in Poland—called system heating. It is convenient for developers because heating costs fall on future owners or tenants, so the developer is not interested in minimizing them. Second way is to implement RES directly in buildings by developer. In this case developers should analyse the available RES and choose the appropriate for the particular project. It can be recommended that in the case of multi-family and multi-storey buildings, the most effective solution will be a combination of heat pumps and photovoltaic panels. The use of only photovoltaic panels is in most cases insufficient due to the need to find a large area for their installation. In the case of buildings in large cities, due to the high land prices, the buildings have several or more storeys to obtain a relatively large number of apartments on a plot of small area. Initially, the introduction of home wind farms was also discussed, but the change in tax regulations from 2017 in Poland made this solution unprofitable.

Furthermore, in the context of the Green Deal, which is assumed to eliminate net greenhouse gases by 55% by 2030 and no emissions by 2050, RES become of prime importance. Reaching this target will require action by all sectors of economy, including decarbonizing the energy sector and ensuring buildings are more energy efficient. Therefore, both heating plants and developers will have to introduce RES in their activity. The sooner they do it the better for them because of the experience and efficiency of introducing RES [53].

The research conducted has shown that residential developers in Poland point at, among others, competitive pressures, concern about the quality of products and services offered, increasing market share or satisfaction and increasing customer requirements. This clearly shows that the implementation of innovation is perceived through the prism of increasing competitiveness.

The conducted research revealed that competition is the basic factor driving the implementation of innovations among developers. The developers operating in the area of the five largest Polish cities attach the greatest importance to implementing innovations in their activities. The area of their activity is characterized by very high competition related to the possibility of quick sale of flats at a high price. However, in order to attract customers and be able to achieve high rates of return on investment, developers must offer a unique and attractive product. Therefore, it is recommended to developers, not only operating on the market of the five largest cities, to introduce innovations, including RES, in order to ensure the highest quality of offered apartments. The process of implementing the RES should, however, be preceded by the study of customer preferences and even a combination of the study of preferences with the simultaneous process of presenting possible innovative solutions, which customers are often unaware of, as shown by the study. The best solution would be the ability to create a specific need among customers for flats with specific innovative solutions, and even fashion for them. Identification of tools that can awaken such a need among customers can be a good direction for further research.

One of the ways to make the project more competitive are certificates. In the largest Polish real estate market, namely in Warsaw, the investment completed in 2019 was pre-certified under BiodiverCity. It is an international certificate that emphasizes biodiversity in newly constructed buildings and their surroundings. Thanks to its unique character, confirmed by the above certificate, the investment achieved very good sales. Therefore, another recommendation already confirmed by the market is the use of pro-ecological innovations confirmed by special certificates, which highlight projects by facilitating their sale and obtaining high rates of return.

Growth of competitiveness through the implementation of innovations, including RES, is motivated not only by market conditions but also by the assumptions of energy policy and development strategies of particular governments. It is these top-down regulations that are the main driving force behind the implementation of RES innovations.

In the study, developers as one of the barriers of introducing innovations, apart from the lack of adequate support, mainly from public entities in the field of, inter alia, appropriate law favouring the implementation of innovations and financial support, indicate difficulties of the client's market manifested by the lack of knowledge and identification of needs in the field of innovation, and the lack of willingness to pay a higher price in regard to the product with higher innovation level. Of course, developers have no influence on the legal solution and the implemented facilitations and preferences that the government should introduce. They can only indicate their preferences and needs in this regard. On the other hand, they should conduct market research to identify customer needs in relation to innovation, while informing and educating about the possibilities and advantages of implementing individual innovations, with particular emphasis on RES. In addition, the abovementioned barrier, which reflects the higher construction costs of innovative buildings, which in turn results in the higher price of the offered apartments, should be justified by the limitation of future operating costs and, what is interesting, the uniqueness and character of the offered apartments. Flats that take into account innovations are currently rare in Poland, especially in smaller real estate markets, which may distinguish them from the competition and allow them to achieve high rates of return on sales at higher-than-average prices. Of course, this forces developers to increase the regime and monitor the project in accordance with the plan, so that the implementation costs do not increase even more. This is all the more important for the Polish realities because it has been shown that the construction costs of buildings taking into account innovations, e.g., heat pumps, recuperation or photovoltaics, increase on average by 10%, while in countries such as Germany or Austria only by 2% [54]. Moreover, markets that are more developed and thus more competitive require the use of innovative solutions, which was also confirmed by the study. Developers operating nationwide and on the market of the five largest Polish cities, usually larger companies, showed a greater willingness and even the need to introduce innovation in their activities.

Taking into account all aspects of innovation in the light of RES from the adoption and support of the Polish Energy Policy until 2030 by the Polish government (sustainable consumption of energy from renewable sources), it has become necessary to develop new management principles ensuring constructive development of management theory and practice [38].

From the perspective of the research, new management principles should concern:

- Adequate financial management (acquiring sources of financing) to meet the need for a large amount of capital to carry out innovation;
- Appropriate project management to introduce innovation, strategic management tailored towards innovation and its introduction and creation in the long-term;
- In the phase of planning the investment appropriate identification of possible to implement innovation with the market survey in the field of determining the customers' needs in innovations;
- Introducing special tools connecting education in possible innovations and its benefits with simultaneous awakening the need for innovation in the housing among customers; and
- Increasing the regime in carrying out the project according to the plan because of the possibility of exceeding the costs related to the implementation of innovation.

Management is a tool to enable the organization to achieve this goal. Therefore, the main task of management is to meet the customer needs, so it is important to know the value and potential needs of the company's customers and adjust the quality of manufactured products (goods and services) to these needs. It should be noted that in the case of RES, such a need has not yet been identified by many customers due to the lack of full awareness of the benefits and, most importantly, future legal guidelines and their consequences (contained, for example, in the Energy Policy until 2030 for the EU).

In conclusion, it should be clearly emphasized that activities in the field of increasing innovation and, in particular, energy efficiency through, inter alia, implementation of RES, driven largely by strategies introduced by national governments or, e.g., the EU, will change the perception of RES as an innovation. Implemented programs, such as the program for the modernization of social housing in Ireland presented in the literature review, are carried out on such a large scale that the use of RES ceases to be an innovation on the market and is slowly becoming a standard.

6. Conclusions

The analysis of the results of the survey allows for drawing the following conclusions. First should be mentioned the factors motivating developers to undertake innovative activities in the real estate market. It has been identified that the most important factors include concerns for the quality of offered products and services, increased market share, reduction of environmental impact and pressure from the competitors. It can be observed that the above-mentioned, most important factors driving the implementation of innovations mostly relate to increasing competitiveness. It is true that pressure from the competitors is in the top 4 factors, but not in the first place, which informs that when analysing the national situation, innovations are not implemented on a large scale by developers.

On the other hand, the situation in the most developed real estate markets in Poland has shown that developers operating there most willingly introduce innovations and believe that introducing innovations is slowly becoming a necessary requirement in their activities. Additionally, in the context of Polish Energy Policy until 2030 and EU Green Deal, introducing innovations, especially RES, will become a necessity. Smaller development companies operating on local markets have shown that they have, so far, largely failed to introduce innovations and currently maintain their average or minimum level in order to avoid bankruptcy. For companies of this type, it is recommended to change the strategy in order to increase the share of innovation in their activities, which will be forced overtime by both the market (competition) and legal regulations.

According to the implementation of innovations in the field of RES, the conducted re-search revealed that almost all developers—with local, voivodeship and 5 biggest polish cities activity, pointed at use of renewable energy sources as the most important innovation which brings the environmental benefits. Only nationwide developers pointed at effective water economy as the most important innovations which brings the environmental benefits. However, this factor partly, also referred to RES because usage of solar panels enabled heating the water. Therefore, it can be concluded that developers should focus on implementation of RES in their innovative activity. Furthermore, the future operation cost limitation will be the best incentive for clients to pay higher price for the flat. Furthermore, the future operation cost limitation will be the best incentive for clients to pay higher price for the flat. In the context of high lack of knowledge about innovations and its benefits, this type of innovation seems to be one of the most important and recognizable.

However, innovation still faces several barriers, which entrepreneurs consider to be the most important. These include lack of financial resources, complex legal regulations, the uncertainty of the results of innovative activities, lack of sufficient qualifications, lack of assessment of market needs for innovation, lack of updated information on available technology, and rush and time pressure. Nevertheless, the main barriers that should be emphasized, besides complex and unfriendly legal regulations, which are the greatest barriers, are a lack of sufficient qualifications, rush and time pressure and poor recognition of the enterprise's market needs for innovation. When analysing the above, apart from changes in the legal framework on which developers have no direct influence, it should be stated that in order to overcome the barriers to implementing innovations, it is necessary to transfer knowledge—knowhow—in relation to the implementation of innovations, which will also minimize the time pressure. When analysing the above, apart from changes in the legal framework on which developers have no direct influence, it should be stated that in order to overcome the barriers to implementing innovations, it is necessary to transfer knowledge—knowhow—in relation to the implementation of innovations, which will also minimize the time pressure. Moreover, as part of carrying out investment projects, in the planning phase, developers should identify the possibilities of implementing innovation along with the resulting benefits and the market needs in this regard.

From the perspective of the research, change in management principles should concern, among others, adequate financial management (acquiring sources of financing) to meet the need for a large amount of capital to carry out innovation. Then appropriate project management to introduce innovation, with emphasis on identification of, in the phase of planning the investment, possible to implement innovation with the market sur-vey in the field of determining the customers' needs. Furthermore, developers should consider introducing of special tools connecting education about possible to implement innovations and its benefits with simultaneous awakening the need for such innovation in the housing among customers. Increasing the regime in carrying out the project according to the plan because of the possibility of exceeding the costs related to the implementation of innovation and strategic management tailored towards innovation and its introduction and creation in the long-term.

The research problems addressed in this study in the context of data obtained in direct research, studies and reports for 2020 were fully confirmed to be topical. Therefore, it can be assumed that they were realized and consistent with global data for the period analysed.

Since the study concerned broadly understood innovations in the activity of housing developers in Poland, and renewable energy sources of RES were only a small part of this study, it seems justified to undertake further research to identify the possibilities of introducing RES in residential development projects, identification of barriers and benefits resulting from the implementation of RES. Such research seems to be needed in the housing market, which was indicated by the respondents of the survey, at the same time indicating the use of RES.

Author Contributions: Conceptualization, M.S.; methodology, M.S.; validation, M.S. and M.T.; formal analysis, M.S. and M.T.; investigation, M.S.; resources, M.S. and M.T.; data curation, M.S.; writing—original draft preparation, M.S. and M.T.; writing—review and editing, M.T.; visualization, M.S.; supervision, M.T. All authors have read and agreed to the published version of the manuscript.

Funding: This research received no external funding.

Informed Consent Statement: Informed consent was obtained from all subjects involved in the study.

Data Availability Statement: Data sharing not applicable.

Conflicts of Interest: The authors declare no conflict of interest.

References

1. European Commission. Press Release, 30 November 2016 Brussels, Clean Energy for all Europeans, i.e., Unleashing Europe's Growth Potential. 2016. Available online: https://ec.europa.eu/commission/presscorner/detail/pl/IP_16_4009 (accessed on 10 May 2020).
2. *Regulation (EU) 2019/941 of the European Parliament and of the Council of 5 June 2019 on Emergency Preparedness in the Electricity Sector and Repealing Directive 2005/89/EC*; EU of 14.6.2019, L 158/1; Official Journal of the European Union, EUR-LEX: Brussels, Belgium, 2019; p. 1.
3. *Regulation (EU) 2019/942 of the European Parliament and of the Council of 5 June 2019 establishing a European Union Agency for the Cooperation of Energy Regulators*; EU of 14.6.2019, L 158/22; Official Journal of the European Union, EUR-LEX: Brussels, Belgium, 2019; p. 1.
4. *Regulation (EU) 2019/943 of the European Parliament and of the Council of 5 June 2019 on the Internal Market in Electricity*; 14.6.2019 L 158/54; Official Journal of the European Union, EUR-LEX: Brussels, Belgium, 2019; p. 1.
5. Regulation (EU) 2019/944 of the European Parliament and of the Council of 5 June 2019 on Common Rules for the Internal Market in Electricity and Amending. Directive. Available online: https://ec.europa.eu/commission/sites/beta-political/files/priorities-report_pl_0.pdf (accessed on 10 May 2020).
6. European Commission. Clean Energy for All Europeans—Unlocking Europe's Potential. 2016. Available online: https://ec.europa.eu/commission/presscorner/detail/en/IP_16_4009 (accessed on 10 May 2020).
7. European Commission. Progress towards the Commission's 10 Priorities European Union. 2019. Available online: https://www.europarl.europa.eu/RegData/etudes/IDAN/2019/637943/EPRS_IDA(2019)637943_EN.pdf (accessed on 10 May 2020).
8. Horizon Europe. *The Next EU Research & Innovation Investment Programme (2021–2027)*; Publications Office of the European Union (European Commission): Brussels, Belgium, 2019.
9. Thompson, B. Innovation in property management. *J. Prop. Invest. Financ.* **2015**, *33*, 436–445. [CrossRef]
10. Hartmann, A. The context of innovation management in construction firms. *Constr. Manag. Econ.* **2006**, *24*, 567–578. [CrossRef]
11. Abusalah, M.; Tait, J. Innovation management in construction—Practical approach. In Proceedings of the 13th Pipeline Technology Conference, Berlin, Germany, 12–14 March 2018. Available online: https://www.pipeline-conference.com/conferences/13th-pipeline-technology-conference-2018 (accessed on 18 February 2021).
12. Grondys, K.; Androniceanu, A.; Dacko-Pikiewicz, Z. Energy management in the operation of enterprises in the light of the applicable provisions of the energy efficiency directive (2012/27/EU). *Energies* **2020**, *13*, 4338. [CrossRef]
13. Sitek, M. Identification of risk factors as an element of the process of risk management in the real estate market. *Pol. J. Manag. Stud.* **2013**, *7*, 255–264.
14. Grabara, J.; Tleppayev, A.; Dabylova, M.; Mihardjo, L.W.W.; Dacko-Pikiewicz, Z. Empirical research on the relationship amongst renewable energy consumption, economic growth and foreign direct investment in Kazakhstan and Uzbekistan. *Energies* **2021**, *14*, 332. [CrossRef]

15. Skibiński, A.; Sipa, M. Sources of innovation of small businesses: Polish perspective. *Procedia Econ. Financ.* **2015**, *27*, 429–437. [CrossRef]
16. Sarma, U.; Karnitis, G.; Zuters, J.; Karnitis, E. District heating networks: Enhancement of the efficiency. *Insights Reg. Dev.* **2019**, *1*, 200–213. [CrossRef]
17. Vlasov, A.I.; Shakhnov, V.A.; Filin, S.S.; Krivoshein, A.I. Sustainable energy systems in the digital economy: Concept of smart machines. *Entrep. Sustain. Issues* **2019**, *6*, 1975–1986. [CrossRef]
18. Dudin, M.N.; Frolova, E.E.; Protopopova, O.V.; Mamedov, A.A.; Odintsov, S.V. Study of innovative technologies in the energy industry: Nontraditional and renewable energy sources. *Entrep. Sustain. Issues* **2019**, *6*, 1704–1713. [CrossRef]
19. Jiang, H.; Payne, S. Green housing transition in the Chinese housing market: A behavioural analysis of real estate enterprises. *J. Clean. Prod.* **2019**, *241*, 118381. [CrossRef]
20. National Development and Reform Commission (NDRC). The US-China Business Council. 2016. Available online: En.ndrc.gov.cn (accessed on 10 May 2020).
21. Maassen, M.A. Correlations between energy economy and housing market prices in the EU-impacts on future sustainability. In *Proceedings of the 11th International Conference on Business Excellence, Bucharest, Romania, 30–31 March 2017*; Dima, A.M., Ed.; Sciendo: Bucharest, Romania, 2017; pp. 45–54.
22. Rossiter, W.; Smith, D.J. Green innovation and the development of sustainable communities: The case of blueprint regeneration's trent basin development. *Int. J. Entrep. Innov.* **2018**, *19*, 21–32. [CrossRef]
23. Sprake, D.; Vagapov, Y.; Lupin, S.; Anuchin, A. Housing estate energy storage feasibility for a 2050 scenario. In *Proceedings of the 2017 7th International Conference Internet Technologies and Applications (ITA), Wrexham, North Wales, UK, 12–15 September 2017*; Picking, R., Ed.; IEEE: Piscataway, NJ, USA, 2017; pp. 137–142.
24. Lane, M.; van der Horst, D.; Tingey, M.; Smith, C.; Webb, J. Social innovation in the shadow of policy failure: Energy efficiency in self-build housing. *Glob. Transit.* **2020**, *2*, 180–189. [CrossRef]
25. Lowe, R.; Chiu, L.F. Innovation in deep housing retrofit in the United Kingdom: The role of situated creativity in transforming practice. *Energy Res. Soc. Sci.* **2020**, *63*, 101391. [CrossRef]
26. Uidhir, T.M.; Rogan, F.; Collins, M.; Curtis, J.; Gallachóir, B.Ó. Improving energy savings from a residential retrofit policy: A new model to inform better retrofit decisions. *Energy Build.* **2020**, *209*, 109656. [CrossRef]
27. Dehmel, C. *Der Einfluss von Progressiven Tarifen Auf den Stromkonsum in Privaten Haushalten in Italien und Kalifornien*; Transpose Working Paper No10; University of Münster: Berlin, Germany, 2011.
28. Zhang, D. Innovation dynamics -what are the housing market uncertainty's impacts. *Int. Rev. Econ. Financ.* **2020**, *70*, 413–422. [CrossRef]
29. Deschermeier, P.; Henger, R. Die Bedeutung des zukünftigen Kohorteneffekts auf den Wohnflächenkonsum. *Iw-Trends Inst. Dtsch. Wirtsch. Köln* **2015**, *3*, 23–37.
30. Ho, A. *The Unlivable Dwellings in Hongkong and the Minimum Living Space*; Hong Kong Free Press: Kennedy Town, Hong Kong, 2015; Available online: https://www.hongkongfp.com/the_unlivable_dwellings_in_Hongkong_and_the_minimum_living_space (accessed on 15 October 2016).
31. Lorek, S.; Spangenberg, J.H. Energy sufficiency through social innovation in housing. *Energy Policy* **2019**, *126*, 287–294. [CrossRef]
32. Killip, G.; Owen, A. The construction industry as agents of energy demand configuration in the existing housing stock. *Energy Policy* **2020**, *147*, 111816. [CrossRef]
33. Brown, D. Business models for residential retrofit in the UK: A critical assessment of five key archetypes. *Energy Effic.* **2018**, *11*, 1497–1517. [CrossRef]
34. Kedir, F.; Hall, D.M. Resource efficiency in industrialized housing construction—A systematic review of current performance and future opportunities. *J. Clean. Prod.* **2021**, *286*, 125443. [CrossRef]
35. Rohn, H.; Pastewski, N.; Lettenmeier, M.; Wiesen, K.; Bienge, K. Resource efficiency potential of selected technologies, products and strategies. *Sci. Total Env.* **2014**, *473–474*, 32–35. [CrossRef]
36. Khasreen, M.M.; Banfill, P.F.G.; Menzies, G.F. Life-cycle assessment and the environmental impact of buildings: A review. *Sustainability* **2009**, *1*, 674–701. [CrossRef]
37. Drucker, P.F. *Zarządzanie XXI Wieku—Wyzwania*; MT Biznes: Warsaw, Poland, 2009.
38. Grudzewski, W.M. Współczesne kierunki rozwoju nauk o zarządzaniu. *Ekon. Organ. Przedsiębiorstwa* **2006**, *3*, 13.
39. Harvard Business Review Polska. *Dystrybucja Niezawodności. Studium Przypadku PKP Energetyka*; ICAN Institute: Warsaw, Poland, 2019.
40. Green Projects. Dekarbonizacja Najważniejszym Zadaniem Ludzkości. 2020. Available online: https://www.green-projects.pl/dekarbonizacja-najwazniejszym-zadaniem-ludzkosci/ (accessed on 15 September 2020).
41. Goldstein, J.S.; Qvist, S.A. *Energia dla Klimatu*; Wydawnictwo PWN: Warszawa, Poland, 2019.
42. *Directive (EU) 2019/944 on Common Rules for the Internal Market for Electricity and Amending Directive 2012/27/EU (Recast)*; L 158/125; Official Journal of the European Union, EUR-LEX: Brussels, Belgium, 2019.
43. Institute of Renewable Energy (IEO). Raport Photovoltaic Market in Poland. 2020. Available online: https://www.ieo.pl/pl/aktualnosci/1471-raport-rynek-fotowoltaiki-w-posce-2021/ (accessed on 10 May 2020).
44. GlobEnergia, News. There Is Already 2537 GW of Installed Capacity from RES in the World. 2020. Available online: https://globenergia.pl/na-swiecie-jest-juz-2-537-gw-mocy-zainstalowanej-z-oze (accessed on 10 May 2020).

45. REN 21. Renewables 2020, Global Status Report. 2020. pp. 81–131. Available online: https://www.ren21.net/wp-content/uploads/2019/05/gsr_2020_full_report_en.pdf (accessed on 10 December 2020).
46. Rosario, N.; Mantegna, R.N.; Stanley, E.E. *Ekonofizyka*; Wydawnictwo Naukowe PWN: Warszawa, Poland, 2001; pp. 23–29.
47. Berenson, M.L.; Levine, D.M. *Basic Business Statistics. Concepts and Applications*; Prentice Hall Inc.: Upper Saddle River, NJ, USA, 1999; pp. 291–335.
48. Tyo, A.; Jazykbayeva, B.; Ten, T.; Kogay, G.; Spanova, B. Development tendencies of heat and energy resources: Evidence of Kazakhstan. *Entrep. Sustain. Issues* **2019**, *7*, 1514–1524. [CrossRef]
49. El Iysaouy, L.; El Idrissi, N.E.; Tvaronavičienė, M.; Lahbabi, M.; Oumnad, A. Towards energy efficiency: Case of Morocco. *Insights Reg. Dev.* **2019**, *1*, 259–271. [CrossRef]
50. Cox, M.G. The evaluation of key comparison data. *Metrology* **2002**, *39*, 589–595.
51. European Innovation Scoreboard 2020. Internal Market, Industry, Entrepreneurship and SMEs. Available online: https://ec.europa.eu/growth/industry/policy/innovation/scoreboards_en (accessed on 10 May 2020).
52. Kwinta, W. Odnawialne Źródła Energii w Polsce i na Świecie. Available online: https://inzynieria.com/energetyka/odnawialne_zrodla_energii/rankingi/58459,odnawialne-zrodla-energii-w-polsce-i-na-swiecie (accessed on 10 January 2021).
53. European Commission. Actions Being Taken by EU. Available online: https://ec.europa.eu/info/strategy/priorities-2019-2024/european-green-deal_en (accessed on 20 February 2021).
54. Rewolucja w Budownictwie—po 31 Grudnia 2020 r. Budynki o Niemal Zerowym Zużyciu Energii. Available online: https://wentylacja.com.pl/news/rewolucja-w-budownictwie-po-31-grudnia-2020-r-budynki-o-niemal-zerowym-zuzyciu-energii-55356.html (accessed on 20 February 2021).

energies

MDPI

Article

Households' Energy Autonomy: Risks or Benefits for a State?

Marko Milojević [1], Paweł Nowodziński [2,*], Ivica Terzić [3] and Svetlana Danshina [4]

1 Department of Accounting and Audit, Singidunum University, 11000 Belgrade, Serbia; mmilojevic@singidunum.ac.rs
2 Faculty of Management, Czestochowa University of Technology, 42-201 Częstochowa, Poland
3 Department of Finance and Banking, Singidunum University, 11000 Belgrade, Serbia; iterzic@singidunum.ac.rs
4 Department of Propaedeutics of Dental Diseases, Sechenov First Moscow State Medical University, 115201 Moscow, Russia; SvetTi.Danshina@yandex.ru
* Correspondence: pawel.nowodzinski@wz.pcz.pl

Abstract: The purpose of this study is to determine the impact of households' energy autonomy on a country's energy independence level, to identify prospects and risks. To assess the economic efficiency of households' energy autonomy, the study used a modeling method based on maximizing the net present value, determining the average notional cost of energy efficiency and the level of energy independence in 20 countries. Based on the analysis of the volumes of electricity consumption by households in the studied countries for the period 2000–2018, it was revealed that in developed and developing countries there is an increase in this indicator. Diagnostics of the investment attractiveness of the installation and operation of energy systems for households makes it possible to determine the boundaries of a possible increase in the level of their energy autonomy. The scientific novelty of the research is represented by the proposed methodological approach, which makes it possible to assess the level of energy dependence of countries, possible deviations, and an increase in households' energy autonomy in relation to the risk limit of energy dependence. The proposed methodological approach allowed the authors to prove the positive impact of increasing households' energy autonomy for most developed countries. The most positive effect is characteristic of the leading countries in fossil energy market.

Keywords: deviation; energy dependence; energy efficiency management; energy saving; cost; risk limit

Citation: Milojević, M.; Nowodziński, P.; Terzić, I.; Danshina, S. Households' Energy Autonomy: Risks or Benefits for a State? *Energies* **2021**, *14*, 2026. https://doi.org/10.3390/en14072026

Academic Editor: Manuela Tvaronavičienė

Received: 18 February 2021
Accepted: 27 March 2021
Published: 6 April 2021

1. Introduction

In the modern world, the main factor in the economic development of any country is energy, the efficiency of which determines the level and rate of improvement in population welfare [1]. Each country chooses its own way of developing the energy sector, which depends both on the availability of natural resources (coal, oil, natural gas) and on the level of well-being of the population. The formation of an energy-efficient model of economic development of countries is a fundamental problem of the global level, which is of a systemic nature and significantly determines the key parameters of energy independence. A reliable and uninterrupted supply of electrical energy is essential for the functioning of any economy. As the economy grows, the demand for electricity increases in line with population growth, industrialization, and income [2].

New technologies contribute to the economical use of energy in households, which leads to greater household independence. Digitalization and automation processes are bringing smart home systems to interact with power systems. Automated systems for managing energy consumption of a household are considered in the context of the transition to smart home and a grid based on contractual energy supply and a scenario of variable payment for electricity [3]. In addition to the development of energy-producing technologies, it is planned to introduce new energy service technologies based on the concept of "smart" power grids [4]. The concept of universal intelligent machines is especially popular to improve the efficiency of services provided to households in the context of information management [5]. Along with this, an important direction is to reduce the costs for the transportation of electricity by optimizing the design of district heating networks [6]. It is energy that today acts as the initiator of the progress of the world economy and occupies one of the first places, performing a key function in the economic, political, and social life of any state [7]. Decentralized energy production offers households significant potential to support the achievement of climate goals [8]. Therefore, this study is aimed at identifying the benefits and risks for countries with different levels of economic development and is valuable in the context of determining the possibilities of influencing the level of households' energy autonomy in the context of the global energy transition.

Households are both producers and consumers in the energy market. Decentralized energy production is becoming a subject of increased relevance, but at the same time, it also signifies an uncertainty in the event of disruptions in the distribution of autonomous energy systems. Energy supply companies, communities, and small businesses and households have access to the energy market today. Thus, increasing the level of energy autonomy of households provides an opportunity for homeowners to function in the energy market. At the same time, the motivational aspects for both households and a state have not been sufficiently studied, including their mutual benefits and risks in the event of an unsuccessful proliferation of autonomous energy systems and the ongoing liberalization of the energy market. This study aims to fill this gap by examining the mutual effects for households and a state in relation to their opportunities and threats. All this contributed to the formation of the goal of this study, which is to determine the impact of households' energy autonomy on the level of energy independence of a state, to identify prospects and risks.

The scientific novelty of the research is represented by the proposed methodological approach, which makes it possible to assess the level of energy dependence of countries, possible deviations, and an increase in households' energy autonomy in relation to the risk limit of energy dependence. This study is based on a methodological approach to modeling the maximization of net present value, which allows identifying the average notional costs of energy efficiency and the level of energy independence in the 20 countries studied. This made it possible to shape their overall dynamics until 2030. Using this methodology, it has been determined that independent energy systems have clear advantages. Decentralized energy systems are emerging against the backdrop of the population's tendency towards energy consumption, and therefore energy autonomy is the main motive for investing in local renewable energy sources.

The approach proposed in the study involves the formation of a set of predictive indicators of the impact of households' energy autonomy on the level of energy independence of the studied developed and developing countries. The motivation for the study is to determine the equality of opportunities for energy autonomy of households in countries with different income levels. In the end, government support for household energy autonomy can be energy efficient, but it can also best benefit higher-income households and thereby contribute to increasing inequality in society.

To achieve this goal, research tasks were formed, which determined the structure of this study:

- identifying the possibility of changing the level of energy autonomy of households and analyzing the volumes of electricity consumption in the studied countries for the period 2000–2018;
- determination of households' energy autonomy based on the level of energy savings in a country and the notional cost of 1% of energy savings for households;
- analysis and comparison of the impact of households' energy autonomy on the level of energy saving in the studied countries.

2. Literature Review

The concept of planning sustainable energy systems is viewed as multi-criteria decision analysis. However, in most of the previous studies, the impact of energy production technologies on public and private sectors has been considered separately. Sustainable planning of energy systems and their components should include both options for possible influence when making decisions within the framework of a multi-criteria decision analysis [9].

The use of information intervention to stimulate energy saving of residents is attracting more and more attention. However, the effect of different information content and intervention strategies remains controversial. In addition, there are not enough studies that assess (using a field experiment) the effectiveness of individual information interventions in motivating energy conservation in urban households. Of the four separate communication strategies, only environmental feedback and cost-benefit feedback had a significant incentive effect on household energy savings, while the impact of regulatory information and information on environmental education was negligible. The energy-saving effect of feedback on environmental contributions was more significant than the effect of feedback on costs and benefits, regardless of whether individual or joint activities were implemented, while the energy-saving effect of environmental education information was negligible [10–12].

It is critical to achieve the carbon emissions reductions set out in the EU 2050 targets, limiting energy consumption [6]. The transition of households to efficient energy consumption in the residential sector proved to be quite difficult, while one of the factors contributing to the regression was determined by human behavior in the field of energy consumption [13,14]. More traditional methods are being used (information campaigns and feedback) to stimulate households to change their behavior. However, these measures tend to be of a short-term nature, as they ignore the underlying causes of such practices [15]. A more efficient solution is a practice-oriented design, where innovative technologies are created jointly with a user. In addition, the emergence and use of automated technologies allow practitioners to act independently of a user. However, the success of automation also depends on understanding the home practice system, the needs and skills of a user who represents a household [16,17].

Contemporary research examines psychological barriers to reducing energy demand in the context of introducing energy-efficient technologies in households and discusses ways to overcome them. At the same time, behavioral approaches to overcoming these limitations are discussed, namely:

- an emphasis on the public choice of "green" technologies, simplification and optimization of this choice;
- reframing benefits;
- changing the time structure of costs and benefits;
- emphasis on the symbolic attributes of new technologies [18];
- behavior change aimed to reduce energy consumption in households [19].

Depending on the level of income, groups of households can be distinguished that react differently to fluctuations in energy prices in residential buildings. At the same time, energy poor households mainly belong to the group of households with the highest elasticity. Income insecurity does not necessarily mean fuel poverty [20]. At the same time, fluctuations in the development of energy companies are possible [21], which also affects the motivation for energy autonomy of households. However, subsidies, which cover a significant portion of total investment, play a significant role in household energy decision-making. Depending on their design, support measures can best benefit certain groups in society and thus can increase inequality. The potential distributional impact of investment subsidies is determined by a combination of targeted technologies and their costs, household income, support intensity, subsidy structure constraints, and additional measures [22].

For developing countries in the energy sector, electricity is not predominant. Biofuels and kerosene are the most common fuels used in the daily life of people in developing countries. The transition from these fuels to more modern forms of energy is already taking place in the 21st century. The government has a great influence on household energy consumption, and different governments have different priorities. For example, a government can subsidize up to 90% of the final electricity for households, thereby significantly increasing energy consumption, which has negative consequences for sustainable development. Effective strategies to reduce energy consumption represent individually tailored information and feedback from users, as well as a clear statement of goals by decision makers for both developed and developing countries [23,24].

Although there is a high level of knowledge about reducing energy demand, scattered information needs to be integrated to develop a combined and inclusive approach to managing energy demand in households. The knowledge gathered will inform decision makers who are involved in the design of residential buildings, energy consumption of households, and planning for sustainable communities to identify activities that can be implemented in a given locality [25–28].

Households' green energy production can make a positive contribution to a state's energy supply, increasing the environmental friendliness of the energy complex, ensuring the transition to the use of renewable energy sources, and reducing the use of fossil fuels [29,30]. At the same time, the pace of development of the private renewable energy sector is not sufficient to make a significant contribution to the achievement of indicators of national plans and programs [31,32]. The main reasons are doubts about the financial feasibility of such projects, requiring state support, and insufficient incomes of a country's population, which do not allow accumulating funds for investment in renewables, along with the high cost of credit resources. In this regard, the main direction for further development of private power plants using renewable energy sources is to strengthen financial state support for such projects [33]. However, with the development of a renewable energy sector, another problem is gradually being identified, which will worsen over time. Namely, an increase in renewable energy volumes at existing high rates of the "green" tariff can lead to a gradual increase in electricity prices, since the compensation for increased "green" tariffs occurs due to an increase in average prices for electricity, obtained from both traditional and from renewable sources. A situation arises when the growth in the production of "green" energy is paid for through the mechanism of increased prices by all energy consumers, and only a few can earn on increased tariffs [34–36]. Further expansion of a renewable energy sector with relatively high "green" tariffs can lead to social problems generated by the stratification of the population in terms of income based on the production of "green" electricity [37–39]. Taking into account the multidirectionality of modern research and the obtained results of the impact of households' energy autonomy, there is a need to develop methods for assessing the efficiency and permissible limits of households' energy autonomy in the context of a country's energy security. Therefore, this study is aimed at identifying possible threats and benefits of increasing households' energy autonomy in countries of different levels. For this, the authors have developed a methodological approach to assessing the impact of households' energy autonomy on the level of energy dependence of the studied developed countries.

3. Materials and Methods

To assess the economic efficiency of households' energy autonomy, the study used a model that maximizes the net present value and the level of energy independence of countries. Therefore, for the study, countries were selected that met certain criteria: the level of GDP per capita, and the second factor was the energy intensity of GDP (Figure 1). These criteria were the main ones for including countries in the study since energy autonomy requires a study of the equality of opportunities for households in countries with different income levels. A key prerequisite for this is that energy autonomy of households can be supported by a state and be quite effective, but at the same time it can be more beneficial for households with higher incomes, which may increase inequality. Besides, countries representing different regions of the world were included in the sample.

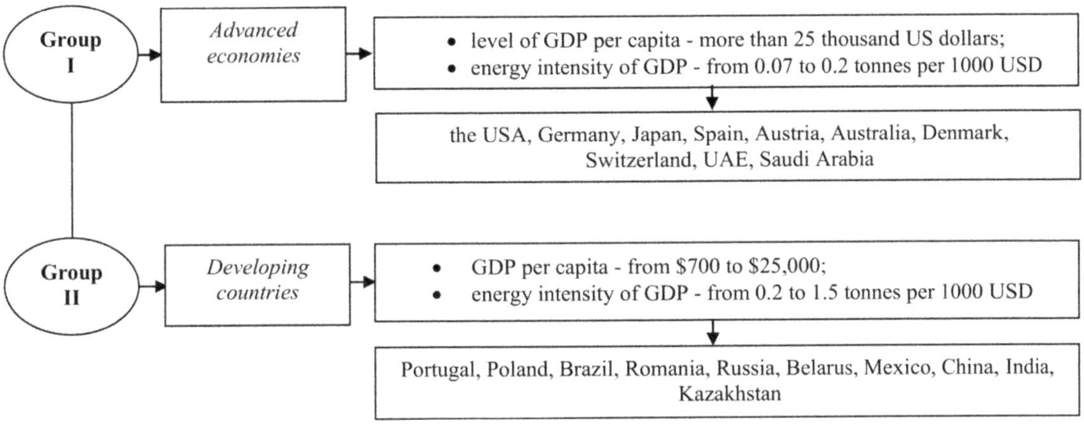

Figure 1. Distribution of the studied countries by groups. Source: generated by the authors.

To determine the level of costs for the transition to autonomous systems in households, a modeling method was used. Based on the formed investment projects to increase the level of households' energy autonomy, an assessment was carried out, and the average notional cost of 1% of energy savings for the autonomous systems under study was determined.

The modeling is based on the assumption that there is an effective amount of investment for the use of each autonomous energy system by a household. Based on econometric diagnostics of the obtained results of modeling, the study analyzed the interdependence of energy saving and energy efficiency of households' autonomous energy systems, namely, the corresponding volume of investments. In this case, the following dependence is assumed:

$$\varphi_\tau(\Delta HS_{aut}) = hs_{0\tau}^{autin} + hs_{1\tau}^{autin}\Delta HS_{aut} \quad (1)$$

where HS_{aut}—the level of energy savings of a household as a result of introducing an autonomous energy system (%);

τ—investment option index according to autonomous system of using renewable energy sources, $\tau = \overline{1.5}$;

$\varphi_\tau(\Delta HS_{aut})$ —the amount of power generation for the investment option τ;

$hs_{0\tau}^{autin}, hs_{1\tau}^{autin}$—parameters of the econometric model for the investment option τ on the implementation of an autonomous energy system.

Based on the proposed methodological approach to the construction of investment projects for the introduction of autonomous energy systems by households, the authors determined the expected value of the energy saving volumes function $\varphi_\tau(\Delta HS_{aut})$ of random variable ΔHS_{aut} with a distribution density $f(\Delta HS_{aut})$. Wherein $f(\Delta HS_{aut})$ has the following value:

$$f(\Delta HS_{aut}) = \frac{1}{\sigma\sqrt{2\pi}}e^{-\frac{(\Delta HS_{aut}-\mathrm{EVS})^2}{2\sigma^2}} \quad (2)$$

where EVS_{res}—the expected value of saving a certain energy resource;

σ—the level of its possible mean deviation.

In this case, the expected value for the energy saving function has the form:

$$ME[\varphi_\tau(\Delta HS_{aut})] = \int_{-\infty}^{+\infty} \varphi_\tau(\Delta HS_{aut})f(\Delta HS_{aut})d\Delta HS_{aut} \quad (3)$$

Based on the substitution within the integration and replacement of certain mathematical models, the equation takes the following form:

$$\begin{aligned} ME[\varphi_\tau(\Delta HS_{aut})] = & \int_0^5 (hs_{0\tau}^{autin} + hs_{1\tau}^{autin}\Delta HS_{aut})\cdot\frac{1}{\sigma\sqrt{2\pi}}e^{-\frac{(\Delta HS_{aut}-EVS_{res})^2}{2\sigma^2}} d\Delta HS_{aut} \\ & = \frac{hs_{0\tau}^{autin}}{\sigma\sqrt{2\pi}}\int_0^5 e^{-\frac{(\Delta HS_{aut}-EVS_{res})^2}{2\sigma^2}} d\Delta HS_{aut} \\ & + \frac{hs_{1\tau}^{autin}}{\sigma\sqrt{2\pi}}\int_0^5 \Delta HS_{aut} e^{-\frac{(\Delta HS_{aut}-EVS_{res})^2}{2\sigma^2}} d\Delta HS_{aut} \\ & = hs_{0\tau}^{autin}\left[\varphi\left(\frac{5-EVS_{res}}{\sigma}\right) + \varphi\left(\frac{EVS_{res}}{\sigma}\right)\right] \\ & + \frac{hs_{1\tau}^{autin}}{\sigma\sqrt{2\pi}}\int_0^5 \Delta HS_{aut} e^{-\frac{(\Delta HS_{aut}-EVS_{res})^2}{2\sigma^2}} d\Delta HS_{aut} \end{aligned} \tag{4}$$

where $\varphi\ (y)$—integral Laplace function taking into account the parameter y.

In order to determine the integral $\frac{1}{\sigma\sqrt{2\pi}}\int_0^5 \Delta HS_{aut} e^{-\frac{(\Delta HS_{aut}-EVS_{res})^2}{2\sigma^2}} d\Delta HS_{aut}$ it is assumed that $x = \frac{\Delta HS_{aut}-EVS_{res}}{\sigma}$ and x is accepted as a new variable. As a result $\Delta HS_{aut} = x\sigma + EVS_{res}$. Wherein $d\Delta HS_{aut} = \sigma dx$. By replacing the variables, one can get:

$$\begin{aligned} & \frac{1}{\sigma\sqrt{2\pi}}\int_0^5 \Delta HS_{aut} e^{-\frac{(\Delta HS_{aut}-EVS_{res})^2}{2\sigma^2}} d\Delta HS_{aut} = \frac{\sigma}{\sigma\sqrt{2\pi}}\int_{-EVS_{res}/\sigma}^{(5-EVS_{res})/\sigma} (x\sigma + EVS_{res}) e^{\frac{x^2}{2}} dx = \\ & \frac{\sigma}{\sqrt{2\pi}}\int_{-EVS_{res}/\sigma}^{(5-EVS_{res})/\sigma} x\sigma e^{\frac{x^2}{2}} dx + \frac{EVS_{res}}{\sqrt{2\pi}}\int_{-EVS_{res}/\sigma}^{(5-EVS_{res})/\sigma} e^{\frac{x^2}{2}} dx = \\ & \frac{\sigma}{\sqrt{2\pi}}\int_{-\frac{EVS_{res}}{\sigma}}^{\frac{(5-EVS_{res})}{\sigma}} e^{\frac{x^2}{2}} d\left(-\frac{x^2}{2}\right) + EVS_{res}\left[\varphi\left(\frac{5-EVS_{res}}{\sigma}\right) + \varphi\left(\frac{EVS_{res}}{\sigma}\right)\right] = \frac{\sigma}{\sqrt{2\pi}}\left[e^{-\frac{EVS_{res}^2}{2\sigma^2}} - \right. \\ & \left. e^{-\frac{(5-EVS_{res})^2}{2\sigma^2}}\right] + EVS_{res}\left[\varphi\left(\frac{5-EVS_{res}}{\sigma}\right) + \varphi\left(\frac{EVS_{res}}{\sigma}\right)\right] \end{aligned} \tag{5}$$

Thus, the expected value of household energy savings as a result of the introduction of an autonomous energy system will have the following form:

$$\begin{aligned} ME_\tau = ME\ & [\varphi(\Delta HS_{aut})] \\ & = hs_{0\tau}^{autin}\left[\varphi\left(\frac{5-EVS_{res}}{\sigma}\right) + \varphi\left(\frac{EVS_{res}}{\sigma}\right)\right] \\ & + + hs_{1\tau}^{autin}\left\{EVS_{res}\left[\varphi\left(\frac{5-EVS_{res}}{\sigma}\right)\right] + \varphi\left(\frac{EVS_{res}}{\sigma}\right)\right. \\ & \left. + \frac{\sigma}{\sqrt{2\pi}}\left(e^{-\frac{EVS_{res}^2}{2\sigma^2}} - e^{-\frac{(5-EVS_{res})^2}{2\sigma^2}}\right)\right\} \end{aligned} \tag{6}$$

The proposed methodological approach to assessing the effectiveness of introducing an autonomous energy system by households allows:

- taking into account energy saving drivers;
- determining directions for increasing energy efficiency;
- promoting favorable investment support for the introduction of autonomous systems, which are to be implemented in households.

The study took into account autonomous energy supply systems, including all costs. It is assumed that the average life of the systems is twenty years since income and expenses during this period remain in the same ratio. Discounted cash flows allow for the overall expected life of the system. At the same time, the self-consumption rate assumes the production of electricity by installations that are autonomous in a household, divided by the total production of electricity by the system.

When predicting the efficiency of investments in autonomous energy systems by households, it is assumed that maintenance costs can be as low as 1% per year of investments [40]. In the period up to 2030, it is assumed that there are no tariffs for the introduction of new systems and the presence of income, taking into account the market price for the duration of the entire study period. Based on this, the profitability of the proposed investment in an autonomous energy system for a household was assessed.

Determining the economic value of introducing an autonomous energy system for a household involves defining the net present value (NPV_{haut}) in the studied period i:

$$NPV_{haut} = \sum_{i=1}^{n} \frac{CF_{hauti}}{(1+d)^i} - \sum_{i=0}^{n} \frac{\text{Inv}_{haut}}{(1+d)^i} \tag{7}$$

When justifying the economic feasibility of introducing autonomous energy systems in households, the possible emergence of new technologies was also taken into account. At the same time, the study took into account a decrease in investments for households' autonomous systems in the period under review. Taking into account that the assessment was formed for a long period (20 years), it should be noted that there is some limit of the proposed methodological approach, namely, the emergence of innovative technologies in the formation of autonomous energy systems. This could result in a slight bias in the projected performance indicators in the context of economic benefits.

In order to determine the dependencies between such factors of energy independence of countries as GDP per capita, the volume of energy production per capita, the volume of energy imports per capita, regression equations were constructed as single cases of multiple relationships. Based on the formed dependencies, it is possible to determine the relationship between the factors under study. The following is the standardized form of the dependence of energy consumption per capita on indicators of energy resources impact on a country's economic development:

$$t_x = 0.68t_{x_1} + 0.14t_{x_2} + 0.04t_{x_3} + 0.81t_{x_4} + 0.35t_{x_5} - 0.18t_{x_6} \tag{8}$$

where t_{x_1}—the volume of energy resources production per capita (tons); t_{x_2}– energy resources imports (USD per capita); t_{x_3}—energy resources exports (USD per capita); t_{x_4}—energy intensity of GDP (tons per 1000 USD); t_{x_5}—GDP volume (USD per capita); t_{x_6}—the value of the import quota of energy resources (%).

Based on the integral coefficient of energy dependence, a forecast of indicators of energy dependence of the studied countries was compiled to assess the effect of an increase in the level of households' energy autonomy by 2030. A realistic (the level of autonomy will increase by 20%), optimistic (the level of autonomy will increase by 30%), and pessimistic (the level of autonomy will increase by 10%) scenarios were formed. The authors proposed the methodological approach to assessing the impact of risks of increasing households' energy autonomy based on such indicators as the probable deviation of the level of a country's energy dependence, as well as the determination of risk limits.

The integral coefficient of energy dependence of each of the studied countries is determined by the Equation:

$$IED = \sqrt{t_x} \tag{9}$$

Deviation of the level of energy dependence (Dev_{ed}) is determined by the Equation

$$Dev_{ed} = \sqrt{\sum_{i=1}^{n} (IED_i - Exp(IED)^2 \times p_i} \tag{10}$$

where IED_i—integral coefficient for optimistic, realistic, and pessimistic scenarios;

$Exp(IED)$—expected total cumulative coefficient of energy dependence of a country;

p_i—the probable value of the integral coefficient of a country's energy dependence according to the optimistic, realistic, and pessimistic scenarios.

The expected total integral coefficient of energy dependence of a country is determined by the Equation:

$$Exp(IED) = \sum_{i=1}^{\infty} IED_i \times p_i. \tag{11}$$

The risk limit of the level of energy autonomy of households in a country is determined by the Equation:

$$EAH_{rl} = \frac{Dev_{ed\ min}}{Exp(IED)} \tag{12}$$

Using the probable deviation of energy dependence and the risk limit, the authors analyzed the impact of changes in the energy market on changes in the energy security of the countries under study as a result of households' energy autonomy. The results were compared to determine future trends.

The initial data for the assessment given in the proposed methodological approach are shown in Table 1.

Table 1. Initial data for determining the risk limit of the level of energy autonomy of households.

Country	Energy Saving Level, %	Notional Cost of 1% of Energy Saving, USD	IED 10%	IED 20%	IED 30%
Australia	18	375	0.3039	0.2271	0.2135
Austria	10	474	0.7473	0.6639	0.5238
Belarus	8	736	1.1417	1.0040	0.8367
Brazil	12	525	0.9700	0.6800	0.6540
China	22	562	0.9329	0.7236	0.6587
Denmark	6	358	0.7511	0.6123	0.5498
Germany	12	421	0.7536	0.5283	0.4966
India	15	625	1.0496	0.8062	0.7632
Japan	13	384	0.4508	0.4069	0.3825
Kazakhstan	11	672	0.5476	0.4740	0.4332
Mexico	12	457	0.8768	0.7790	0.6543
Poland	6	598	0.7696	0.5972	0.5185
Portugal	5	589	0.7288	0.6475	0.5287
Romania	16	829	1.0367	0.9117	0.6598
Russia	15	680	0.4907	0.3576	0.3376
Saudi Arabia	21	459	0.4700	0.3786	0.3559
Spain	12	338	0.8269	0.6417	0.5543
Switzerland	7	295	0.5045	0.4554	0.4281
UAE	14	425	0.3701	0.3017	0.2836
USA	22	359	0.3703	0.2694	0.2532

Source: compiled by the authors based on statistical data [41].

4. Results

To study the possibilities of changing the level of energy autonomy of households, the volumes of their electricity consumption in the studied countries for the period 2000–2018 were determined (Figure 2). It is electricity that is the main energy resource, based on which a household can increase its energy independence from a state based on renewables, including solar and wind energy.

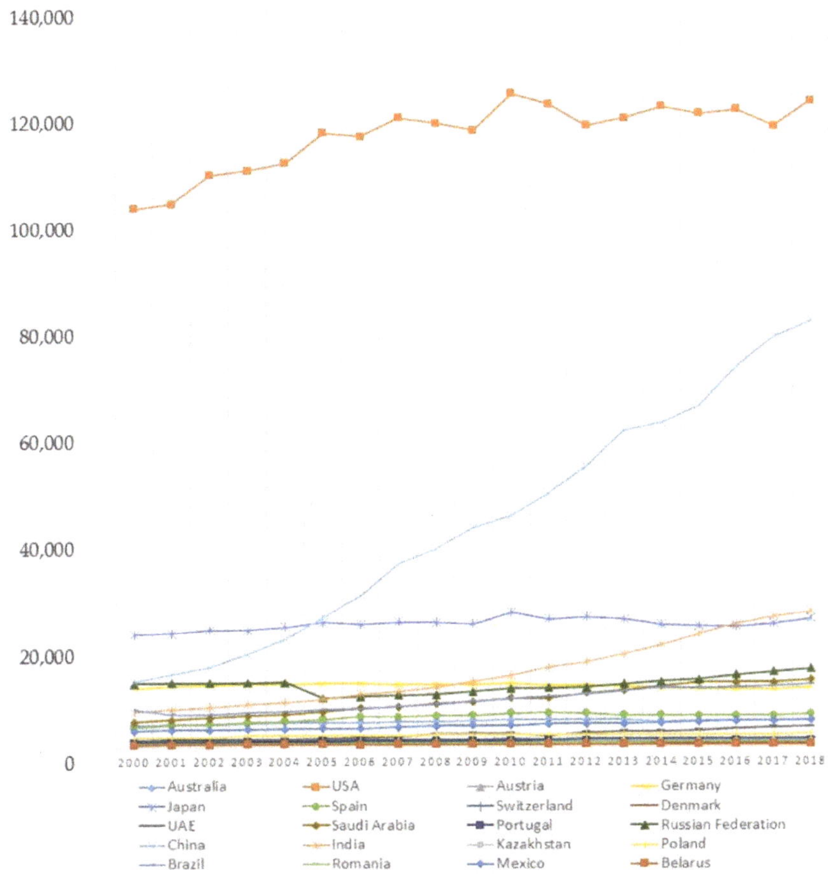

Figure 2. Final electricity consumption by households in the studied countries. Source: developed by the authors based on data [41].

Throughout the study period, the highest level of electricity consumption by households was recorded in the United States. It has grown by 22% compared to 2000. The most pronounced increase in the volume of electricity consumption by households is in China, where the indicator has increased almost seven times. A significant increase in the level of consumption is typical for India (about four times), the United Arab Emirates (3.5 times), and Kazakhstan (about three times). The lowest increase in consumption was recorded in Germany (6%) and Denmark (4%). In general, in all the studied countries, there is an increase in the volume of electricity consumption by households, resulting from the development of science and technology. The indicators do not depend on whether it is a developed or a developing country. At the same time, in each country, there are certain conditions for the introduction of the same technologies of renewable sources by households, the cost of equipment, conditions, and terms of operation. Taking into account all these determinants, the investment attractiveness of the installation and operation of autonomous energy systems for households has been assessed. The forecast is based on the assumption that 50% of households will gain energy independence. Based on this, factors for increasing the level of households' energy autonomy were determined on the basis of country's energy saving and the notional cost of 1% of households' energy saving (Figure 3).

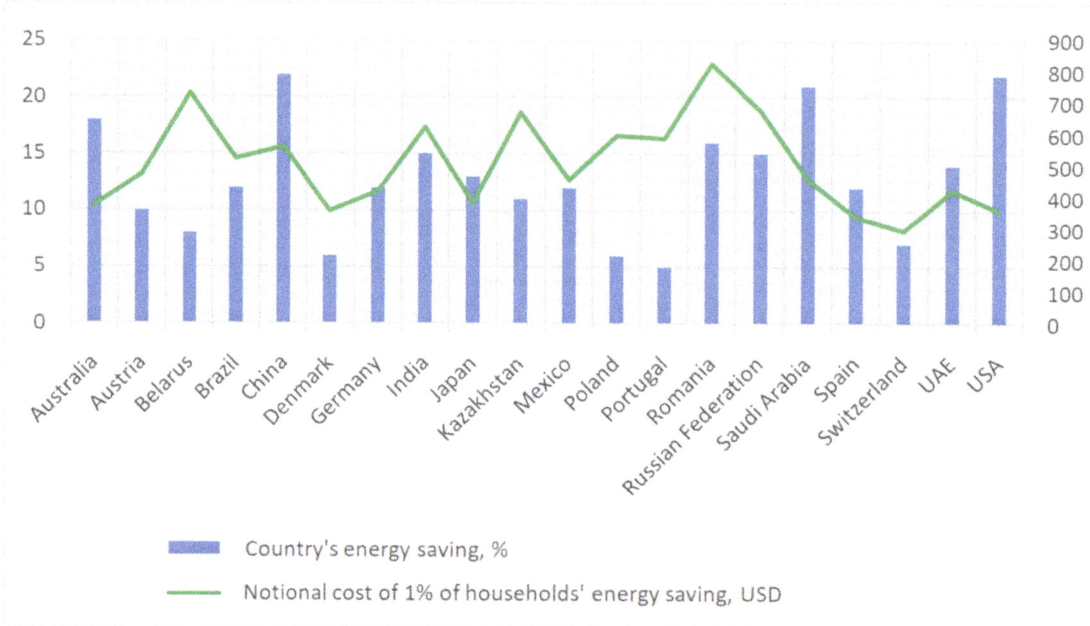

Figure 3. Motivational benchmarks for increasing state's and household's energy autonomy. Source: generated by the authors.

Households' energy autonomy is the most attractive in the USA, China, and Saudi Arabia, where a possible increase in the level of energy savings is expected by 21–22%. Most developed countries are characterized by the prospect of increasing energy savings above 10%, with the exception of Denmark and Switzerland. This is primarily due to the fact that Denmark and Switzerland already have a fairly high level of households' energy autonomy. Among developing countries, a high level of energy saving is possible in Russia, India, and Romania. This is due to the effective government policy in recent years to develop solar energy and to stimulate the introduction of renewable energy technologies in households. However, despite the prospects for a state, it is necessary to take into account households' interests as well. For example, the lowest notional cost of 1% energy savings for households was recorded in developed countries such as Switzerland, Spain, and Denmark, which is a consequence of effective government policies in previous years. The highest notional cost of energy savings for households is found in Romania, Belarus, Russia, and Kazakhstan. Given the prospect of benefits for Romania, the state should reconsider its strategic priorities and methods of stimulating household energy independence. In Belarus, there is a confrontation of state interests with households, expressed in a need for significant investments from households, which, as a result, affects the payback period and reduces investment attractiveness. For Russia and Kazakhstan, the main reason for this situation is the absence of an urgent need to motivate households since there is a priority of providing fossil energy resources.

Taking into account the opposition of the interests of a state and households, the forecast of energy dependence indicators was modeled (Index of country's energy dependence—IED) according to three scenarios, possible deviations (Dev_{ed}), and the impact of households' energy autonomy in relation to the risk limit of energy dependence of the studied countries (EAH_{rl}) until 2030. At the same time, three scenarios have been formed that imply an increase in the energy autonomy of households in a country by 10, 20, and 30% (Figure 4).

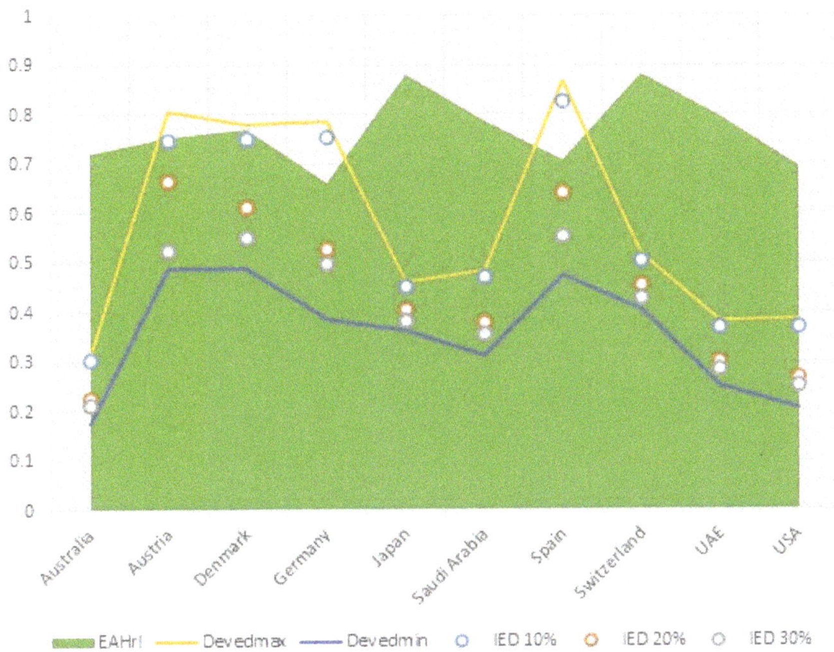

Figure 4. Forecast indicators of the impact of households' increasing energy autonomy on the level of energy dependence of the studied developed countries until 2030. Source: generated by the authors.

For most developed countries, there is a positive impact of increased energy autonomy of households. This is manifested in the fact that in all three scenarios the index of energy independence is within the risk of possible deviations. Moreover, the lower the index of energy dependence in comparison with the risk limit, the more effectively households' autonomy is manifested. The most positive effect in this context is typical for Japan, Australia, UAE, the USA, and Saudi Arabia. At the same time, one should pay attention to the paradoxical situation, which consists in the fact that most of these countries are leaders in the fossil energy market. Although, at first glance, households' energy autonomy is not a priority for these states, increasing its level can contribute to the development of additional energy resources to increase the level of their development. Despite the sufficient level of energy independence, an increase in households' energy autonomy for Austria, Germany, and Spain can lead to an excess of the risk limit and negatively affect the energy efficiency of countries since they are characterized by a deviation that exceeds the level of risk limit of energy dependence.

By analogy, the assessment was carried out for developing countries (Figure 5).

The forecasted indicators confirm the effectiveness of expanding the household's energy autonomy in Russia, Kazakhstan, and practically in Portugal. This process can ensure the further development of these countries based on energy savings and the development of additional energy resources. Russia and Kazakhstan have an advantage in the context of the availability of fossil energy resources that provide for the needs of households, but at the same time harm the environment. Energy autonomy of households in these countries has both economic and sustainable effects. Romania and most other developing countries can get a positive effect only as a result of a total increase in the energy autonomy of households up to 30%. For Belarus and India, so far, the energy autonomy of households is not feasible, since the energy dependence of these countries increases due to a rather high cost of improving energy efficiency. At the same time, deviations of possible results have exorbitant values in relation to the risk limit, which indicates that these countries are not ready for such a transition.

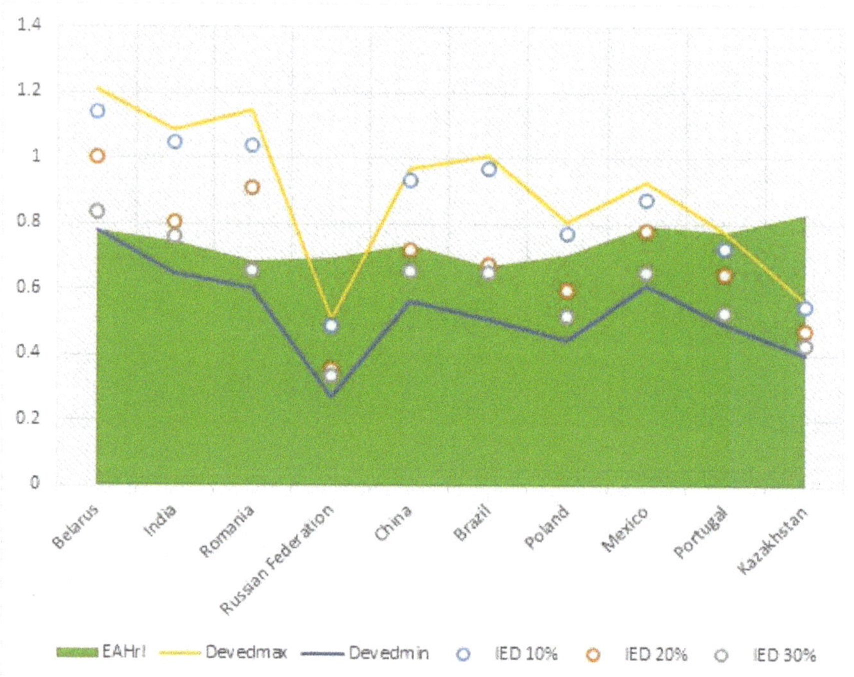

Figure 5. Forecast indicators of the impact of households' energy autonomy on the level of energy dependence of the studied developing countries until 2030. Source: generated by the authors.

For developed countries with a high level of energy independence and developing countries that own fossil energy resources, it is imperative to increase households' energy autonomy, which will mitigate the pressure on the environment. For developing countries, it is not advisable to prioritize this autonomy, since sufficient conditions have not been created for its successful and effective implementation.

5. Discussion

The household share is significant in global energy consumption, however, it is difficult to characterize it since consumer characteristics are determined by several factors, ranging from geographic location to the behavior of each of the users of a particular residential building. In addition, it is important to take into account the significant amount of energy used in this sector due to the use of traditional fuels and the implementation of modern energy policies [42]. Based on this specificity, the current study is relevant and can serve as a basis for further developments in this direction.

The advantage of the proposed methodological toolkit is a set of predictive indicators of the impact of households' energy autonomy regarding the degree of energy independence of the studied developed and developing countries, which made it possible to form their trend until 2030. This study confirms that while government support for household autonomy is energy efficient, it provides the greatest benefits to higher income households and thus contributes to increasing inequality [43]. Subsidies do not help reduce energy poverty because of the low investment capacity of households in developing countries. To achieve a more even distribution of benefits when developing policies to increase households' energy autonomy, the current situation of low-income households and issues of energy availability should be taken into account [22]. Considering that the development and construction of a functional autonomous intelligent network require deep knowledge, high costs, and availability of opportunities to use modern technologies, as well as the fact that the first installations have been produced in highly developed countries, the study confirms the benefits of households' energy independence for these countries. Today, a significant proportion of autonomous energy systems for households in Europe have smart grid characteristics. Developing countries should use rich practical experience of other countries for the construction of modern equipment [44].

Based on the study, it is assumed that benefits for a state in the form of load switching and savings are possible, but taking into account the differences in socio-economic aspects. That being said, developing a strategy, in turn, can affect the net economic cost of energy and the total amount of energy that each household consumes [45].

The conducted research indicates that giving additional energy autonomy of households cannot be achieved solely by legislative and regulatory acts. Instead, it is necessary to form a motivational mechanism for homeowners towards the material benefits of more efficient energy use. In order to improve the energy efficiency of one's home, it is critical to provide appropriate, helpful guidance. At the same time, state policy at the level of developed and developing countries must take into account specific cultural and ethnic circumstances. Therefore, this study can be complemented by a study of the relationship between cultural values and household energy consumption [46]. Cultural differences are possible. For example, Japan and European countries show clear cultural differences that affect household energy consumption [23].

Based on the proposed methodological approach, it was determined that decentralized energy systems have clear advantages, but one of the main disadvantages is the lack of economic scale effect. Decentralized energy systems are developing in parallel with the trend towards the use of energy resources by the population. Energy autonomy is the main driving force behind investing in local renewable energies, according to many of these communities [47].

A limitation of the study is a certain degree of averaging of the initial indicators for assessing the level of energy independence of households since each of them has its own specifics and can establish different types of autonomous systems, as well as their integration. At the same time, a household can introduce the newly created autonomous system earlier than the period suggested in the study. This can affect the final results of deviations of the obtained indicators [48]. This study can also be expanded towards identifying potential opportunities for improving the quality of housing maintenance and modernization to increase the energy autonomy of households. In addition, there is a need to consider the possibilities of co-production with household participation, including opportunities for households to revise the parameters of tasks related to research and project implementation [49].

The results of the study confirm that energy consumption in the residential sector continues to increase, and there is reason to believe that these dynamics will continue in the future. In the conditions of the economic crisis, as well as within the framework of the current policy on environmental protection and energy saving, a significant number of households are forced to live in conditions of a decrease in income and an increase in electricity prices. In this regard, an increase in the number of energy-poor households is expected. In addition to the study, a regression model of panel shutdown can be used to experimentally diagnose the sensitivity of households to fluctuations in energy prices due to their flexibility [20]. Therefore, in the long term, this study should be aimed at identifying the factors that determine the demand for energy in the residential sector, while special attention is to be paid to the analysis of income distribution and household vulnerability to the impact of changes in energy prices. The study is relevant in terms of periodic discussions on energy efficiency and reducing energy dependence.

6. Conclusions

Based on the assessment and comparison of the volumes of electricity consumption by households in the studied countries for the period 2000–2018, the highest level of electricity consumption by households is characteristic of the United States. The highest rates of increase in electricity consumption by households among the studied countries were recorded in China, India, the United Arab Emirates, and Kazakhstan. At the same time, a low level of increase in consumed volumes is typical for Germany and Denmark. In all the countries studied, both developed and developing, there is an increase in the volume of electricity consumption by households, as a result of the development of scientific and technological progress.

Diagnostics of the investment attractiveness of the installation and operation of energy systems for households makes it possible to determine the boundaries of a possible increase in the level of their energy autonomy. The most indicative countries of households' energy autonomy are the United States, China, and Saudi Arabia. In most developed countries, there is a prospect of increasing energy savings above 10%. Among developing countries, a high level of energy saving is possible in Russia, India, and Romania based on effective government policies. A low level of notional cost of energy savings for households is typical for Switzerland, Spain, and Denmark. The highest notional cost of energy savings for households was recorded in Romania, Belarus, Russia, and Kazakhstan. Therefore, states should reconsider their strategic priorities and methods of stimulating household energy independence. A conflict of interests between a state and households has been identified, which manifests itself in the need for significant investment from households, which, as a result, affects the payback period and reduces the level of investment attractiveness. For countries rich in fossil fuels, the main reason for this situation is insufficient motivation of households.

The modeled indicators of a country's energy dependence for three scenarios, possible deviations, and determination of households' energy autonomy in relation to the risk limit of energy dependence demonstrate the positive impact of households' energy autonomy for most developed countries. The most positive effect in this context is characteristic of the leading countries in the fossil energy market. Despite the absence of a clear political priority of households' energy autonomy for these countries, an increase in its level can contribute to the development of additional energy resources to increase the level of development. Increasing households' energy autonomy for Austria, Germany, and Spain may lead to an excess of the risk limit and negatively affect the energy efficiency of countries. There is a positive economic and sustainable effect of increasing households' energy autonomy in Russia, Kazakhstan, and Portugal. For most developing countries, a positive effect is possible as a result of a total increase in household energy autonomy up to 30%. In Belarus and India, household energy autonomy is not sufficiently expedient. The reason for this is a rather high notional cost of improving energy efficiency, which affects the increase in the level of energy dependence. On the other hand, the exorbitant deviations of indicators in relation to the risk limit indicate an insufficient readiness of these countries for the energy transition.

For countries that have a high level of energy independence, households' energy autonomy is focused on effective sustainable development. An increase in households' energy autonomy in developing countries at the present stage is possible based on sufficient provision for its successful and effective implementation without threats to economic development.

The study is valuable in the context of determining the possibilities of influencing the energy autonomy of households, identifying the benefits and risks of taking measures in the process of transforming the global energy for countries with different economic statuses. In the long term, within the framework of this study, it is possible to analyze the determinants of energy consumption in the housing sector, namely, the significance of income and the vulnerability of households in relation to changes in energy tariffs.

Author Contributions: Conceptualization, M.M., P.N. and S.D.; methodology, I.T.; software, M.M.; validation, M.M., I.T. and S.D.; formal analysis, S.D.; investigation, I.T. and S.D.; resources, S.D.; data curation, I.T.; writing—original draft preparation, M.M.; writing—review and editing, P.N. and I.T.; visualization, S.D.; supervision, M.M.; project administration, I.T.; funding acquisition, M.M. and P.N. All authors have read and agreed to the published version of the manuscript.

Funding: This research received no external funding.

Institutional Review Board Statement: Not applicable.

Informed Consent Statement: Not applicable.

Data Availability Statement: Not applicable.

Conflicts of Interest: The authors declare no conflict of interest.

References

1. Bezpalov, V.V.; Lochan, S.A.; Fedyunin, D.V. The Signs of Economic Disparity in Russia Following the Implosion of the USSR. *Vis. Anthropol.* **2020**, *33*, 116–127. [CrossRef]
2. Tvaronavičienė, M.; Ślusarczyk, B. *Energy Transformation towards Sustainability*; Elsevier Ltd.: Amsterdam, The Netherlands, 2019; pp. 1–333. [CrossRef]
3. Gonçalves, I.; Gomes, Á.; Antunes, C.H. Optimizing the management of smart home energy resources under different power cost scenarios. *Appl. Energy* **2019**, *242*, 351–363. [CrossRef]
4. Dudin, M.N.; Frolova, E.E.; Protopopova, O.V.; Mamedov, A.A.; Odintsov, S.V. Study of innovative technologies in the energy industry: Nontraditional and renewable energy sources. *Entrep. Sustain. Issues* **2019**, *6*, 1704–1713. [CrossRef]
5. Vlasov, A.I.; Shakhnov, V.A.; Filin, S.S.; Krivoshein, A.I. Sustainable energy systems in the digital economy: Concept of smart machines. *Entrep. Sustain. Issues* **2019**, *6*, 1975–1986. [CrossRef]
6. Sarma, U.; Karnitis, G.; Zuters, J.; Karnitis, E. District heating networks: Enhancement of the efficiency. *Insights Reg. Dev.* **2019**, *1*, 200–213. [CrossRef]
7. Tyo, A.; Jazykbayeva, B.; Ten, T.; Kogay, G.; Spanova, B. Development tendencies of heat and energy resources: Evidence of Kazakhstan. *Entrep. Sustain. Issues* **2019**, *7*, 1514–1524. [CrossRef]
8. Karjalainen, S.; Ahvenniemi, H. Pleasure is the profit-the adoption of solar PV systems by households in Finland. *Renew. Energy* **2019**, *133*, 44–52. [CrossRef]
9. Zhang, C.; Wang, Q.; Zeng, S.; Baležentis, T.; Štreimikienė, D.; Ališauskaitė-Šeškienė, I.; Chen, X. Probabilistic multi-criteria assessment of renewable micro-generation technologies in households. *J. Clean. Prod.* **2019**, *212*, 582–592. [CrossRef]
10. Katsuba, S.; Shestak, V.; Kvasnikova, T.; Bokov, Y. Liability for Violation of Environmental Legislation in the EU. *Eur. Energy Env. Law Rev.* **2021**, *30*, 9–19.
11. Hussain, H.I.; Slusarczyk, B.; Kamarudin, F.; Thaker, H.M.T.; Szczepańska-Woszczyna, K. An investigation of an adaptive neuro-fuzzy inference system to predict the relationship among energy intensity, globalization, and financial development in major ASEAN economies. *Energies* **2020**, *13*, 850. [CrossRef]
12. Chehabeddine, M.; Tvaronavičienė, M. Securing regional development. *Insights Reg. Dev.* **2020**, *2*, 430–442. [CrossRef]
13. Iweka, O.; Liu, S.; Shukla, A.; Yan, D. Energy and behaviour at home: A review of intervention methods and practices. *Energy Res. Soc. Sci.* **2019**, *57*, 101238. [CrossRef]
14. El Iysaouy, L.; El Idrissi, N.E.; Tvaronavičienė, M.; Lahbabi, M.; Oumnad, A. Towards energy efficiency: Case of Morocco. *Insights Reg. Dev.* **2019**, *1*, 259–271. [CrossRef]
15. He, K.; Zhang, J.; Zeng, Y. Rural households' willingness to accept compensation for energy utilization of crop straw in China. *Energy* **2018**, *165*, 562–571. [CrossRef]
16. Eon, C.; Breadsell, J.; Morrison, G.; Byrne, J. Shifting home energy consumption through a holistic understanding of the home system of practice. In *Decarbonising the Built Environment*; Palgrave Macmillan: Singapore, 2019; pp. 431–447.
17. Kim, J.; Moon, C. A Robot System Maintained with Small Scale Distributed Energy Sources. *Energies* **2019**, *12*, 3851. [CrossRef]
18. Rogalev, A.; Komarov, I.; Kindra, V.; Zlyvk, O. Entrepreneurial assessment of sustainable development technologies for power energy sector. *Enterp. Sustain. Issues* **2018**, *6*, 429–445. [CrossRef]
19. Hafner, R.J.; Elmes, D.; Read, D. Promoting behavioural change to reduce thermal energy demand in households: A review. *Renew. Sustain. Energy Rev.* **2019**, *102*, 205–214. [CrossRef]
20. Charlier, D.; Kahouli, S. From residential energy demand to fuel poverty: Income-induced non-linearities in the reactions of households to energy price fluctuations. *Energy J.* **2019**, *40*, 101–137. [CrossRef]
21. Boichenko, K.S.; Tepliuk, M.A.; Rekova, N.Y.; Stashkevych, I.I.; Morkunas, M. Management of fluctuation of financial and economic integrated development of innovative enterprise. *Financ. Credit Act. Probl. Theory Pract.* **2019**, *3*, 62–69. [CrossRef]
22. Afanasyev, M.P.; Shash, N. Russian Federation Cross-Border Investments and Bank Expansion. *Public Admin. Issues* **2019**, *6*, 105–120.

23. Johansson, T.; Pirouzfar, P. Sustainability challenges in energy use behaviour in households: Comparative review of selected survey-based publications from developed and developing countries. *Probl. Ekorozw.* **2019**, *14*, 33–44.
24. Tishkov, S.; Shcherbak, A.; Karginova-Gubinova, V.; Volkov, A.; Tleppayev, A.; Pakhomova, A. Assessment the role of renewable energy in socio-economic development of rural and Arctic regions. *Entrep. Sustain. Issues* **2020**, *7*, 3354–3368. [CrossRef]
25. Karunathilake, H.; Hewage, K.; Sadiq, R. Opportunities and challenges in energy demand reduction for Canadian residential sector: A review. *Renew. Sustain. Energy Rev.* **2018**, *82*, 2005–2016. [CrossRef]
26. Eddelani, O.; El Idrissi, N.E.; Monni, S. Territorialized forms of production in Morocco: Provisional assessment for an own model in gestation. *Insights Reg. Dev.* **2019**, *1*, 6–18. [CrossRef]
27. Nasr, A.K.; Kashan, M.K.; Maleki, A.; Jafari, N.; Hashemi, H. Assessment of Barriers to Renewable Energy Development Using Stakeholders Approach. *Entrep. Sustain. Issues* **2020**, *7*, 2526–2541. [CrossRef]
28. Mazzoni, F. Circular economy and eco-innovation in Italian industrial clusters. Best practices from Prato textile cluster. *Insights Reg. Dev.* **2020**, *2*, 661–676. [CrossRef]
29. Taghizadeh-Hesary, F.; Yoshino, N. The way to induce private participation in green finance and investment. *Financ. Res. Lett.* **2019**, *31*, 98–103. [CrossRef]
30. Pobedinsky, V.; Shestak, V. Improving Environmental Legislation in Central Asia. *Environ. Policy Law* **2020**, *50*, 69–79. [CrossRef]
31. Thøgersen, J. Frugal or green? Basic drivers of energy saving in European households. *J. Clean. Prod.* **2018**, *197*, 1521–1530. [CrossRef]
32. Sabishchenko, O.; Rębilas, R.; Sczygiol, N.; Urbański, M. Ukraine energy sector management using hybrid renewable energy systems. *Energies* **2020**, *13*, 1776. [CrossRef]
33. Haseeb, M.; Kot, S.; Iqbal Hussain, H.; Kamarudin, F. The natural resources curse-economic growth hypotheses: Quantile–on–Quantile evidence from top Asian economies. *J. Clean. Prod.* **2021**, *279*, 123596. [CrossRef]
34. Sangroya, D.; Nayak, J.K. Factors influencing buying behaviour of green energy consumer. *J. Clean. Prod.* **2017**, *151*, 393–405. [CrossRef]
35. Bae, J.H.; Rishi, M. Increasing consumer participation rates for green pricing programs: A choice experiment for South Korea. *Energy Econ.* **2018**, *74*, 490–502. [CrossRef]
36. Hvelplund, F.; Djørup, S. Consumer ownership, natural monopolies and transition to 100% renewable energy systems. *Energy* **2019**, *181*, 440–449. [CrossRef]
37. Cloke, J.; Mohr, A.; Brown, E. Imagining renewable energy: Towards a Social Energy Systems approach to community renewable energy projects in the Global South. *Energy Res. Soc. Sci.* **2017**, *31*, 263–272. [CrossRef]
38. Palm, J. Household installation of solar panels–Motives and barriers in a 10-year perspective. *Energy Policy* **2018**, *113*, 1–8. [CrossRef]
39. Kaiser, M.; Bernauer, M.; Sunstein, C.R.; Reisch, L.A. The power of green defaults: The impact of regional variation of opt-out tariffs on green energy demand in Germany. *Ecol. Econ.* **2020**, *174*, 106685. [CrossRef]
40. Delgado, D.; Carvalho, M.; Junior, L.M.C.; Abrahão, R.; Chacartegui, R. Photovoltaic solar energy in the economic optimisation of energy supply and conversion. *IET Renew. Power Gener.* **2018**, *12*, 1263–1268. [CrossRef]
41. International Energy Agency. Data and Statistics. 2020. Available online: https://www.iea.org/data-and-statistics?country=WORLD&fuel=Energy%20consumption&indicator=Carbon%20intensity%20of%20industry%20energy%20consumption (accessed on 13 December 2020).
42. Suciu, R.; Stadler, P.; Kantor, I.; Girardin, L.; Maréchal, F. Systematic integration of energy-optimal buildings with district networks. *Energies* **2019**, *12*, 2945. [CrossRef]
43. Plutshack, V.; Sengupta, S.; Sahay, A.; Viñuales, J.E. New and renewable energy social enterprises accessing government support: Findings from India. *Energy Policy* **2019**, *132*, 367–378. [CrossRef]
44. Kott, J.; Kott, M. Generic ontology of energy consumption households. *Energies* **2019**, *12*, 3712. [CrossRef]
45. Ngoma, R.; Tambatamba, A.; Oyoo, B.; Mulongoti, D.; Kumwenda, B.; Louie, H. How households adapted their energy use during the Zambian energy crisis. *Energy Sustain. Dev.* **2018**, *44*, 125–138. [CrossRef]
46. Satish, B.K.; Brennan, J. Understanding the energy use behaviour of British Indian households to shape optimised sustainable housing strategies in existing housing stock. *Sustain. Cities Soc.* **2019**, *48*, 101542. [CrossRef]
47. McKenna, R. The double-edged sword of decentralized energy autonomy. *Energy Policy* **2018**, *113*, 747–750. [CrossRef]
48. Bertheau, P.; Dionisio, J.; Jütte, C.; Aquino, C. Challenges for implementing renewable energy in a cooperative-driven off-grid system in the Philippines. *Environ. Innov. Soc. Transit.* **2020**, *35*, 333–345. [CrossRef]
49. Skjølsvold, T.M.; Throndsen, W.; Ryghaug, M.; Fjellså, I.F.; Koksvik, G.H. Orchestrating households as collectives of participation in the distributed energy transition: New empirical and conceptual insights. *Energy Res. Soc. Sci.* **2018**, *46*, 252–261. [CrossRef]

energies

MDPI

Article

The Dilemma of Long-Term Development of the Electric Power Industry in Kazakhstan

Elena Vechkinzova [1,*], Yelena Petrenko [2], Yana S. Matkovskaya [3] and Gaukhar Koshebayeva [4]

1 Department of Business Administration, Higher School of Management, Financial University under the Government of the Russian Federation, 49 Leningradsky Prospekt, 125993 Moscow, Russia
2 Academic Department of Management and Business Technologies, Faculty of Management, Plekhanov Russian University of Economics, 36 Stremyanny Lane, 117997 Moscow, Russia; petrenko_yelena@bk.ru
3 V.A. Trapeznikov Institute of Control Sciences, Russian Academy of Sciences, 65 Profsoyuznaya Street, 117997 Moscow, Russia; yana.s.matkovskaya@gmail.com
4 Department of «Economics of the Enterprise», Karaganda State Technical University, 56 Nursultan Nazarbayev Ave., Karaganda 100027, Kazakhstan; gauhark@bk.ru
* Correspondence: kvin07@list.ru; Tel.: +7-965-520-2615

Abstract: This article discusses the current state and trends in the development of the electric power industry in Kazakhstan. An analysis of the energy supply and energy intensity of Kazakhstan's GDP is provided in this paper. The results of the foresight of the risks and opportunities of the energy industry are described. This study identifies the relationship between the traditional development of the energy industry and the development of alternative energy sources. In addition, the work examines the risks and consequences of various trends in the development of national and global energy. Previous studies have shown that government efforts are insufficient in developing an alternative energy sector in Kazakhstan. The research results show that there is a need to transform energy production from traditional sources towards greater efficiency and environmental friendliness, as well as the active involvement of the business community in the development of an alternative energy market. This is expected to attract more investments and transfer technologies to maintain the country's position in the energy market of the future.

Keywords: energy industry; energy supply; energy intensity; energy efficiency; energy industry risks; dilemma; development; Kazakhstan

Citation: Vechkinzova, E.; Petrenko, Y.; Matkovskaya, Y.S.; Koshebayeva, G. The Dilemma of Long-Term Development of the Electric Power Industry in Kazakhstan. *Energies* **2021**, *14*, 2374. https://doi.org/10.3390/en14092374

Academic Editor: Manuela Tvaronavičienė

Received: 6 April 2021
Accepted: 19 April 2021
Published: 22 April 2021

1. Introduction

In 1996, Kazakhstan became one of the first states of the former Soviet Union to begin reforming the existing energy sector and transition to a market economy. Using the British and Norwegian electricity markets as models, the government dissolved the existing vertically integrated state monopoly, splitting it into separate electricity management and business management companies. Since then, more than 85% of the electricity sector has been privatized [1]. Although the privatization of Kazakhstan's energy sector is almost complete, the sale of regional distribution companies has been slower. Kazakhstan's energy sector continues to be under tight government regulation and price controls, and relations between the Eurasian Natural Resources Corporation and international investors are unsatisfactory.

Despite the incoming investments and the development of the power grid infrastructure, there are still no direct electric connections at a voltage of 500 kW in the three regions of the Southern Zone and the Western zone with the unified power system of Kazakhstan. There is also a high degree of wear and tear of the main equipment of power transmission organizations and regional power grid companies, as well as a significant number of unattended power grid facilities.

According to the estimates provided by the Ministry of Energy of Kazakhstan, the average annual growth rate of electricity production in 2020–2025 will be 3% [2]. At the same time, consumption is expected to grow by 1.9% per year and increase from 110.1 billion kWh in 2020 to 120.9 billion. kWh in 2025. By 2025, 28% of all electricity production is expected to come from stations commissioned between 2020 and 2025, which indicates the need for additional capital investments in this industry. At the same time, 19% of the stations planned for commissioning will relate to renewable energy sources (RES) [3].

Today, the cost of electricity in Kazakhstan is one of the cheapest in the world [1]. This situation is related to the fact that energy is produced in coal-fired power plants. A feature of the use of coal is its cheapness, availability, and storage capability. However, given the high deterioration of coal-fired power facilities, in the conditions of maintaining a high energy supply, Kazakhstan faces a difficult choice—to direct its efforts to the modernization of the coal-fired power industry or to develop renewable energy sources.

The dilemma in determining the directions of the country's long-term energy development lies, not only in the choice between traditional and alternative sources, but also includes a whole range of problems of an economic, political, and social nature. We can talk about the complexity of the dilemma, since choosing one of the two opposites is equally difficult, generates risks, but also opens up certain opportunities.

The purpose of this study is to identify alternatives for the development of the national energy sector, based on forecasting the main development trends and assessing development directions in the long term.

The dilemma of the long-term development of the electric power industry in Kazakhstan is also part of ensuring the country's energy security and part of the broader concept of national security, which gives rise to additional important aspects in addressing the problem.

In this regard, representatives of the scientific community note the importance of the concept of energy security for the development of the country's energy system [4–6]. Energy security is seen not only as an uninterrupted supply of electricity to enterprises and end-users [7,8], but also an indicator of environmental safety [9], energy efficiency [10] and sustainable development [11]. Building an energy system and the features of energy security of various countries were considered in Kuik et al. [12], Johnson and Boersma [13], Lei and Xuejun [14], Shiroyama et al. [15]. The prospects for the implementation of projects of renewable energy sources were also studied in detail [16–18].

Within Kazakhstan, these issues are devoted to works [19,20] that consider the potential for the development of the bioenergy sector and the main political, economic, and technological problems that impede the large-scale use of renewable energy sources.

The author's hypothesis assumes that the country cannot abandon the use of traditional raw materials soon, but should instead focus on improving the efficiency and environmental friendliness of its use. In the long term, it is necessary to join the transition to alternative energy, so as not to be technologically behind. In the short term, it is necessary to ensure the maintenance of the level of energy supply based on traditional fuels. The scientific novelty largely lies in the fact that the combination of the development of a pragmatic approach to a more efficient and environmentally friendly use of traditional raw materials, and the promotion of alternative energy production models. These factors will enable Kazakhstan to accumulate resources for a timely transition to innovative energy markets and ensure the position of an energy-secure and economically developed country. The formation of alternative energy in the country should be based on the development and support of modern technologies and the creation of conditions for the development of a free energy market, while ensuring adequate guarantees for the population through appropriate social institutions and mechanisms.

The presented work is based on the study of world experience in researching the problems of the development of the energy industry and is a continuation of the original author's developments in the field of energy in Kazakhstan. We have previously conducted research on certain issues of energy efficiency of Kazakhstani enterprises [21], in particular, analysis of losses and inefficient distribution of electricity in Kazakhstan [22,23]. In this study, we aimed to study the directions of development of the electric power industry in

Kazakhstan to understand possible alternatives for the development of the industry in the coming decades. In Section 2, we consider the views of the modern scientific community on the problems of the development of modern energy, which characterizes the multi-faceted manifestations of the electric power industry in the socio-economic life of any region and country. Section 3 is devoted to the research methodology and the description of the initial data. Section 4 contains an analysis of the results that are discussed in Section 5. Section 6 draws general conclusions and identifies prospects for further research.

2. Literature Review

Studying regional and national energy systems are highly relevant for countries of all levels of development. The efficiency of the country's production activities and the quality of life of households directly depend on the reliability and efficiency of the energy system. At the geopolitical level, the global energy system is seen as the basis for ensuring national security, the most important part of which is energy security. Initially, if energy security was considered because of the functioning of the energy supply chain [4], then later, the concept of energy security acquired different meanings with a few internal contradictions. Researchers from various countries come to the conclusion that the concept of energy security is traditionally used to justify state control over energy and unwillingness to deal with energy problems at the global level [12,13]. Emphasizing the multidimensionality of the key concept of energy security, Soysal and Soysai [24] state that energy security is an "umbrella term" that encompasses many of the issues linking energy, economic growth, and political power.

Under the auspices of the state, energy security has often meant excessive, inefficient, and sometimes dangerous production and consumption of energy. Researchers have found a way to mitigate the various multilevel contradictions in energy security in their need to develop reliable, continuous, affordable, and environmentally efficient energy services. Directly studying the impact of energy efficiency on the need to invest in new technologies, the so-called "avoided power costs" were carried out by York et al. [25], Bilton et al. [26], York et al. [27], Nguyen and Tran [28], Igaliyeva et al. [29]. Improving energy efficiency depends on the ultimate equilibrium impact on energy security through demand, price, investment, and spare capacity in the short- and long-term [4,10,25,30–32].

A few authors focus attention on economic attributes (for example, the amount of energy required at a reasonable price) [31–36], Yergin [37], Luft et al. [38], Hughes [39], Mansson et al. [40], consider the cultural, political and economic aspects of energy security.

Comparative analysis shows that the dilemma of energy development in different regions contains serious contradictions. Considering the evolution of energy policy in East Asia, Shiroyama et al. [15] suggest that the golden principle of "increasing your energy self-sufficiency ratio and not depending on external suppliers" cannot be the most suitable for this region. East Asia, as the region with the highest energy consumption, must strive to design and build a reliable energy supply in the context of changing production and consumption patterns reflecting the global power shift. In a study assessing the level of energy security of 30 countries Du et al. [41] conclude that the level of energy security in these countries varied significantly during 2001–2012. This is closely related to each country's resources, energy technologies and national policies. Therefore, each country needs a differentiated system of energy policy in accordance with the economic, social, political and resource situation. In fact, each country faces its own energy dilemma and solves it based on its own criteria for choosing well-being and independence. The multifaceted nature of this problem is also revealed in Böhringer and Keller [42], Checchi et al. [43], Kruyt et al. [44], Mitchell [45], Chester [46], Cherp and Jewell [47], Joskow [48], Chehabeddine and Tvaronavičienė [49], Alkhateeb et al. [50], Shumilo et al. [51].

Focusing on the developing countries of the Asian region, one can also see that Cronshaw et al. [52]. In the study of energy systems of the countries of the Asia-Pacific region over 30 years draw conclusions about the growing demand for improving environmental performance and at the same time improving energy services [52]. Therefore, the long-term development of the electric power industry in the regions requires approaches that support reliable and sustainable energy networks and promote the flexibility and resilience of energy systems and markets. The focus is on reliability, efficiency, environmental safety and social acceptability of energy services for end users [7,9].

Consideration of various aspects of the reliability of the energy system leads researchers to the idea of avoiding risks affecting the continuity of energy supplies relative to demand. Ecofys [4], Jansen and Seebregts [5], Winzer [6], Levèfre [53], Plėta et al. [54] identify sources of risks associated with energy security. At the same time, the problems of stability of power systems are considered through the prism of the impact of the experience of destruction and the perception of risks. Many researchers have focused on identifying the key threats to a particular energy system. Ecofys [4] defined such categories as extreme events, inadequate market structures and supply shortages associated with resource concentration. Winzer [55] identified the types of risks and their sources, associated risks (threats) with their impact on the energy supply chain, as well as threats, and also identified those risks that can be predictable, probabilistic, heuristic and unknown. Mansson et al. [40] and Gracceva and Zeniewski [8] propose a classification of threats to energy security according to the following criteria: Place in the energy supply chain; time frame; origin.

Modern research is expanding the focus of attention from the sources of destruction to the ability to respond to them (that is, the resilience of energy systems). In later concepts, the emphasis has shifted from the issue of measuring energy security to the characteristics of the energy system, allowing control and immediate response to risks [56]. Blum and Legey [56] see energy system vulnerability as a combination of risk exposure and resilience. At the same time, the differentiation of energy systems makes it possible to better measure their vulnerability and purposefully orient the policy of long-term development [57–60].

The study of various aspects of the energy system of Kazakhstan has been especially active over the past five years. In a study by Xiong et al. [61] analyzes the factors affecting carbon emissions in Kazakhstan in 1992–2014. The results of the analysis showed the largest amount of carbon emissions associated with the production of energy. This is due to the high power consumption of energy resources, high energy consumption and low economic effect of productivity. Kerimra et al. [62] address the problem of unsustainable household energy use and lack of access to energy infrastructure.

In the study of Kazakh scientists, the complexity of the problems of the development of domestic energy is noted, so the authors conclude that the energy system of Kazakhstan requires a significant amount of water resources to cool the power plant, and in the future these needs will increase [63]. A comprehensive overview of the current energy situation in Kazakhstan, including fossil energy sources and renewable resources is presented in the book by Kalyuzhnova et al., showing the possibilities and implications of a global transition to cleaner energy for Kazakhstan [64] and articles [65–67].

The reviewed publications helped the authors of this article in the development of original research, assessing the dilemma of the long-term development of the electric power industry in Kazakhstan.

3. Materials and Methods

The priority, system-forming role of modern energy for the economy and all life activities have determined the basic theoretical approach to the study of the problems of its development. We considered the electric power industry not only as a specific industry, but rather as a unique institution in which development is determined not only and not so much by technological factors, but primarily by accepted norms and rules. In analyzing the patterns of choosing the paths for the long-term development of the energy industry, the authors, based on the institutional approach, since they did not limit themselves to analyzing economic categories but considered non-economic factors. In particular, the most important factors are international norms and obligations, to which Kazakhstan joins and which force it to make a certain choice.

An important methodological approach is Path Dependency, which assumes that we cannot consider the choice of an industry development model based only on current conditions and future opportunities. It is necessary to consider further decisions, depending on the sequence of past actions, each of which led to a certain result.

The assessment of the state of the energy industry in Kazakhstan is considered based on systematization of comparative analysis, which allows obtaining a description and explanation of the similarities and differences in the development of energy industries.

The study used the foresight method, which allows improving decision-making and managing the choice of technologies, creating alternative directions for future development, increasing preparedness for unforeseen circumstances and motivating participants to make and implement decisions to achieve the desired future. Foresight is a system of methods for expert assessment of strategic directions of socio-economic and innovative development, identification of technological breakthroughs that can have an impact on the economy and society in the medium and long term.

In 2020, a "Study of the image of the future and demanded professions in the energy sector of the Republic of Kazakhstan" was carried out in Kazakhstan [68]. The authors were part of the development team of the industry foresight methodology and were directly involved in conducting foresight sessions, interviewing experts, and processing the results. A total of 136 Industry forecasts expert advice from several leading industry experts, academics and educators were involved. The interviewed experts are highly qualified specialists—more than half of them have been working in the industry for more than 15 years, and another 13% have experience from 10 to 15 years. The average work experience in the energy industry of the interviewed experts was 13 years, which is optimal for understanding the current problems and prospects for the development of the industry. Forecasts made by experts combined knowledge of technological innovations with an understanding of the specifics of the real situation in the country and in the industry.

Obtaining expert assessments was carried out by the Delphi method—a technology used to predict and assess development trends. The method consists in structuring the group communication process aimed at creating the conditions for effective team work on a complex problem.

The survey of experts by the Delphi method was supplemented by the method of in-depth interviews with CEOs of KEGOC JSC, Kazakhstan Solar Energy Association, Burnoye Solar-1 LLP, Kazatomprom National Atomic Company JSC, Kentau Transformer Plant JSC, professors of Almaty University of Energy and Communications.

In August 2020, foresight sessions of the energy industry stakeholders were also held in which 56 participant of industry foresight sessions formed forecasts and alternative scenarios for the development of the energy industry in Kazakhstan. Within the framework of these sessions, issues of risks and opportunities for the development of the energy industry in Kazakhstan were also considered.

For a quantitative assessment of risks and opportunities, a map was modeled as an integrated information-graphic method of nature, which allows the risks to be monitored based on their classification and ranking, assessment of the likelihood of occurrence and possible damage, in order to analyze the development of the situation in the industry and further impact on it. The method of mapping the risks and opportunities for the development of the industry applied for its construction makes it possible to comprehensively consider the totality of the opinions expressed and evaluate them on the basis of quantitative data. The mapping method was applied in stages: Factors (risks or opportunities) are classified, the strength of their influence and the likelihood of occurrence are determined, and then the degree of managerial impact is assessed. At the final stage, a map is modeled, which makes it possible to determine the patterns and priorities of the development of factors. The resulting map contributes to the adoption of adequate decisions aimed at compensating risks or realizing opportunities and allows you to display these decisions as a general management system.

4. Results

4.1. Analysis of the Current Situation in the Industry

Despite the active progress of Kazakhstan on the path of diversification, energy resources will remain of primary importance for its economy. At the moment, the oil and gas sector provide 1/5 of GDP (21.3% in 2018), about 2/3 of total export revenue (70% in 2018) and almost half of the country's state budget revenues (44% in 2018). The fuel and energy complex also occupies a leading position in attracting foreign direct investment to Kazakhstan [69].

Energy has always been the "fuel" for the economy and its importance grows with the intensification of industrialization. Most of the territory of Kazakhstan is located in unfavorable climatic conditions, therefore, both enterprises and ordinary people also need fuel in the literal sense. For the country, the provision of energy directly means ensuring the living conditions of the population. Also, the task of power engineering is to meet the needs of the national economy for heat and electric energy and provide an opportunity to export electricity to the countries of near and far abroad. The energy industry is a set of systems that convert primary energy: minerals, natural energy, artificial raw materials into secondary energy: electrical and thermal. The share of the energy sector in the country's GDP is 1.6%.

Energy production involves successive stages:

- transportation of resources and energy carriers to generating power plants,
- processing of the energy carrier into secondary energy,
- distribution of energy and its transportation to the end user.

At each stage of production, specific problems accumulate that require a balanced solution.

In 2019, 138 power plants with an installed capacity of 22,936 MW were operating in Kazakhstan (in 2018—21,902 MW). About 82.7% of all electricity generated in Kazakhstan is produced at thermal power plants. The country has large reserves of energy resources (oil, gas, coal, uranium) and is an energy power. The balance of generated energy in Kazakhstan is as follows: about 70% of electricity is generated from coal, a little more than 10%—from hydro resources, a little more than 10%—from gas and 5%—from oil. The current structure reflects the first dilemma of the energy sector—should coal prevail as the main raw material resource, or is it possible to replace it? Let's consider in more detail the state of each energy resource.

Coal-fired power generation is the most widespread in Kazakhstan. The coal industry and coal generation have historically been and remain an important source of economic development in many countries around the world. A feature of the use of coal is its cheapness, availability, and storage capability. According to some data, today the 13 largest countries in the world, including China, the United States, India, South Africa and others, account for up to 90% of the world's electricity generation from coal. In Kazakhstan, the coal industry is one of the most important resource industries, its contribution to GDP today is about 1.5%. In 2018, the country produced 113.7 million tons of coal, which is 6.5% more than in 2017. The main volume of electricity in Kazakhstan is generated by about six dozen power plants operating on coal (Ekibastuz, Maikubensky, Turgay and Karaganda basins), gas, and fuel oil.

The high share of coal in energy production is due to the orientation of end consumers in Kazakhstan precisely to inexpensive coal-fired generation, while gas in the country is more expensive than coal, although its cost is still low relative to world prices. This makes it difficult for power plants to switch to gas while maintaining price competitiveness. For example, the transfer of the thermal power plants in Nur-Sultan from coal to gas increase the cost of electricity generation by about 50%.

In second place in terms of electricity generation is the hydropower industry in Kazakhstan. Economically efficient hydro resources are concentrated mainly in the east (Gorny Altai) and in the south of the country in the Irtysh, Ili and Syrdarya rivers. The largest hydroelectric power stations in the country are: Bukhtarma, Shulba, Ust-Kamenogorsk (on the

Irtysh River) and Kapchagai (on the Ili River). They provide 10% of the country's electricity needs. Kazakhstan plans to increase the use of water resources in the medium term.

Earlier decisions have influenced the current situation in the energy sector of Kazakhstan. Despite the presence of significant hydrological resources, preference was given to raw coal. The use of oil and gas is limited for energy production in the republic due to the high demand for these resources in the external market. Therefore, when choosing the path of long-term development, the national energy sector fell into the so-called path dependence, when the choice of further development depends on previous decisions and past experience of operating the industry.

The global energy sector is shifting towards renewable energy sources (RES), but the pace of this process is not fast enough to compensate for the growth of the global economy and population. The problems of energy saving, and alternative energy sources are relevant for the Republic of Kazakhstan, the issues of "green" energy have become one of the strategic directions of the national economy, as a component of energy resources supplied to the domestic market and as an additional source of income. They form a second strategic dilemma—can Kazakhstan, when replacing coal energy with an alternative one, ensure the need for energy and security?

Priority sources are solar power plants, wind farms, small hydro power plants with a capacity of less than 25 kW, biomass power plants. The country has launched 100-megawatt solar stations in the Kapchagai region near Almaty, in the city of Saran, Karaganda region, in the village of Burnoye, Zhambyl region. Today there is a 50-megawatt wind station in the Zhambyl region, a 45-megawatt station in the city of Ereimentau, and a 42-megawatt station in the Mangistau region. The first phase of a 100-megawatt wind station has been commissioned near the capital. Alternative energy power plants are less powerful than conventional power plants. Renewable energy also brings imbalances to the country's energy systems. In Kazakhstan, there is no flexibility in capacities that can quickly provide or reduce unplanned electricity.

The development of nuclear energy is promising for Kazakhstan, but social acceptance is an important condition for its development. For nuclear power to arise in any country, society must accept it. The only nuclear power plant in Kazakhstan was in the city of Aktau with a fast neutron reactor with a capacity of 350 MW. Nuclear power plants (NPP) operated in 1973–1999. Currently, nuclear energy is not used in Kazakhstan, even though the country's uranium reserves (according to the IAEA) are estimated at 900 thousand tons. The main deposits are in the south of Kazakhstan (South Kazakhstan region and Kyzylorda region), in the west of Mangystau, in the north of Kazakhstan (Semizbay field). The prospect of nuclear power in Kazakhstan raises many doubts, primarily related to the culture of maintenance, operation, and reliability of equipment. The public has no confidence that it will be safe and secure. Since the south of Kazakhstan is quite in a high seismic zone, the construction of nuclear power plants there is unsafe, and in the northern regions this project is quite feasible. NPP will generate a significant amount of electricity at an affordable price. Although the construction of a nuclear power plant is not a cheap option, but later getting electricity from there will be much cheaper than from the traditional one.

Nuclear, solar, wind, hydropower and even energy from biofuels pose not one dilemma for the republic, but a whole "bunch of dilemmas" when it is necessary to make a choice from difficult options and none of them is definitely advantageous.

Infrastructure issues are added to the choice of energy sources. Electric networks in Kazakhstan consist of substations, switchgears, and power transmission lines with a voltage from 0.4 kV to 1150 kV. The national electric grid of Kazakhstan provides a connection between the regions within the country and the energy systems of neighboring states (Russia, Kyrgyzstan, Uzbekistan). In addition, the national electricity grid transfers electricity from producers to wholesale consumers. The largest organization that carries out the transmission of electricity in Kazakhstan is KEGOS JSC. It serves interstate power lines and transmission lines that provide electricity from power plants with a voltage of

220 kV or more. The distribution of electricity at the regional level is handled by 21 regional energy companies and almost 300 small energy transmission organizations. Regional energy companies transfer energy to retail consumers and provide connections within the region. Energy transmission organizations also transmit electrical energy through electrical networks to wholesale and retail consumers or supply organizations. Energy supply organizations buy electricity from energy-producing companies and resell it to the final retail consumers.

Previous studies by the authors show that it is in the electric grids that the price of energy increases, including the administrative burden of a significant corruption component [21]. The increase in energy prices from the source of production to the source of consumption is a serious problem for economic development. As our previous calculations show [22,23], the costs of heat and electric energy form "cost nodes" of production costs and generate their growth both in production and in the service sector. Thus, the following dilemma arises: An increase in energy tariffs ensures the modernization of the industry but increases the total costs of the economy and worsens the situation of the population. In Kazakhstan, there are 2.5 thousand heat supply sources providing the production of about 90 million Gcal. The received heat energy is distributed to end users through networks with a length of 11,357.9 km in two-pipe calculation. From these networks, 27.5% are in a dilapidated state, 27.9% are in need of replacement. This leads to the fact that the losses amounted to 16.9% of the total volume of energy supplied to the consumer.

In 2025, it is planned to create a common electricity market of the Eurasian Economic Union. Kazakhstan's electric power industry will also have to seriously prepare for this. In the country's strategic development plan until 2020, one of the main goals in the energy sector is the creation of a vertically integrated company with a nuclear fuel cycle. This means that dozens of specialists will be required in nuclear reactors and power plants, protection and non-proliferation of nuclear materials, electronics, and automation of physical installations.

4.2. Energy Supply and Energy Intensity of Kazakhstan

Kazakhstan is one of the top ten energy-rich countries in the world. Energy supply indicator (the ratio of primary energy production/extraction to the volume of gross consumption of fuel and energy resources) of Kazakhstan is 228%. The ratio of gross consumption and production of fuel and energy resources is shown in Figure 1.

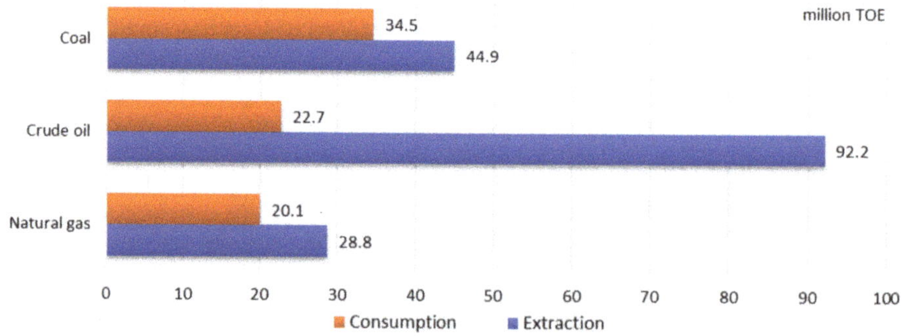

Figure 1. Dynamics of gross consumption and production of fuel and energy resources in Kazakhstan in 2019, million tons of oil equivalent (TOE). Source: compiled by the authors based on data from [69,70].

The gross consumption of fuel and energy resources in 2019 in Kazakhstan was 73 million TOE, and the final energy consumption was 41.6 million TOE. The largest energy consumption was shown by the industrial sector, the housing sector, and the transport sector (Figure 2).

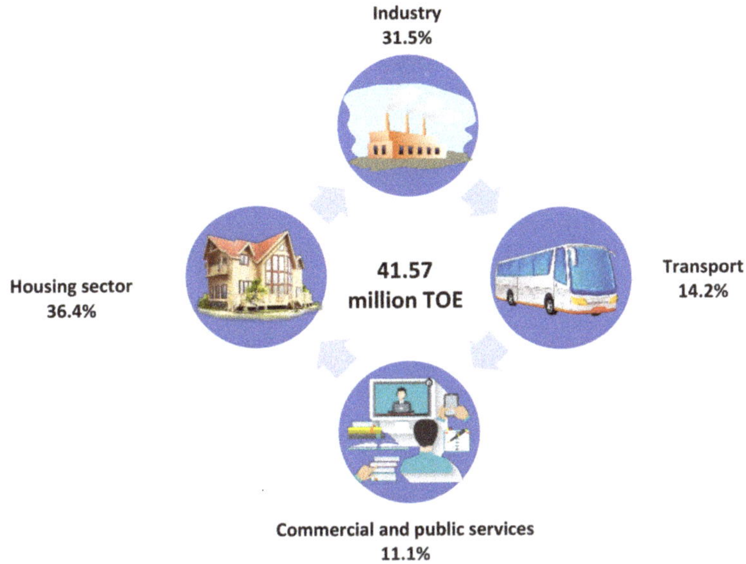

Figure 2. Structure of the main items of final consumption of fuel and energy resources in Kazakhstan in 2019, million TOE. Source: Compiled by the authors based on data from [69,70].

An indicator of the economic efficiency of consumption of fuel and energy resources is the indicator of the energy intensity of GDP. It is defined as the ratio of the volume of gross consumption of fuel and energy resources for all production and non-production needs in tons of oil equivalent to the value of GDP. The energy intensity of the economy is a particularly strong indicator of the nature intensity. This is a key indicator that characterizes the sustainability of the development of both the country as a whole and the energy sector. This indicator is one of the basic indicators in most systems of sustainability indicators.

According to statistics, with a high energy supply of the economy, the indicator of energy intensity of GDP in Kazakhstan is one of the highest (Figure 3). This situation demonstrates a very low efficiency of the economy of Kazakhstan.

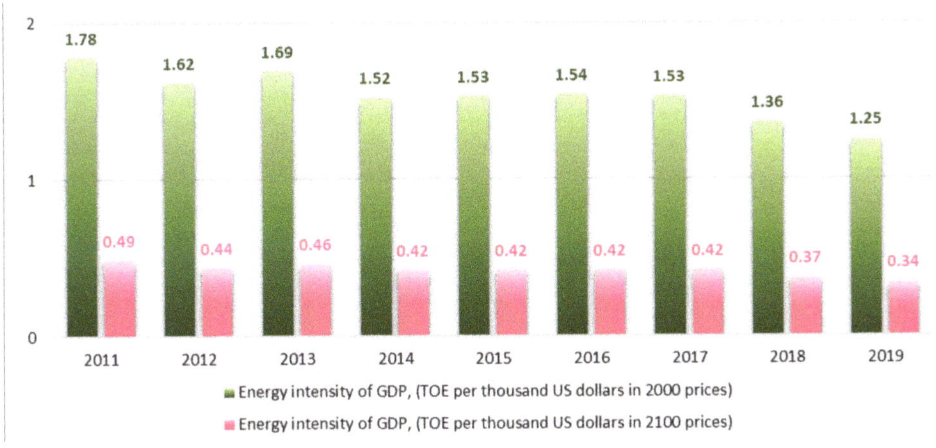

Figure 3. Dynamics of energy intensity of Kazakhstan's GDP in 2011–2019. Source: compiled by the authors based on data from [70].

The energy intensity of Kazakhstan's GDP in comparison with the world average exceeds 2 times, with the OECD countries-4 times, among the CIS countries is on the 4th place. The strategic plan for the development of the Republic of Kazakhstan until 2025 and the Concept for the transition to a "green economy" set goals to reduce the energy intensity of the country's GDP by 25% by 2025 and by 50% by 2050. Another indicator of the modernization that has begun in recent years is the boom in street lighting at night. Today, about one third of street lamps in Kazakhstan have been converted to LED lamps. Annual savings in this segment alone amounted to more than 500 million tenge (about 1 million euros) [69].

The development of energy efficiency in Kazakhstan is regulated by the law "On Energy Saving and Energy Efficiency Improvement" (adopted in 2012). At the same time, in 2015, the Ministry of Investment and Development of the Republic of Kazakhstan developed an energy efficiency map. As part of the implementation of the energy efficiency map in September 2019, the Institute for the Development of Electric Power and Energy Saving JSC published the results of an energy audit of the country's main industrial enterprises. It turned out that, on average, enterprises can reduce energy consumption by 10%, and budget organizations—up to 40%. However, the heads of institutions are not interested in the introduction of energy-efficient technologies due to the long payback period of such energy projects [71].

One of the tools that stimulate the introduction of energy-efficient technologies in Kazakhstan is the State Energy Register. State institutions (kindergartens, schools, medical organizations, theaters) that consume more than 100 tons of conventional fuel per year and industrial enterprises that consume more than 1500 tons of conventional fuel per year are added to the register. Organizations that are included in the register are required to conduct an annual energy audit and follow the auditor's recommendations for five years, otherwise they will be fined. But the fines system is designed in such a way that in some cases it is much cheaper to pay a fine than to invest in energy efficiency. Nevertheless, according to the Ministry of Industry and Infrastructure Development of Kazakhstan, over 5 years the share of enterprises implementing the energy efficiency system increased from 9.7% to 41.7%, and the effect of the introduction of energy-saving technologies in Kazakhstan in 2018 amounted to 27 billion tenge (about 54 million euros), and in 2019—28.2 billion tenge (about 54 million euros) [72].

The main reasons for the high energy intensity of Kazakhstan's GDP:

- the current structure of the economy with a predominance of energy-intensive types of industries—extractive industries, mining and metallurgical complex, oil and gas sector, coal energy.
- climatic conditions when the heating season lasts 9 months in the northern parts of Kazakhstan. This leads to an energy-spending and energy-consuming heat supply sector.
- the general technological backwardness of many sectors of the economy, which leads to a high energy intensity of products (in some industries, it exceeds the similar European indicator by several times).
- relatively low cost of energy prices, which does not encourage many consumers to be energy-efficient.
- the deterioration of networks and equipment in the housing and communal services, the associated significant losses, fuel, and energy consumption.

Summarizing the review of energy availability and energy intensity, it should be noted that increasing the level of energy management remains an important task. Because whichever of the development alternatives the industry chooses, the indicators of efficiency of production and energy consumption should be significantly improved.

4.3. Alternative Energy Sources

Despite the still significant role of coal in the energy sector, global challenges are leading to a reduction in the share of coal generation in the global energy mix. To ensure sustainable energy development, one of the priorities is the environmental friendliness of energy sources. Immediate and complete rejection of the use of coal is not possible. The latest technological solutions make it possible to ensure coal mining and the operation of coal stations more environmentally friendly and maneuverable. On this basis, the development of coal seams becomes "green", and the operation of modern coal plants is almost as clean as gas plants: carbon dioxide emissions at such facilities can be reduced or captured and used effectively.

A large-scale transition from coal to gas in the economy, as well as improving energy efficiency and further increasing the use of renewable energy sources are the most important factors that allow Kazakhstan to fully reach the unconditional emission reduction target (15% of 1990 levels by 2030) under the Paris Agreement.

Another direction in the development of the coal industry is the production of liquid fuel and various chemical products by hydrogenation, liquefaction, and coal extraction methods. This is one of the important directions in the coal chemical industry of the future. In several countries, public and private companies are conducting intensive research into the production of synthetic liquid fuels from coal. In addition, today it is relevant to study the issues of obtaining synthetic liquid fuel and humic preparations from Kazakhstani coals, which will make it possible in the future to bring the processing of local solid hydrocarbon raw materials closer to a comprehensive one.

According to international experts [73], by 2100 the share of oil and coal in the world fuel and energy balance will be 2.1%, and 0.9%, respectively, thermonuclear energy will reach 10% of the market, and more than 25% of all world electricity will be generated by the sun. There will be a gradual decrease in the use of hydrocarbons and a reorientation to the production of cleaner energy industries.

In 2009, Kazakhstan passed a law "On the support of renewable energy sources". The share of renewable energy sources is still small—just over 10% (including hydropower) (Figure 4, Table 1) [2,74–76]. But the existing difficulties in its implementation call into question that by 2050 the share of the use of renewable energy sources in Kazakhstan will reach 50%.

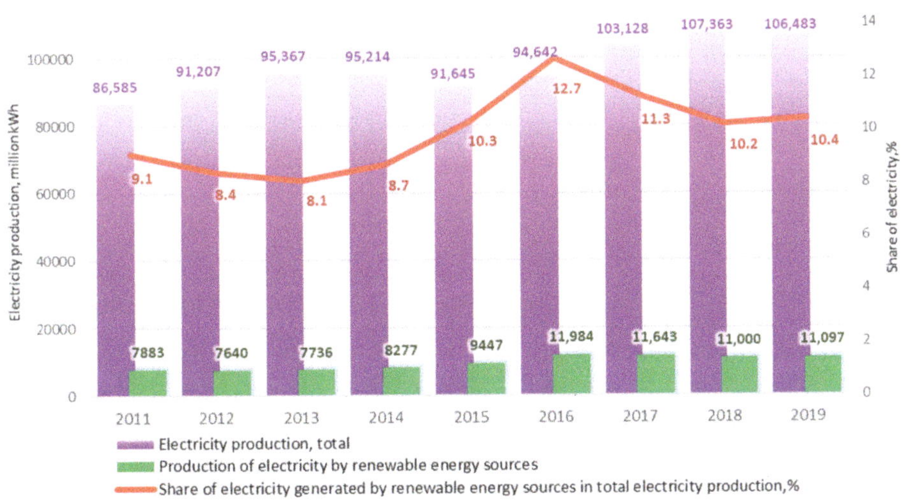

Figure 4. Dynamics of electricity production in Kazakhstan in 2011–2019, million kWh. Source: compiled by the authors based on data from [2,74–76].

Table 1. Structure of production of renewable energy sources in Kazakhstan, thousand kWh.

	2011	2012	2013	2014	2015	2016	2017	2018	2019
produced by hydroelectric	7,883,323.0	7,637,265.9	7,730,762.5	8,262,830.9	9,269,190.4	11,620,763.9	11,210,191.7	10,395,354.1	9,993,658.8
produced by wind farms	147.0	2665.0	4546.9	13,300.8	131,722.3	274,982.8	339,840.3	460,583.1	707,135.1
produced by solar power plants		21	775.8	1268.3	46,171.0	88,403.1	93,038.8	141,311.1	391,229.6
produced using biogas							200.0	2832.1	4967.1

Source: compiled by the authors based on data from [2,74–76].

The use of biological fuel has a certain reserve. Due to the processing of agricultural waste, up to 35 billion kWh of electricity and 44 million Gcal. of thermal energy can be obtained annually. However, apart from a part of hydropower, these resources have not been widely used until now (Table 1).

The potential of renewable energy resources (hydropower, wind, and solar energy) in Kazakhstan is quite significant. The Ministry of Energy of Kazakhstan has estimated the potential of the republic at more than 2.7 trillion. kW, of which:

- wind energy potential—920 billion kWh,
- potential of solar energy—15 billion kWh (1300–1800 kW per 1 sq. m of area per year),
- potential of hydropower—170 billion kWh.

The calculated indicators demonstrate significant potential to enable Kazakhstan to make the transition from traditional to alternative energy sources. However, the choice of an alternative path of development is not carried out only based on calculations, it is accepted and carried out by people. The authors proceed from the assumption that the decision in the dilemma of long-term development of the electric power industry in Kazakhstan will be made under the influence of the industry stakeholders and will be implemented in the overall result. Therefore, it will reflect the common vision of the most interested and influential participants.

4.4. Industry Risks and Trends

The considered tendencies of the electric power industry in Kazakhstan pose a difficult choice for the country to modernize the existing production base and improve coal processing methods or direct its efforts to a more expensive but promising direction for the development of renewable energy sources.

The authors of the article were directly involved in a series of events to determine the forecast for the development of the energy industry in Kazakhstan in August 2020. In particular, the authors were the moderators of the foresight sessions, which allowed further analysis to be carried out using the original author's methodology. The Foresight was attended by 136 experts representing the manufacturing sector of the energy sector of Kazakhstan, industry science and education. The assessment of the future development of the industry was an important point in studying the opinion of experts, enabling us to determine, in a first approximation, the difference and coincidence in the vision of stakeholders. (Figure 5).

The weighted average assessment of the industry's prospects on a 10-point scale was 8.03 points—stable development of the industry. Moreover, the weighted average assessment of the possibility of an expert's personal influence on the development of the industry on a 10-point scale was 5.89 points, i.e., experts assess their influence as average, and the influence of companies in which experts work—6.75 points, which is slightly higher than the assessment of personal influence.

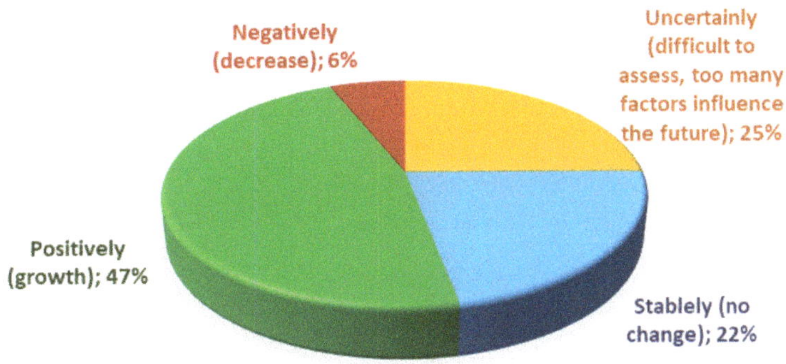

Figure 5. Expert assessment of the future development of the energy industry in Kazakhstan. Source: compiled by the authors based on the results of the Foresight.

According to experts, the following subjects have the greatest influence on the development of the industry (in descending order of the power of influence):

- Government of Kazakhstan,
- line ministry,
- owners and shareholders of energy companies,
- multinational companies,
- resident top managers,
- production personnel of energy companies,
- banks and credit institutions,
- functional management of energy companies,
- local executive authorities,
- universities and colleges.

It is obvious that it these participants not only form the future development agenda of the industry, but also have the greatest influence on the choice of the development model. The influence of the state and big business is still decisive, the role of the population and the public on such an important issue as the future of energy is, unfortunately, small.

All experts showed restrained caution in assessing the prospects for the development of the industry. Experts consider the development of monitoring and data processing technologies and new technologies for transmission and distribution of energy as growth zones for the domestic energy industry. This is where the main investment and management efforts of the next period should be directed.

Like any production system, the energy sector of Kazakhstan is subject to risks, not only of a technical and technological nature, but also of an environmental, social and labor nature. The authors of the article also took part in the development of the foresight methodology and in the analysis of its results. The experts were asked to assess 12 types of risks in the energy sector of Kazakhstan (Figure 6).

To further use the views of stakeholders as a basis for making management decisions, the weighted average probability of each risk and the severity of their consequences was determined, which made it possible to map the risks of the energy industry in Kazakhstan (Figure 7). Figure 7 uses the same risk numbering as in Figure 6.

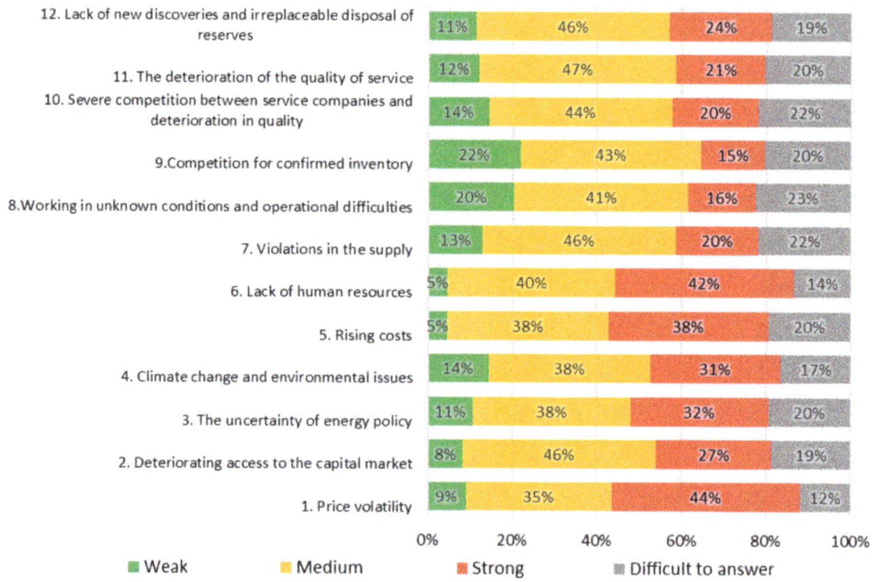

Figure 6. Expert assessment of the risks of negative development of the electric power industry in Kazakhstan. Source: compiled by the authors based on the results of the Foresight.

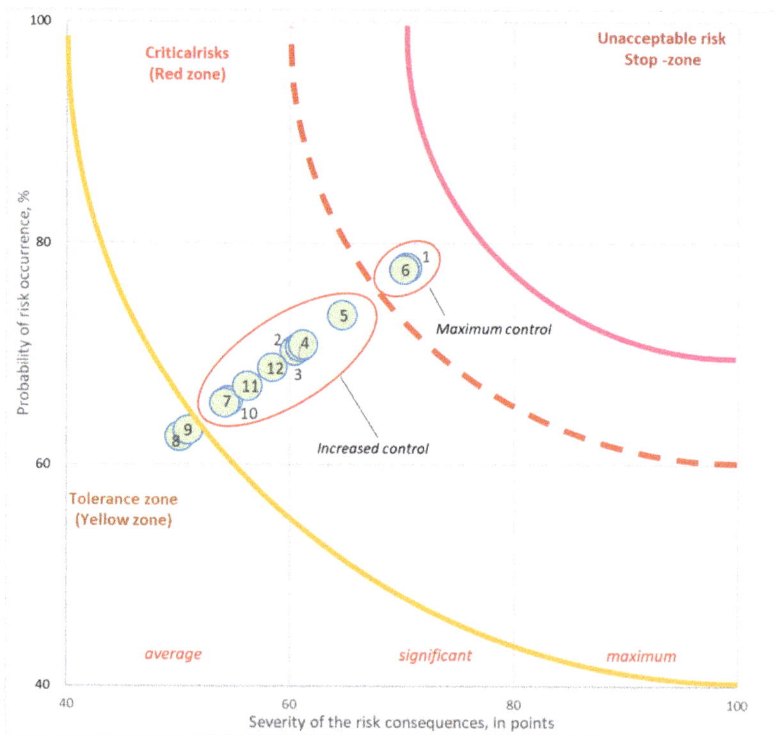

Figure 7. Risk mapping of the energy sector in Kazakhstan. Source: compiled by the authors based on the results of the Foresight.

When mapping risks, four zones are distinguished: Unacceptable risk, critical risk, tolerant risks, acceptable risks. Given that, base on the calculations, the risks were concentrated mainly in one zone, the scale of the map was increased by the authors and the zone of acceptable risks was not shown (it did not include the risks of the energy industry).

As can be seen from the figure, according to experts, there are currently no unacceptable risks in the industry. However, there are two risks that require maximum control: 1. Price volatility and 6. Lack of human resources. Eight more risks fall into the critical control zone. And two risks—8. Working in unknown conditions and operational difficulties and 9. Competition for confirmed inventory—the experts recognized as tolerant.

The risks selected by the foresight participants reflect not only their expert assessment, but also the conviction of specialists working in the industry. In every day practice, it is necessary to confront the noted threats that arise from the risks of the implementation of a particular development scenario. Therefore, the authors consider it possible to use the collective opinion of experts to assess the development dilemma. We have built a comparative matrix (Table 2), which allows us to evaluate the alternatives for new construction and operation of various power generation facilities according to four characteristics. The points in the table were distributed as follows: 4—the maximum negative assessment for this characteristic, 1—the minimum negative assessment. The matrix makes it possible to formalize the foresight experts' assessments, taking into account the strength of the influence of each characteristic when choosing the use of energy sources.

Table 2. Matrix for assessing the choice of using energy sources.

Assessment Characteristics	Nuclear Power Plants	RES	Coal Power Plants
Construction cost including equipment manufacturing	4	2	2
Lack of human resources, including the availability of educational infrastructure on the relevant technologies of the electric power industry	4	2	1
Electricity price	2	3	1
Environmental impact	4	1	4
Total	14	8	8

Source: compiled by the authors based on the results of the Foresight.

The cost of building a nuclear power plant is the highest compared to renewable energy sources and coal-fired power plants. In addition, a project for Kazakhstan, requiring significant investment and the involvement of experienced contractors.

The lack of personnel is also the most typical for the nuclear power industry due to its absence in the mill. Colleges and universities in Kazakhstan do not train specialists in nuclear energy. Traditionally, personnel are trained in the specialties of coal energy, and in the last 5 years, renewable energy sources.

Due to the low price of coal, the stability and availability of technological processes, the price of coal energy is minimal. The price of nuclear power is slightly higher since it is necessary to take into account the return-on-investment costs. However, since nuclear power plants are designed for a long service life and a large volume of electricity production, the cost of 1 unit of electricity is relatively low. The highest price for renewable energy sources is associated with high investment costs with unstable electricity production, as well as the need to maintain shunting capacities.

The assessment of the environmental impact is the highest in the nuclear power industry due to the catastrophic consequences in case of possible accidents and the need to dispose of nuclear waste. The maximum assessment in coal power engineering is associated with both emissions into the atmosphere and disturbance of the landscape and microclimate during coal mining.

Thus, having calculated the scores for each type of power plant, we see the maximum negative (risk) potential for nuclear energy and the same values for the assessment of RES and coal energy. This result once again confirms the presence of a dilemma in choosing a further path for the development of Kazakhstan's energy sector.

The choice of management methods for identified risks directly depends on industry development trends that are expected soon and on actors and stakeholders, whose influence is most significant in this sector of the economy of Kazakhstan.

Among the strong and dominant (in total), experts identified the following trends:

- introduction of new technologies for generation, transmission, and distribution of energy (65% of respondents),
- digitalization, data collection and big data analytics (60% of respondents),
- growth of environmental requirements (51% of respondents).

The rest of the trends have a low dominant influence (less than 50% of respondents), or rather an indirect influence (Figure 8).

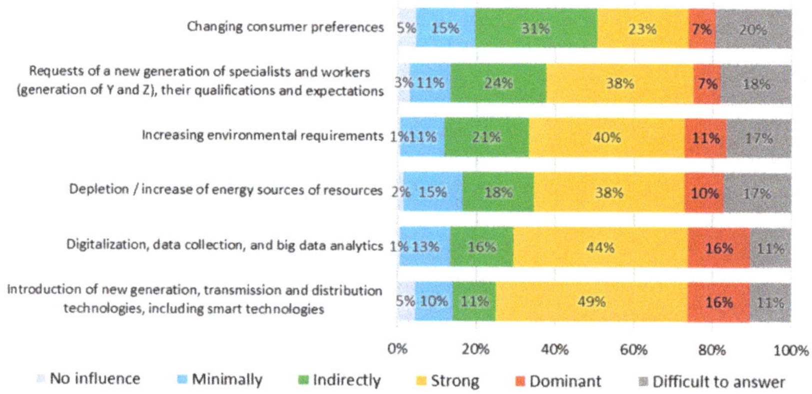

Figure 8. Expert assessment of the impact of major trends on the future of the energy industry in Kazakhstan. Source: compiled by the authors based on the results of the Foresight.

Experts predict different activity of implementation of the identified trends and the likelihood of risks. The participation of experts in building the desired future of the industry depends on how much they themselves position their influence on the development of the industry, as well as new opportunities for the development of the industry experts see.

As shown by the answers of experts, the main opportunities in the industry in the future open up primarily in the construction of generating stations in alternative energy (2 in Figure 9) and coal, gas, oil (1 Figure 9).

Further, in the group of dominant influence, there are opportunities (in order of decreasing probability of occurrence and strength of influence):

6—Significant reduction in energy losses,

7—Significant reduction in personnel due to the introduction of telemetry.

The following opportunities fell into the zone of strong influence with a 50% probability of attack (in order of decreasing strength and possibility of influence):

5—Consolidation of companies, optimization of processes and costs.

3—Stabilization of the price market and their stability in the next 5 years,

4—Reduced energy transportation costs.

Considering the almost equal dominant prospects that the experts identified—the construction of classical power plants (fueled by coal, gas, and oil) and renewable energy sources, we can state that the dilemma of choosing a development path for experts.

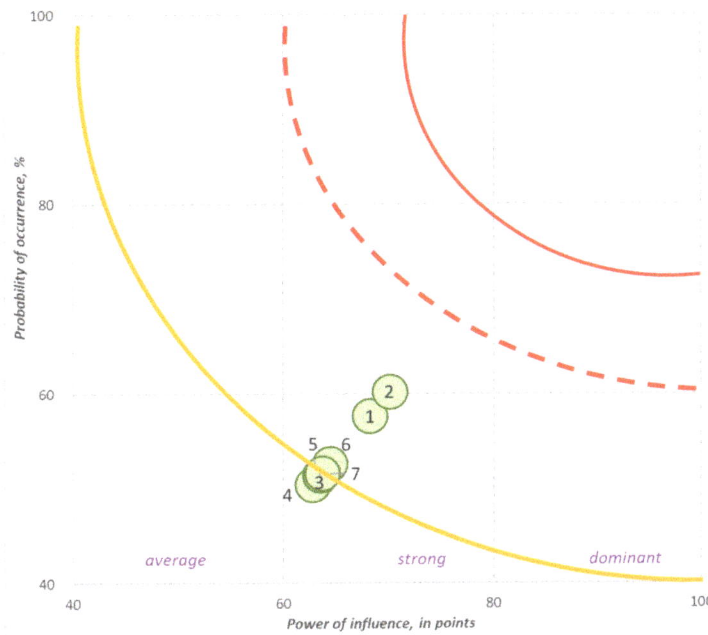

Figure 9. Expert assessment of opportunities for the development of the electric power industry in Kazakhstan. Source: compiled by the authors based on the results of the Foresight.

5. Discussion

The existing electric power facilities in Kazakhstan were built in Soviet times and are quite worn out. They require serious modernization or complete replacement of equipment that is not produced in Kazakhstan. In this regard, we are witnessing a dilemma facing the country—where to direct financial resources: Modernize existing predominantly coal-fired thermal power plants or build renewable energy facilities?

On the one hand, the country has vast coal resources, and thus, low prices for traditional coal energy. According to the Internet portal GlobalPetrolPrices [77], the cost of electricity in Kazakhstan for one kWh is $0.04, while the average price in the world is $0.14. In Belarus, one kWh of electricity costs $0.07, in Georgia $0.06, and in Ukraine $0.05. Modernization projects for existing coal-fired power generation facilities are an order of magnitude lower than projects for new construction of renewable energy sources. Traditional heat producers will resist the invasion of alternative green energy competitors.

On the other hand, the global environmental situation dictates the need for a transition to green electricity based on renewable energy sources. The Government of Kazakhstan has set a strategic goal that by 2050, half of the country's total energy consumption should be generated by renewable energy sources (RES), using water, wind, sun. Large investments are required for the construction of renewable energy sources. The unstable position of the state currency in the world market and an opaque operating environment are a deterrent for foreign investment in Kazakhstan. Therefore, the main obstacles to the development of renewable energy sources in Kazakhstan are the lack of access to long-term money and taxes on imported equipment. It is also necessary to understand that an increase in the share of renewable energy sources in the energy sector requires a proportional volume of modern shunting capacities, which is absent in Kazakhstan. In addition, green energy will increase the tariff burden on end users by 2.5–3 times.

Since the issues of long-term development of the electric power industry are closely related to the energy security policy, we consider it expedient to consider the choice of a development model with the provision of an appropriate level of energy security for the country (Table 3). The authors rely on the international classification of energy security, depending on the period of the forecast assessment [24].

Table 3. Methodological approaches to solving the dilemma.

Period	Energy Security	Choice Dilemma
short	associated with the risks of power outages	*Social approach*—ensuring the preservation of moderate costs of energy costs for the population and enterprises, especially during the period of coming out of the pandemic lockdown and overcoming the crisis generated by them
mid-term	uninterrupted, reliable, sufficient power supply ranging from years to decades	*A pragmatic approach*—a transition to effective energy management, reduction of harmful effects on the environment, accumulation of funds and their investment in alternative sources, gradual integration into the global energy system
long term	ensures the availability of energy sources in the future	*An innovative approach* is to ensure such a share of innovative (alternative) energy, which will allow the country to fully participate in global economic and energy cooperation and ensure high efficiency in the production and use of energy

Source: compiled by the authors based on the source [24] and their own developments.

It should also be noted that the problem of choosing an energy model affects the interests of various territories and groups of society in the republic, which are contradictory in nature and content. The authors previously considered regional features in the study of energy management problems and noted the difference in its effectiveness both by region and by industry [21–23]. The works of Kazakh authors also note the need to differentiate approaches to energy management, models of energy systems are proposed with a breakdown by subnational regions, types of buildings and urban/rural areas, which in their opinion, will increase the energy security of households [59]. The various choices in the way electricity is developed in the republic, combined with significant social obligations, reduce the importance of rational factors, especially in the short term.

The authors consider that a controversial position about the need to introduce a ban on the use of coal and move to cleaner alternatives for generating energy in the shortest possible time. The dilemma of choosing the long-term development of the electric power industry in Kazakhstan should be solved using different approaches for solutions of different implementation periods.

6. Conclusions (Prospects for the Development of the Industry)

It is necessary to consider the directions of development of the energy industry in Kazakhstan from different time perspectives. Soon, the development of environmentally friendly and efficient technologies for the processing of fossil fuels (oil, coal, gas) based on steam-gas plants and methods of deep processing of coal, is of greatest relevance. At the same time, fossil fuel will remain a priority energy source. It should provide an exit for the economy from the crisis generated by the need for a lockdown and ensure the accumulation of investments for the transition to alternative energy sources.

In the future, there will be an active introduction of renewable energy sources and the development of effective methods for converting and storing energy, including fuel cells. These technologies have already begun to be implemented, but a radical change in the structure of the world energy because of the displacement of coal and its replacement by carbon-free sources will come after 2050. Nuclear energy plays a key role, in combination with solar energy, hydropower, and environmentally friendly biofuels. Although it is nuclear energy, which has indisputable economic advantages for Kazakhstan, may be not chosen for political and social factors.

In conclusion, it should be noted that modern energy is already in the process of digital transformation. Digital technologies are actively penetrating the energy sector, making it possible to analyze and manage the production, transportation, and consumption of energy more efficiently. As a result, technological innovations shape the future of the energy sector and form the basis for economic and social choice.

The staffing dilemma remained outside the scope of this study. The issues of staffing the electric power industry in Kazakhstan are important, but require separate consideration, which will be presented in subsequent publications of the authors.

Author Contributions: Conceptualization, E.V. and Y.P.; methodology, E.V.; validation, E.V., Y.P. and Y.S.M.; formal analysis, E.V.; investigation, E.V.; resources, G.K.; data curation, Y.P.; writing—original draft preparation, E.V.; writing—review and editing, Y.P.; visualization, E.V.; supervision, Y.S.M.; project administration, Y.P.; funding acquisition, G.K. All authors have read and agreed to the published version of the manuscript.

Funding: This research received no external funding.

Conflicts of Interest: The authors declare no conflict of interest.

References

1. Overview of the electricity market in Kazakhstan. Halyk Global Markets JSC. Available online: https://kazks.kz/ru/analytics/macroeconomics/obzor-rynka-elektroenergii (accessed on 14 April 2021).
2. National Energy Report Kazenergy 2019. p. 149. Available online: http://kazenergy.com/upload/document/energy-report/NationalReport19_ru.pdf (accessed on 14 April 2021).
3. Order of the Ministry of Energy of the Republic of Kazakhstan No. 6 Dated 12 January 2021 "On approval of the forecast balances of Electric Power of Capacity for 2021–2027". Available online: https://kegoc.kz/ru/prognoznyy-balans-moshchnosti-i-elektroenergii-ees-kazahstana-na-period-2021-2027gg (accessed on 14 April 2021).
4. Ecofys, Redpoint, ERAS. Analysis of Impacts of Climate Change Policies on Energy Security. European Commission DG Environment, 2009. Available online: https://ec.europa.eu/environment/integration/energy/pdf/cces.pdf (accessed on 21 April 2021).
5. Jansen, J.C.; Seebregts, A.J. Long-term energy services security: What is it and how can it be measured and valued? *Energy Policy* **2010**, *38*, 1654–1664. [CrossRef]
6. Winzer, C. Conceptualizing Energy Security. *Energy Policy* **2012**, *46*, 36–48. [CrossRef]
7. Tziogas, C.; Georgidas, P. Sustainable Energy Security: Critical Taxonomy and System Dynamics Decision-Making Methodology. *Chem. Eng.* **2015**, *43*, 1951–1956.
8. Gracceva, F.; Zeniewski, P. A systemic approach to assessing energy security in a low-carbon EU energy system. *Appl. Energy* **2014**, *123*, 335–348. [CrossRef]
9. Sovacool, B.K.; Sodortsov, R.V. *Jones, B.R. Energy Security, Equality and Justice*; Routledge, Oxon: London, UK, 2014. [CrossRef]
10. Kim, J.-H.; Shcherbakova, A. Common failures of demand response. *Energy* **2011**, *36*, 873–880. [CrossRef]
11. Bohi, D.R.; Toman, M.A.; Walls, M.A. *The Economics of Energy Security*; Kluwer Academic Publishers: Norwell, MA, USA, 1996; ISBN 978-94-009-1808-5.
12. Kuik, O.J.; Lima, M.B.; Gupta, J. Energy security in a developing world. *Wires Clim. Chang.* **2011**, *2*, 627–634. [CrossRef]
13. Johnson, C.; Boersma, T. The politics of energy security: Contrasts between the United States and the European Union. *WIREs Energy Environ.* **2015**, *4*, 171–177. [CrossRef]
14. Lei, W.; Xuejun, L. China or the United States: Which threatens energy security? *OPEC Rev.* **2007**, *31*, 215–234. [CrossRef]
15. Shiroyama, H.; Yoshikawa, H.; Orsi, R.; Sato, C. The Future of Energy Policy in East Asia: Prospects and Risks for a Sustainable Framework. In *Handbook of Clean Energy Systems*; John Wiley & Sons: Hoboken, NJ, USA, 2015. [CrossRef]
16. McNally, A.; Magee, D.; Wolf, A.T. Hydropower and sustainability: Resilience and vulnerability in China's powersheds. *J. Environ. Manag.* **2009**, *90*, 286–293. [CrossRef]
17. Meyer, N.I. Learning from wind energy policy in the EU: Lessons from Denmark, Sweden and Spain. *Environ. Policy Gov.* **2007**, *17*, 347–362. [CrossRef]
18. Creţan, R.; Vesalon, L. The Political Economy of Hydropower in the Communist Space: Iron Gates Revisited. *J. Econ. Hum. Geogr.* **2017**, *108*, 688–701. [CrossRef]
19. Karatayev, M.; Hall, S.; Kalyuzhnova, Y.; Clarke, M.L. Renewable energy technology uptake in Kazakhstan: Policy drivers and barriers in a transitional economy. *Renew. Sustain. Energy Rev.* **2016**, *66*, 120–136. [CrossRef]
20. Koshim, A.; Karatayev, M.; Clarke, M.L.; Nock, W. Spatial assessment of the distribution and potential of bioenergy resources in Kazakhstan. *Adv. Geosci.* **2018**, *45*, 217–225. [CrossRef]
21. Petrenko, Y.; Denisov, I.; Koshebayeva, G.; Biryukov, V. Energy Efficiency of Kazakhstan Enterprises: Unexpected Findings. *Energies* **2020**, *13*, 1055. [CrossRef]
22. Velinov, E.; Petrenko, Y.; Vechkinzova, E.; Denisov, I.; Siguencia, L.; Ródek-Szostak, Z. "Leaky Bucket" of Kazakhstan's Power Grid: Losses and Ineffcient Distribution of Electric Power. *Energies* **2020**, *13*, 2947. [CrossRef]
23. Petrenko, Y.; Vechkinzova, E.; Antonov, V. Transition from the industrial clusters to the smart specialization of the regions in Kazakhstan. *Insights Reg. Dev.* **2019**, *1*, 118–128. [CrossRef]

24. Soysal, O.A.; Soysal, H.S. *Energy for Sustainable Society: From Resources to Users*; John Wiley & Sons: Hoboken, NJ, USA, 2020; p. 526, Online ISBN 9781119561330, Print ISBN 9781119561309. [CrossRef]
25. York, D.; Kushler, M. *Exploring the Relationship Between Demand Response and Energy Efficiency: A Review of Experience and Discussion of Key Issues*; Report Number U052; American Council for an Energy-Efficient Economy: Washington, DC, USA, 2005.
26. Bilton, M.; Ebeanu, N.C.; Woolf, M.; Djapic, P.; Wilcox, M.; Strbac, G. *Impact of Low Voltage Connected Low Carbon Technologies on Network Utilization*; Report B4 for the "Low Carbon London" LCNF Project; Imperial College: London, UK, 2014.
27. York, D.; Kushler, M.; Witte, P. Examining the peak demand impacts of energy efficiency. *Manag. Q.* **2007**, *48*, 2–15.
28. Nguyen, V.P.; Tran, T.K. Explicating Energy Saving Intention from the Prospect of Small Medium Enterprises. *Entrep. Sustain. Issues* **2020**, *8*, 716–734. [CrossRef]
29. Igaliyeva, L.; Niyazbekova, S.H.S.; Serikova, M.; Kenzhegaliyeva, Z.; Mussirov, G.; Zueva, A.; Tyurina, Y.; Maisigova, L. Towards environmental security via energy efficiency: A case study. *Entrep. Sustain. Issues* **2020**, *7*, 3488–3499. [CrossRef]
30. CEER. *Advice on Ensuring Market and Regulatory Arrangements help deliver Demand-Side Flexibility*; Ref: C14-SDE-40-03; Council of European Energy Regulators: Brussels, Belgium, 2014.
31. Torriti, J.; Hassan, M.G.; Leach, M. Demand response experience in Europe: Policies, programmes and implementation. *Energy* **2010**, *35*, 1575–1583. [CrossRef]
32. Tolstolesova, L.L.; Yumanova, N.; Mazikova, E.; Glukhikh, I.; Vorobieva, M. Sustainability issues of territorial power systems in market conditions. *Entrep. Sustain. Issues* **2019**, *7*, 263–277. [CrossRef]
33. Sarma, U.; Karnitis, G.; Zuters, J.; Karnitis, E. District heating networks: Enhancement of the efficiency. *Insights Reg. Dev.* **2019**, *1*, 200–213. [CrossRef]
34. Rogalev, A.; Komarov, I.; Kindra, V.; Zlyvk, O. Entrepreneurial assessment of sustainable development technologies for power energy sector. *Enterp. Sustain. Issues* **2018**, *6*, 429–445. [CrossRef]
35. NREL. *Chapter 10: Peak Demand and Time-Differentiated Energy Savings Cross-Cutting Protocols, The Uniform Methods Project: Methods for Determining Energy Efficiency Savings for Specific Measures (Frank Stern, Navigant Consulting)*; Subcontract Report NREL/SR-7A30-53827; 2013. Available online: https://www.energy.gov/sites/prod/files/2013/05/f0/53827-10.pdf (accessed on 21 April 2021).
36. Le Coq, C.; Paltseva, E. Measuring the Security of External Energy Supply in the European Union (22 May 2009). *Energy Policy* **2009**, *37*, 4474–4481. [CrossRef]
37. Yergin, D. Energy Security in the 1990s. *Foreign. Aff.* **1988**, *67*, 110–132. [CrossRef]
38. Luft, G.; Korin, A.; Gupta, E. Energy security and climate change: A tenuous Link. In *The Routledge Handbook of Energy Security*; Sovacool, B.K., Ed.; Routledge: New York, NY, USA, 2011; pp. 43–55.
39. Hughes, L. A generic framework for the description and analysis of energy security in an energy system. *Energy Policy* **2012**, *42*, 221–231. [CrossRef]
40. Mansson, A.; Johansson, B.; Nilsson, L.J. Assessing energy security: An overview of commonly used methodologies. *Energy* **2014**, *73*, 1–14. [CrossRef]
41. Du, H.B.; Chen, Z.N.; Brown, M.A.; Yin, Y.Y.; Zuo, J.; Chai, L.H. How secure are national energy systems: A dynamic assessment approach. *Ecol. Indic.* **2020**, *108*, 105666. [CrossRef]
42. Böhringer, C.; Keller, A. *Energy Security: An Impact Assessment of the EU Climate and Energy Package*; No V-335-11, Working Papers; Department of Economics, University of Oldenburg: Oldenburg, Germany, 2011. Available online: https://uol.de/f/2/dept/wire/fachgebiete/vwl/DP_V-335_11.pdf (accessed on 21 April 2021).
43. Checchi, A.; Behrens, A.; Egenhofer, C. *Long-Term Energy Security Risks for Europe: A Sector-Specific Approach*; CEPS Working Paper No. 309; Centre for European Policy Studies: Brussels, Belgium, 2009. Available online: https://ssrn.com/abstract=1334620 (accessed on 21 April 2021). [CrossRef]
44. Kruyt, B.; van Vuuren, D.P.; de Vries, H.J.M.; Groenenberg, H. Indicators for energy security. *Energy Policy* **2009**, *37*, 2166–2181. [CrossRef]
45. Mitchell, J.V. Renewing Energy Security. Royal Institute of International Affairs. 2002. Available online: https://www.chathamhouse.org/sites/default/files/public/Research/Energy,%20Environment%20and%20Development/renewing_energy_security_mitchell_july_2002.pdf (accessed on 21 April 2021).
46. Chester, L. Conceptualising energy security and making explicit its polysemic nature. *Energy Policy* **2010**, *38*, 887–895. [CrossRef]
47. Cherp, A.; Jewell, J. The concept of energy security: Beyont the four As. *Energy Policy* **2014**, *75*, 415–421. [CrossRef]
48. Joskow, P. The U.S. Energy Sector: Prospects and Challenges, 1972–2009. *Dialogue* **2009**, *17*, 7–11.
49. Chehabeddine, M.; Tvaronavičienė, M. Securing regional development. *Insights Reg. Dev.* **2020**, *2*, 430–442. [CrossRef]
50. Alkhateeb, T.T.Y.; Mahmood, H.; Altamimi, N.N.; Furqan, M. Role of education and economic growth on the CO2 emissions in Saudi Arabia. *Entrep. Sustain. Issues* **2020**, *8*, 195–209. [CrossRef]
51. Shumilo, O.; Lytvyn, I.; Shablystyi, V.; Kornyakova, T.; Popovich, I. Legal mechanism to ensure national security in the field of use of natural resources. *Entrep. Sustain. Issues* **2021**, *8*, 455–470. [CrossRef]
52. Cronshaw, I.; Grafton, Q. Reflections on Energy Security in the Asia Pacific. *Asia Pac. Policy Stud.* **2014**, *1*, 127–143. [CrossRef]
53. Levèfre, N. Measuring the energy security implications of fossil fuel resource concentration. *Energy Policy* **2010**, *38*, 1635–1644. [CrossRef]
54. Plėta, T.; Tvaronavičienė, M.; Della Casa, S.; Agafonov, K. Cyber-attacks to critical energy infrastructure and management issues: Overview of selected cases. *Insights Reg. Dev.* **2020**, *2*, 703–715. [CrossRef]

55. Winzer, C. Conceptualizing Energy Security, EPRG Working Paper 1123, University Cambridge, July 2011. Available online: http://www.econ.cam.ac.uk/research-files/repec/cam/pdf/cwpe1151.pdf (accessed on 21 April 2021).
56. Cotella, G.; Cassen, C.; Graccdva, F. The Energy Security Trends and Strategies in the European Territorial Cohesion. Presentation at MILESECURE-2050 Energy Day, Brussels, Belgium, 25 June 2014.
57. Blum, H.; Legey, L.F.L. The challenging economics of energy security: Ensuring energy benefits in support to sustainable development. *Energy Econ.* **2012**, *34*, 1982–1989. [CrossRef]
58. Cherp, A.; Jewell, J. The three perspectives on energy security: Intellectual history, disciplinary roots and the potential for integration. *Curr. Opin. Environ. Sustain.* **2011**, *3*, 202–212. [CrossRef]
59. Cherp, A.; Jewell, J. Energy security assessment framework and three case studies. In *International Handbook of Energy Security*; Dyer, H., Trombetta, M.J., Eds.; Edward Elgar: Cheltenham, UK; Northampton, MA, USA, 2013; pp. 146–173.
60. Jewell, J.; Cherp, A.; Riahi, K. Energy security under de-carbonization scenarios: An assessment framework and evaluation under different technology and policy choices. *Energy Policy* **2014**, *65*, 743–760. [CrossRef]
61. Xiong, C.H.; Chen, S.; Gao, Q.; Xu, L.T. Analysis of the influencing factors of energy-related carbon emissions in Kazakhstan at different stages. *Environ. Sci. Pollut. Res.* **2020**, *27*, 36630–36638. [CrossRef]
62. Kerimray, A.; Suleimenov, B.; De Miglio, R.; Rojas-Solorzano, L.; Torkmahalleh, M.A.; Gallachoir, B.P.O. Investigating the energy transition to a coal free residential sector in Kazakhstan using a regionally disaggregated energy systems model. *J. Clean. Prod.* **2018**, *196*, 1532–1548. [CrossRef]
63. Rivotti, P.; Karatayev, M.; Mourão, Z.S.; Shah, N.; Clarke, M.L.; Konadu, D.D. Impact of future energy policy on water resources in Kazakhstan. *Energy Strategy Rev.* **2019**, *24*, 261–267. [CrossRef]
64. Kalyuzhnova, Y.; Pomfret, R. (Eds.) *Sustainable Energy in Kazakhstan: Moving to Cleaner Energy in A Resource-Rich Country*; Taylor & Francis: Routledge, London, 2019; p. 304, ISBN 9780367885106. [CrossRef]
65. Zhunussova, G.Z.; Ratner, S.V.; Nurmukhanova, G.Z.; Shaihutdinova, A.K. Renewable Energy in Kazakhstan: Challenges and Prospects. *Int. Energy J.* **2020**, *20*, 311–324.
66. Mouraviev, N. Renewable energy in Kazakhstan: Challenges to policy and governance. *Energy Policy* **2021**, *149*, 112051. [CrossRef]
67. Tyo, A.; Jazykbayeva, B.; Ten, T.; Kogay, G.; Spanova, B. Development tendencies of heat and energy resources: Evidence of Kazakhstan. *Entrep. Sustain. Issues* **2019**, *7*, 1514–1524. [CrossRef]
68. Atlas of New Professions and Competencies in the Republic of Kazakhstan (2020). Available online: https://www.enbek.kz/atlas/en/industry/2 (accessed on 6 April 2021). (In Russian).
69. Kazakhstan Energy Week/ XII Eurasian Forum Kazenergy "The Future of Energy Sources: Innovative Growth" Final Report. Available online: http://kazenergyforum.com/wp-content/uploads/files/KEW-2019-summary-report-ru.pdf (accessed on 6 April 2021).
70. Bureau of National Statistics of the Agency for Strategic Planning and Reforms of the Republic of Kazakhstan. Energy Statistics. Available online: https://stat.gov.kz/official/industry/30/statistic/6 (accessed on 6 April 2021).
71. Informburo.kz, Kazistayev Ye. Kak Stimulirovat' Razvitiye Energoeffektivnosti v Kazakhstane. [Kazistaev E. How to Stimulate the Development of Energy Efficiency in Kazakhstan]. Available online: https://informburo.kz/stati/chtoby-vnedryat-energosberezhenie-nuzhno-podnimat-tarify-kak-stimulirovat-razvitie-energoeffektivnosti-v-kazahstane.html (accessed on 6 April 2021). (In Russian).
72. Annual Effect from the Implementation of Energy Saving Measures. Website of the Ministry of Industry and Infrastructure Development of the Republic of Kazakhstan. Available online: https://www.gov.kz/memleket/entities/miid/press/news/details/116442?lang=ru (accessed on 10 April 2021). (In Russian)
73. *Evolution of World Energy Markets and Its Consequences for Russia*; Makarova, A.A.; Grigorieva, L.M.; Mitrova, T.A.M. (Eds.) ERI RAS-AC under the Government of the Russian Federation: Moscow, Russia, 2015; p. 400, ISBN 978-5-91438-019-6. Available online: https://www.eriras.ru/files/evolyutsiya-mirovyh-energeticheskih-rynkov-i-ee-posledstviya-dlya-rossii.pdf (accessed on 10 April 2021). (In Russian)
74. Kazakhstan Electricity and Coal Market Analysis Report January-December 2020. p. 9. Available online: https://www.samruk-energy.kz/ru/press-center/analytical-review#2020 (accessed on 10 April 2021). (In Russian).
75. Shayakhmetova, D. Monitoring the 2030 Sustainable Development Goals. Webinar-Waste Management Indicators and Policies, UNDA Project "Improved Environmental Monitoring and Assessment in Support of the 2030 Agenda in South-Eastern Europe, Central Asia and the Caucasus", 10 March 2021 9:15–15:30. Available online: https://unece.org/sites/default/files/2021-02/10_Shayakhmetova_Kazakhstan_%D1%8D%D0%BD%D0%B5%D1%80%D0%B3%D0%BE%D0%B5%D0%BC%D0%BA%D0%BE%D1%81%D1%82%D1%8C.pdf (accessed on 10 April 2021).
76. Renewable Energy Data, Bureau of National Statistics. Available online: https://stat.gov.kz/official/industry/30/statistic/8 (accessed on 10 April 2021).
77. Electricity prices. Internet portal GlobalPetrolPrices. Available online: https://www.globalpetrolprices.com/electricity_prices/ (accessed on 21 April 2021).

energies

MDPI

Article

Composed Index for the Evaluation of Energy Security in Power Systems within the Frame of Energy Transitions—The Case of Latin America and the Caribbean

Sergio Fuentes, Roberto Villafafila-Robles *, Joan Rull-Duran and Samuel Galceran-Arellano

Centre d'Innovació Tecnològica en Convertidors Estàtics i Accionaments (CITCEA-UPC), Departament d'Enginyeria Elèctrica, Universitat Politècnica de Catalunya, ETS d'Enginyeria Industrial de Barcelona, Avinguda Diagonal, 647, 2nd Floor, 08028 Barcelona, Spain; sergio.fuentes@upc.edu (S.F.); joan.rull@upc.edu (J.R.-D.); samuel.galceran@upc.edu (S.G.-A.)

* Correspondence: roberto.villafafila@upc.edu; Tel.: +34-934-137-308

Abstract: Energy transitions are transforming energy systems around the globe. Such a shift has caused the power system to become a critical piece of infrastructure for the economic development of every nation on the planet. Therefore, guaranteeing its security is crucial, not only for energy purposes but also as a part of a national security strategy. This paper presents a multidimensional index developed to assess energy security of electrical systems in the long term. This tool, named the Power System Security Index (PSIx), which has been previously used for the evaluation of a country in two different time frames, is applied to evaluate the member countries of the Latin American Energy Organization, located within the Latin America and the Caribbean region, to measure its performance on energy security. Mixed results were obtained from the analysis, with clear top performers in the region such as Argentina, while there are others with broad areas of opportunity, as is the case of Haiti.

Keywords: energy security; energy transitions; Latin America; power system; sustainability

Citation: Fuentes, S.; Villafafila-Robles, R.; Rull-Duran, J.; Galceran-Arellano, S. Composed Index for the Evaluation of Energy Security in Power Systems within the Frame of Energy Transitions—The Case of Latin America and the Caribbean. *Energies* **2021**, *14*, 2467. https://doi.org/10.3390/en14092467

Academic Editors: Tomonobu Senjyu and Gianfranco Chicco

Received: 4 March 2021
Accepted: 22 April 2021
Published: 26 April 2021

1. Introduction

Energy transitions are motivated, as stated by [1], by global efforts to efficiently increase competitiveness while respecting the environment and guaranteeing energy supply. The transitions are changing the way energy is produced, consumed, stored, and transmitted, not only by boosting the presence of renewable energy technologies but also by improving the system's flexibility through innovative infrastructure solutions, at the same time that enhancing energy productivity has become a state priority worldwide [2,3]. This new paradigm also presents new challenges, the security of energy supply being an utmost important matter for the efficient functioning of modern economies [4].

Despite its importance and its wide presence in different national energy strategies, the concept of energy security is highly context-dependent and, consequently, differs significantly from one policymaker to another [5], requiring different authorities to determine their own approach to the concept for creating solutions for the procurement of energy supply in their respective populations. With the objective of maintaining uniformity of the concept and for the development of the present paper, the definition proposed by [6] has been taken on board, i.e., energy security is understood as the sustainable supply of energy.

Latin America is an energy-rich region, not only in fossil-fuel reservoirs, but also in renewable energy potential. At the same time, some nations in the region do not possess strong economies, requiring them to undermine their electricity systems, independently of their possession—or lack—of fuels basins. Due to this diversity of circumstance, it is pertinent to evaluate how policies of different countries in the region are translated into improvements to energy security in their respective power systems.

As stated by [7], to be analytically helpful, a measure for evaluating energy security must be quantifiable. One instrument to do this is composed indexes, which are useful to identify benchmark performances and trends, focusing on particular issues and, by those means, setting policy priorities [8]. In the present document, the composed index developed by [9], named Power System Security Index (PSIx), is applied for the evaluation of different countries in the Latin America and Caribbean region. The developed tool consists of a multidimensional index aimed at evaluating policies regarding energy security in the power sector. Nevertheless, the instrument has been applied to one single country in different time frameworks. The path that every country in the region has determined for transitioning to a sustainable energy system is different, making it pertinent to analyze how each country is positioned in its own transition. The PSIx, which, as shown in Figure 1, has been constructed based on the framework of energy transitions, offers a common framework to evaluate development.

Figure 1. PSIx Approach [9].

The novelty of this paper lies in the application of the developed methodology, based on the PSIx, consisting of the evaluation of energy security policies of multifold economies in a given time frame, as Figure 1 shows, and its further use for the assessment of the case of Latin America. The suitability of the multidimensional index will be evaluated by executing a statistical analysis of different countries' development, as well as their comparison and ranking, according to their performance on achieving energy security in their respective power systems.

To achieve the abovementioned objectives, the present paper is organized into five sections. Section 2 describes the composed index. The third one describes mathematical model. In Section 4, the outcomes of the composed index application are presented and discussed. Finally, the respective conclusions are presented in Section 5.

2. PSIx

The dimensions included in the PSIx for the characterization of energy security are availability, infrastructure, economy, environment, governance, and research, development, and innovation (R + D + i). Each one of these six dimensions possesses multifold indicators grouped, in turn, into different categories. How the index is structured is presented in Figure 2, in which dimensions, categories, and indicators are shown, each of them possessing an alphanumeric code for easing its identification throughout the document.

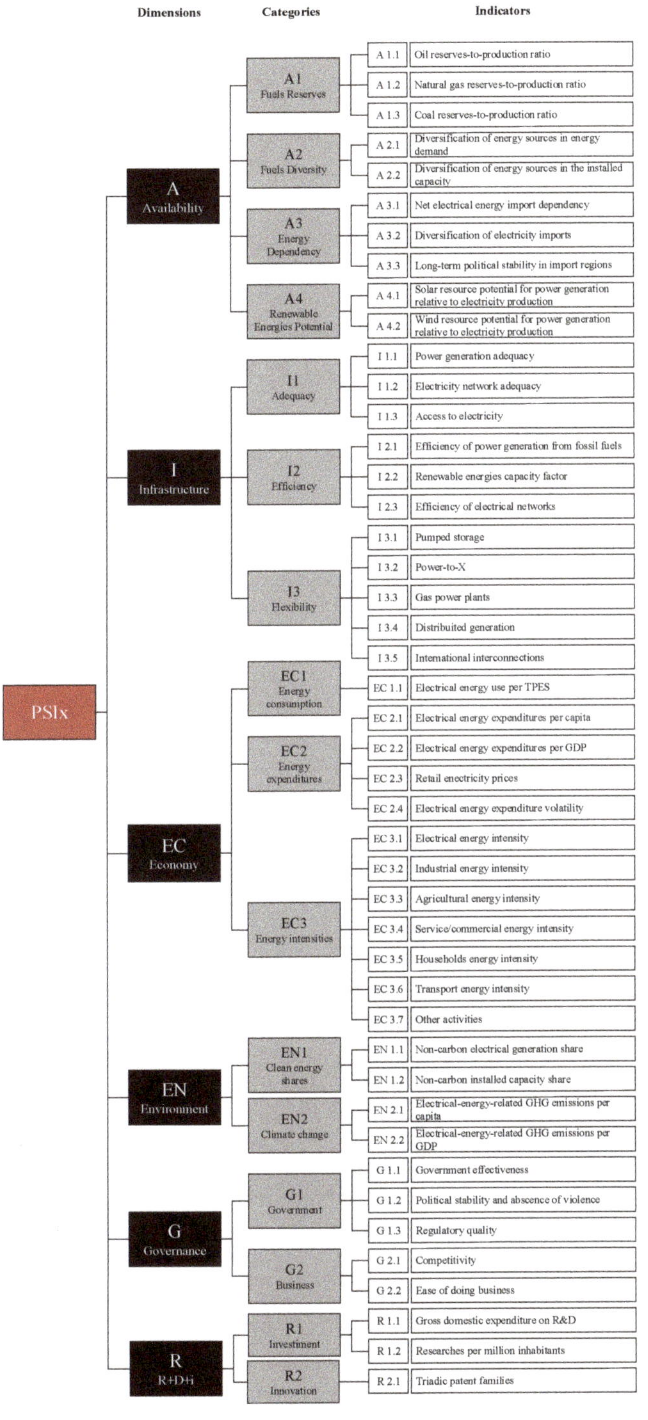

Figure 2. PSIx structure. Reprinted from ref. [9].

The dimension of availability (A) is directly related to energy independence [10]. It evaluates the geological presence of energy resources within a determined area, as well as the degree of their replacement by alternative energy resources [11,12]. The dimension also evaluates the diversification of energy technologies and sources for fulfilling the energy needs of a specific region.

The infrastructure dimension (I), also known as accessibility [4,13], evaluates the ability to access energy resources to provide a stable and uninterrupted supply of electrical energy, i.e., the reliability of the power system.

The economy dimension (EC), also called affordability, measures energy prices as well as their volatility, since, as stated by [14], these two factors have a great influence on the overall economy, as well as on industrial competitivity and trade balance.

The indicators of the environmental dimension (EN) aim to measure the repercussions of energy-generation technologies on the environment, so that they do not represent a menace for sustainable development. Climate change has acquired a very high importance for energy policymakers during the current century, particularly greenhouse gas (GHG) emissions emitted into the atmosphere, as proxy measures of the pollution of human activities. This tendency has been translated into strict restrictions to conventional energy technologies, which has spurred several countries to transform their power systems towards more sustainable models.

Governments are responsible entities for effectively planning infrastructure development to ensure long-term energy security [15]. They also pledge to establish lasting relationships with other countries, so it is possible to assure energy supplies in a politically stable scenario. On their part, there are also the competent bodies for creating an attractive environment for attracting investments, which are lifeblood of the energy system [16].

Finally, research, development, and innovation (R + D + i) play a central role for the enhancement of energy security, since they improve the capacity to adapt and respond to disruption challenges through innovation [10]. The R + D + i dimension (R) has the aim of, as proposed by [17], assessing new technologies in the energy field, as well as the development of intellectual capital as a factor to assess risks on energy security.

The respective formulas and objectives of each indicator are summarized in Table A1 of Appendix A, while the corresponding description and units of formulas are presented in Table A2.

The source from which data was obtained for its analysis is the Latin America and the Caribbean Energy Information System, developed by the Latin America and the Caribbean Energy Organization [18], and the chosen year is the last one with a complete dataset, namely 2018.

3. Mathematical Model

The mathematical model consists of the imputation of missing data, normalization, a multivariate analysis, weighting, and aggregation. Although most of the model has been originally developed for the study of different economies and presented in this paper, the normalization process used here is the same as the one proposed by [9].

3.1. Imputation of Missing Data

Some economies, particularly the smallest ones, have not provided complete datasets on energy information, either to international entities or through their own responsible authorities, which is translated into missing values for the indicators within the composed index. Therefore, it is necessary to complete these values by means of a suitable analytical method.

As defined by [19], missing data are unobservable values, which, if observed, would have a meaningful implication in the analysis. According to [20], there are three types of missing values depending on their predictability of non-appearance in the studied dataset, i.e., missing completely at random (MCAR), missing at random (MAR), and not missing at random (NMAR). MCAR values are independent of the variable of interest or any other observed variable; MAR values are independent of the variable of interest, but other variables in the dataset condition their missingness; and NMAR values are dependent on the missing data.

The indicators containing missing values are I1.2, I3.1, I3.3, EC2.4, EC3.3, EC3.6, EC3.7, G2.1, R1.1, R1.2, and R2.1. Two indicators possess NMAR values, since the availability of data is scarce for every country, not only those gathered by international institutions, but also those collected by each responsible national entity. These indicators are I3.1 and I3.3. It is inferred that these values are unavailable in most cases due to, precisely, the scarcity of data. Moreover, these indicators are relatively new compared to the rest of them, and policies of the covered countries do not consider them as priorities yet. Therefore, their measurement at national level is, in most cases, rather low or nonexistent.

For the NMAR values present in the index, an implicit modeling method has been selected for completing the corresponding datasets, i.e., hot deck imputation. This method is used to impute missing values within a data matrix by using available values from the same matrix with similar figures [21]. The countries are considered to have a similar behavior in the deployment of power-to-x and distributed generation installations. For these two specific indicators, in the case of missing values, they are set to zero, considering, therefore, that the measured value is negligible for its study.

The rest of the indicators with missing values correspond, in general, to small economies, particularly to those in the Caribbean. To achieve a more reliable imputation, the countries of the index have been divided into four categories, depending on the size of their economies and their geographical locations, with the purpose of considering them more equal in energy terms. These categories are:

A. Big continental economies: Argentina, Brazil, Chile, Colombia, Mexico, and Peru
B. Small continental economies: Bolivia, Ecuador, Guatemala, Paraguay, Uruguay, and Venezuela
C. Caribbean and the Guianas: Barbados, Cuba, Grenada, Guyana, Haiti, Jamaica, Dominican Republic, Suriname, and Trinidad and Tobago
D. Central America: Belize, Costa Rica, El Salvador, Honduras, Nicaragua, and Panama

To impute the missing values of indicators, an explicit method, based on a formal statistical model, was selected, specifically the unconditional mean imputation method. This approach consists of the substitution of missing values by means of the sample series. Consequently, such a procedure leads to estimates similar to those found by weighting, provided the sampling weights are constant within weighting classes [19].

3.2. Multivariate Analysis

With the objective of assessing the underlying structure of the gathered data, a multivariate analysis was conducted. This approach is also helpful for assigning weights to the indicators, a crucial step for, according to [22], determining their influence within the index, as well as their trade-off values.

Among the different methodological techniques present in the literature, a data-driven approach was selected, since it depends entirely on the data themselves. A factorial analysis approach, specifically the principal component analysis, was chosen, since this statistical approximation allows the determination of interrelations among a great number of variables, at the same time as also allowing an explanation of their behavior in terms of their subjacent common dimensions [23]. For conducting the statistical analysis, Minitab® software was used (Minitab® and all other trademarks and logos for the Company's products and services are the exclusive property of Minitab, LLC. All other marks referenced remain the property of their respective owners. See minitab.com for more information).

The treated variables have been considered initially neither dependent nor independent from each other. Therefore and, according to [24], an interdependency study can be executed. As the methodology dictates, the statistical study must cover all the variables simultaneously, so an underlying structure can be identified for the whole set of indicators. For performing the principal components analysis, a covariance matrix of the data was employed, containing 44 indicators for the 27 analyzed countries within the composed index.

For analyzing the correlations of the indicators, an item analysis was performed, and the most significant values of the resulting correlation matrix is shown in Table 1. The matrix confirms the existence of a subjacent structure among the gathered data. In the table the most significant correlations among the variables, those equal to or above 0.70, are highlighted.

Table 1. Correlation matrix showing the most significant correlations among variables.

	A1.3	A2.1	A3.2	I1.1	I1.3	I2.2	I3.1	I3.3	I3.4	EN1.1	EN2.1	G1.1	G1.3	G2.1	R1.1
A2.2	0.16	**0.97**													
A3.3	0.00	0.11	**1.00**												
I1.3	0.24	0.03	0.20	**0.83**											
I2.3	0.03	0.13	0.00	0.69	**0.73**	0.10									
I3.3	0.53	0.07	0.28	−0.04	0.12	−0.01	**0.82**								
EC1.1	−0.21	−0.22	−0.18	**−0.76**	−0.63	−0.12	−0.19	−0.15	−0.24						
EN1.1	0.06	0.51	0.45	0.11	0.20	**0.76**	−0.15	−0.12	**0.75**						
EN1.2	0.20	0.33	0.45	0.11	0.12	**0.70**	−0.07	−0.01	0.64	**0.90**					
EN2.2	−0.04	0.33	0.44	−0.16	0.02	0.45	−0.05	0.02	0.64	**0.75**	**0.71**				
G1.3	0.14	0.07	−0.09	0.36	0.41	0.03	0.04	0.01	0.05	0.15	−0.20	**0.78**			
G2.1	0.30	0.23	0.02	0.51	0.63	0.19	0.16	0.14	0.25	0.29	−0.13	**0.76**	**0.80**		
G2.2	0.02	0.21	0.16	0.41	0.45	0.13	0.02	0.06	0.30	0.24	0.04	0.69	**0.79**	**0.87**	
R1.1	**0.78**	0.17	0.00	0.09	0.21	0.15	0.47	0.56	0.05	0.09	−0.04	0.30	0.38	0.53	
R1.2	**0.74**	0.14	0.08	0.06	0.19	0.16	0.64	**0.72**	0.10	0.09	−0.01	0.28	0.30	0.43	**0.95**

With the purpose of evaluating the internal consistency of the analyzed data, the Cronbach's alpha parameter was employed and, since its value overpasses the benchmark of 0.7, specifically 0.7347, it is considered that the analyzed data measures the same characteristic, namely energy security in the power system, in the case of the present study.

To determine the principal components of the data, the methodology proposed by [25] has been followed, in which the correlation matrix serves as the basis for such purpose. A principal component is defined as

$$z = A'x^* \tag{1}$$

where A consists of columns formed by the eigenvectors of the correlation matrix, while x^* is composed by the arrangement of standardized variables. The objective of this approach is to identify the principal components of the standardize version of x^* with regard to x, where x^* possesses the jth element $x_j/\sigma_{jj}^{1/2}$, $j = 1, 2, \ldots, p$, x_j is the jth element of x, and σ_{jj} is the variance of x_j. Therefore, the covariance matrix for x^* is the correlation matrix for x, and the principal components of x^* are determined by Equation (1).

For the selection of the factors to be considered to be relevant for a further analysis, and which give rise to the determination of principal components, an a priori criterion has been chosen, i.e., it will be considered that those factors that contribute for explaining 90% of the variance of the data are those that will be kept.

After the execution of the principal component analysis, the variance values of the principal components with a considerable influence were obtained and they are shown in Table 2. They are 10 of the total sample of 44 values, which explain 91.2% of the variance of the dataset:

Table 2. Values of the factors of the covariance matrix.

Factor	Eigenvalue	Proportion	Accumulated
1	0.54515	0.277	0.277
2	0.37498	0.191	0.468
3	0.18972	0.097	0.565
4	0.17292	0.088	0.653
5	0.14021	0.071	0.724
6	0.11698	0.06	0.783
7	0.08296	0.042	0.826
8	0.07199	0.037	0.862
9	0.0537	0.027	0.889
10	0.04414	0.022	0.912

The scree plot of the total number of factors vs. their corresponding eigenvalues in a descending order, as shown in Figure 3. The considerable high value of the first two components can be observed, while from the 15th value the curve presents practically a flat behavior.

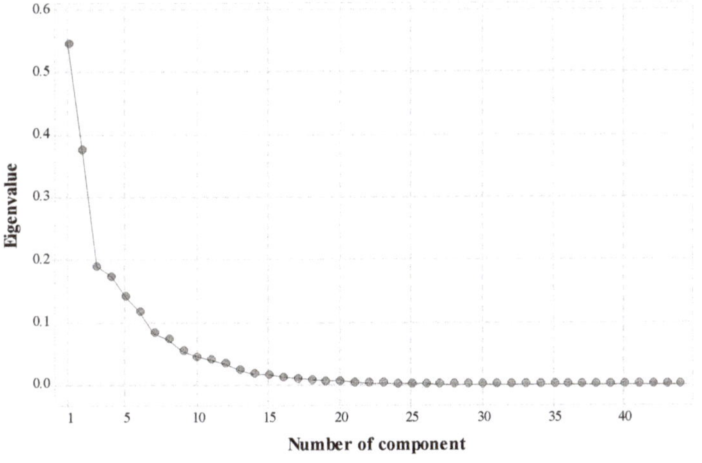

Figure 3. Data scree plot.

The first two principal components, named PC1 and PC2 and which account for 46.8% of the total variance in the data, are presented in Table 3, jointly with the PSIx variables and the corresponding factorial loads or eigenvectors. The load values higher than 0.25 are highlighted, as they are considered significant for each component.

Table 3. Eigenvectors for the first two principal components.

Variable	PC1	PC2	Variable	PC1	PC2	Variable	PC1	PC2	Variable	PC1	PC2
A1.1	**0.27**	**0.44**	I1.2	−0.13	0.16	EC2.2	0.04	−0.03	EN1.2	**−0.28**	0.16
A1.2	**0.29**	**0.36**	I1.3	−0.01	0.08	EC2.3	0.01	−0.01	EN2.1	0.00	0.00
A1.3	0.08	**0.46**	I2.1	−0.05	−0.04	EC2.4	−0.06	0.11	EN2.2	−0.07	0.03
A2.1	−0.15	0.12	I2.2	−0.12	0.07	EC3.1	0.00	0.00	G1.1	0.04	0.05
A2.2	−0.15	0.11	I2.3	0.00	0.05	EC3.2	0.03	0.04	G1.2	0.02	−0.12
A3.1	0.04	−0.01	I3.1	0.00	0.00	EC3.3	0.01	0.06	G1.3	−0.03	0.03
A3.2	**−0.32**	0.09	I3.2	**0.29**	**0.43**	EC3.4	0.02	−0.11	G2.1	−0.01	0.04
A3.3	**−0.24**	0.07	I3.3	0.00	0.00	EC3.5	0.07	0.02	G2.2	−0.04	0.02
A4.1	0.03	−0.01	I3.4	**−0.52**	**0.27**	EC3.6	0.00	0.00	R1.1	0.01	0.05
A4.2	−0.09	0.04	EC1.1	0.05	−0.07	EC3.7	0.00	0.00	R1.2	0.00	0.05
I1.1	−0.01	0.04	EC2.1	0.02	−0.07	EN1.1	**−0.38**	0.18	R2.1	0.00	0.00

To picture these results graphically, Figure 4 shows the loading plot of the data:

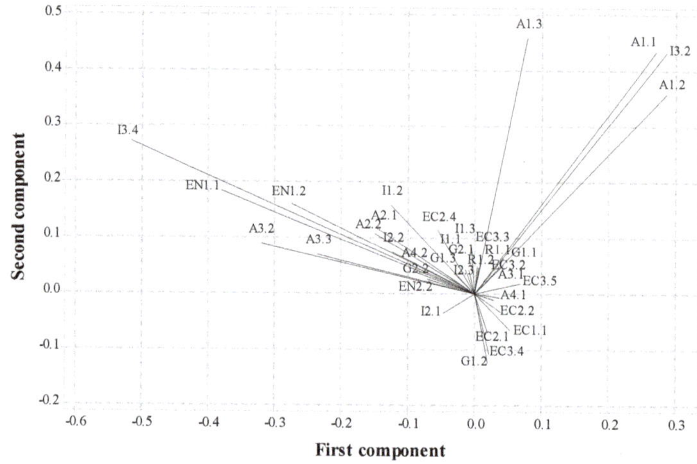

Figure 4. Loading plot for the first two principal components.

PC1 has a large positive influence of loads coming from variables belonging to the availability dimension, particularly A1.1 and A1.2, which measure reserves-to-production ratios of oil and gas fuels, respectively. Therefore, it can be inferred that this component is an indicator related the availability of energy sources. By contrast, indicators I3.4 of international electrical interconnections and availability indicators related to the diversification of sources have a strong negative load in the component. It can be deduced that the larger the ratio of production of fossil fuels compared to the reserves, the lower the diversification of other sources of energy. PC2 has a considerable load of values corresponding to the infrastructure dimension, jointly with other availability indicators.

3.3. Weighting and Aggregation

3.3.1. Weighting

Despite the fact that the relative importance of different indicators for sustainable energy development vary from country to country, depending on country-specific conditions, national energy priorities, sustainability development criteria and their inherent objectives [26], it is necessary to establish a groundwork that assigns weights as importance coefficients to the indicators of the PSIx, so that the analyzed countries can be evaluated, compared, and ranked within a common framework.

A data-driven approach has been determined for assigning weights to the PSIx indicators. For that aim, the outcomes obtained through the principal component analysis in Section 3.2 are highly advantageous, since they offer a statistical approach for comparing the variables of the index and, since a large amount of data is being analyzed, the risk of double-weighting the indicators of the index is avoided [27].

From the correlation matrix, also presented in Section 3.2, new intermediate composites have been obtained by selecting the indicator with the highest correlation to each significant factor, whose value is expressed by:

$$\widetilde{w}_j = arg\max_i\left(\frac{a_{ij}^2}{\sum_{k=1}^{m} a_{ik}^2}\right) \tag{2}$$

In which:

$j = 1, \ldots, m$: index indicators

i: analyzed country

a_{ij}: factor load for country i of j indicator

Therefore, the weight of each jth variable is obtained as follows:

$$w_j = \frac{\widetilde{w}_j \left(\frac{\sum_{k=1}^{m} a_{ik}^2}{\sum_{j=1}^{m-q} \sum_{k=1}^{m} a_{ik}^2} \right)}{\sum_{j=1}^{m} \widetilde{w}_j \left(\frac{\sum_{k=1}^{m} a_{ik}^2}{\sum_{j=1}^{m-q} \sum_{k=1}^{m} a_{ik}^2} \right)} \quad (3)$$

In which q is the last significant factor to be considered for the analysis according to the scope described in Section 3. Table 4 shows the weights assigned to each indicator of the index according to the described methodology. As a result of such procedure, several indicators lack a significant value, with only 18 variables being considered to be significant. Furthermore, from the original six dimensions of the index, only three are of statistical interest, which are availability, infrastructure, and economy, summarizing a weight of 0.24, 0.44, and 0.32, respectively.

Table 4. Weights assigned to each indicator.

Dimension	Variable	Domain Weight	Weight of the Respective Factor	Weight Score (ωi)	Resulting Weight (Σωi = 1)	Dimension Weight (Σωi = 1)
Availability	A1.1	0.1247	0.0040	0.0005	0.0024	0.24
	A1.2	0.3407	0.0370	0.0126	0.0604	
	A1.3	0.2108	0.0910	0.0192	0.0918	
	A3.1	0.1339	0.0020	0.0003	0.0013	
	A3.2	0.2381	0.0710	0.0169	0.0809	
Infrastructure	I2.1	0.1902	0.0010	0.0002	0.0009	0.44
	I2.3	0.2634	0.0010	0.0003	0.0013	
	I3.2	0.1964	0.0220	0.0043	0.0207	
	I3.4	0.2664 0.2277	0.2770 0.0600	0.0874	0.4187	
Economy	EC1.1	0.1822	0.0170	0.0031	0.0148	0.32
	EC2.1	0.2477	0.0710	0.0176	0.0842	
	EC2.2	0.3906	0.0420	0.0164	0.0786	
	EC2.3	0.2854	0.0080	0.0023	0.0109	
	EC2.4	0.2246	0.0270	0.0061	0.0290	
	EC3.2	0.1747 0.0180	0.0050 0.0030	0.0009	0.0044	
	EC3.3	0.5825	0.0170	0.0099	0.0474	
	EC3.4	0.3893	0.0200	0.0078	0.0373	
	EC3.5	0.5186	0.0060	0.0031	0.0149	

3.3.2. Aggregation

Although it is true that according to the impossibility theorem of [28], there does not exist a perfect aggregation method, it is necessary design a frame that fits the needs of the desired scope for the PSIx application. In this process, the use of rules implying additive or multiplicative principles, i.e., linear or geometric aggregation methods, could be possible. Even though the use of any of these techniques implies that weights become able to be substituted by themselves, meaning that a poor development on one variable might be compensated by an over-standing development in another one. The compensability property leads linear and geometric aggregation methods to minimize the importance of the associated indicators. Therefore, the use of a method is necessary which does not allow or restrain compensability according to the scope of the built index.

As stated by [24,29], for weights to be construed as importance coefficients, a non-compensatory framework must be adopted in the aggregation process. The non-compensatory multi-criteria approach (MCA) is the selected method, since it restrains compensability by setting arrangements between two or more legitimate goals.

The elasticity of substitution between indicators j and j', understood, according to [22], as how much one variable has to give up of one achievement to get an extra unit of a second indicator while keeping the level of energy security, is expressed by:

$$\delta_{jj'} = \frac{1}{(1-\beta)} \tag{4}$$

From this expression, it is noticeable that the smaller the value of β, the smaller the allowed substitutability between indicators. Depending on if the values correspond to the same dimension or not, the value of β is considered distinctly in the aggregation process. For intra-dimensional indicators, the value assigned to β is set to 1, therefore $\delta \to \infty$, meaning that all the indicators of one particular dimension are completely substitutable with each other. On the other hand, it is desired that the possibility of substitutions among indicators of different dimensions is zero, so β is set to $-\infty$ and the elasticity of substitution δ is null.

With the purpose of assigning scores to each dimension, the following one-digit classification has been stablished:

The score on each dimension is determined by evaluating the development of each individual country. Since there is no interdimensional substitutability, there will be a grade for each relevant dimension within the index.

4. Results and Discussion

The results obtained from the analysis are, as could be expected from a region with such diversity of countries in energy terms, quite divergent.

From Section 3, and with most of the variance in the data gathered, the score plot, shown in Figure 5, allows the clustering of the analyzed countries depending on their results. It can be observed that all the big economies in the continent, Group A according to the classification presented in Section 3.1, are located in the upper part of the graph, deducing, therefore, according to the principal components defined in Section 3.3, that their infrastructure is more developed than other countries, compared, for instance, with the case of the Caribbean countries and the Guianas. Central American countries can be easily grouped due to their close location in the plot; therefore, their energy security can be considered to be very alike. The plot shows that the geographical location of the covered countries does have a strong influence on the development of their power systems, as well as the size of the respective economies. The first of the characteristics, geographical location, plays a central role in determining the energy security of countries with a similar level of economic development, which is consistent with the geological basins of fuels and similar renewable energy potential.

The score of each country is determined by the multiplication of the performance on each dimension multiplied by its respective weight, as indicated in Table 5. The resulting outcomes from the evaluation of countries in Latin America and the Caribbean region are summarized in Table 6, where the results of each dimension are shown, as well as the overall score of the index.

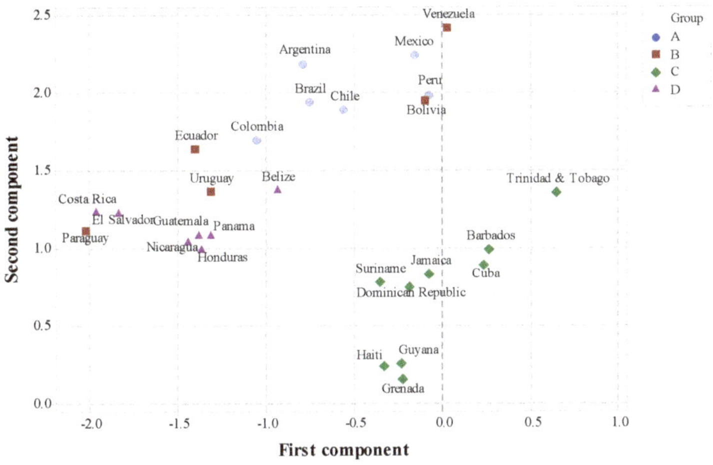

Figure 5. Score plot of the first two components.

Table 5. Dimensions grading system.

Performance	Grade
$X > 90$	1
$80 \leq X < 90$	2
$70 \leq X < 80$	3
$60 \leq X < 70$	4
$50 \leq X < 60$	5
$40 \leq X < 50$	6
$30 \leq X < 40$	7
$20 \leq X < 30$	8
$10 \leq X < 20$	9
$X < 10$	0

Table 6. Resulting scores of the composed index for countries of Latin American and the Caribbean.

		A	I	EC	Score			A	I	EC	Score
1st	Argentina	3	1	0	0.67	15th	Mexico	5	5	0	0.40
2nd	Ecuador	7	1	0	0.58	16th	Venezuela	4	7	0	0.33
3rd	Costa Rica	8	1	8	0.55	17th	Peru	8	8	8	0.24
4th	El Salvador	7	1	0	0.54	18th	Brazil	5	0	0	0.24
5th	Paraguay	7	1	0	0.54	19th	Trinidad and Tobago	8	0	8	0.15
6th	Colombia	7	1	0	0.54	20th	Cuba	8	0	8	0.15
7th	Panama	0	1	8	0.50	21st	Barbados	0	0	8	0.13
8th	Nicaragua	0	1	0	0.49	22nd	Grenada	0	0	8	0.07
9th	Bolivia	8	2	0	0.49	23rd	Guyana	0	0	8	0.07
10th	Uruguay	0	1	0	0.48	24th	Suriname	0	0	0	0.06
11th	Honduras	0	1	0	0.47	25th	Dominican Republic	0	0	0	0.04
12th	Guatemala	0	1	0	0.47	26th	Jamaica	0	0	0	0.04
13th	Chile	6	4	8	0.45	27th	Haiti	0	0	0	0.03
14th	Belize	0	1	0	0.44						

It can be observed that the countries within the region have mixed values in their energy security performance. The country with the highest overall score is Argentina, mainly due to its performance on infrastructure and availability dimensions, even though it does not have an outstanding development in the economic dimension. Indeed, the country has very important reserves of fossil fuels, it has a noticeable energy self-sufficiency, and its electrical interconnections provide an important flexibility capacity to the Argentinean power system. On the other hand, Haiti is the country with more areas of improvement, being weak in all the three evaluated dimensions; the Caribbean country has no fossil-fuel reserves in its territory, has a feeble energy infrastructure and possesses a fragile and inefficient economy.

By dimension, most of the studied countries have an improvable behavior in availability, with Venezuela, Argentina, Brazil, and Mexico being the countries best positioned, in this order. In infrastructure, the gap among countries with relatively good energy infrastructure and those lacking it is deep, with Argentina, Colombia, Ecuador, Paraguay, Uruguay, and the Central American nations as the best performers in this dimension. No country has shown outstanding performance in the economic dimension; most have mediocre behavior. Barbados, Chile, Costa Rica, Cuba, Grenada, Guyana, Panama, Peru, and Trinidad and Tobago are the countries that performed the best in this dimension.

The results show a picture of the current situation of the region. As the path towards an energy transition of every country is defined by each nation state, also each state is also in charge of determining the priority dimension or dimensions it wants to focus on. Notwithstanding, the most efficient way to do it, at least mathematically and as explained in Section 3.3, is to improve the areas in which the country scores the lowest. Nevertheless, no one single dimension nor indicator should be considered to be completely irrelevant, since on they all, as a whole, rely on the possibility of succeeding in achieving a secure and sustainable energy system.

5. Conclusions

Energy transitions are reshaping the global energy system, causing electricity to occupy a predominant role in modern infrastructure. This new paradigm also represents new challenges, and, among them, guaranteeing energy security of the power system has become a priority for policymakers. The path that each nation adopts in this line depends on its own needs, interests, and possibilities; therefore, a single approach on energy security does not exist, but instead a series of divergent strategies.

Latin America and the Caribbean is a very diverse region in energy terms, in which countries range from possessing the largest crude oil reserves in the world to extensive energy poverty; therefore, the analysis of its strategies on how efficient it is for procuring energy security is of outstanding usefulness for the enhancement of power systems on the continent.

The PSIx was conceived as a tool for policymakers for issuing strategies focused on reaching sustainable development through energy security enhancement. The tool offers the possibility to assess energy security in the power system using a multidimensional approach, covering availability, infrastructure, economy, environment, government, and R + D + i spheres. Through the analysis of elements, the internal uniformity of the index has been verified, asserting that the tool measures the same characteristics, i.e., the energy security performance of a nation. The composed index constitutes, therefore, a comprehensive frame in which strategies addressed to enhance energy security in the power system can be evaluated, according to their effectiveness for achieving that purpose. Through the study, it is confirmed that the PSIx is useful not only for tracking the development of a single country regarding its energy security, but it is also suitable for the study of multifold economies.

Three of the six dimensions are of statistical relevance, i.e., availability, infrastructure, and economy. It is pertinent to notice that this does not mean that the rest of the dimensions are not important for energy security, but that variance of data among countries is explained mostly by those dimensions considered statistically significant.

The evaluated countries, as expected, perform very distinctly in the relevant dimensions of the index. Countries that possess considerable fuel reservoirs have higher evaluation results in the energy availability dimension. There exists a wide division between countries with an adequate electrical infrastructure and those that lack one, mainly due to the existence of international interconnections and the presence of gas-fueled power plants, which, additionally, are measures that greatly enhance the flexibility of the electrical network. No country presents distinguished results on the economic dimension. On the contrary, they all have rather lackluster performance. The country with the highest overall score is Argentina, with 0.67 points, followed by Ecuador and Paraguay with 0.58 and 0.54 points, respectively. The first two countries, Argentina and Ecuador, have important fossil-fuel reservoirs, while Paraguay is a net electricity exporter thanks to its large hydropower plants. These three countries are very well interconnected with their neighbors, and Ecuador and Paraguay have experienced important improvements to their economies lately.

The developed multidimensional index constitutes a tool addressed to help policymakers to assess energy security strategies in the power system. Through its application in the case of Latin America and the Caribbean, and after the subsequent statistical analysis, it can be confirmed that this tool can, by means of the betterment of energy security, help national systems to reach sustainable development. Future work shall include the application of the index to other regions at a supranational level to assess the suitability of policies aimed to improve energy security, as well as the incorporation of more indicators aimed to achieve a sustainable energy system.

Author Contributions: Conceptualization, S.F. and R.V.-R.; methodology, S.F. and R.V.-R.; validation, J.R.-D. and S.G.-A.; formal analysis, S.F.; investigation, S.F.; resources, R.V.-R.; data curation, S.F. and R.V.-R.; writing—original draft preparation, S.F. and R.V.-R.; writing—review and editing, S.F. All authors have read and agreed to the published version of the manuscript.

Funding: This research was funded by CONSEJO NACIONAL DE CIENCIA Y TECNOLOGÍA of Mexico, grant number 314487.

Institutional Review Board Statement: Not applicable.

Informed Consent Statement: Not applicable.

Data Availability Statement: The gathered data was obtained from the Energy Information System of Latin America and the Caribbean, available publicly at http://sielac.olade.org/default.aspx (accessed on 1 March 2020).

Acknowledgments: Portions of information contained in this publication are printed with permission of Minitab, LLC. All such material remains the exclusive property and copyright of Minitab, LLC. All rights reserved.

Conflicts of Interest: The authors declare no conflict of interest.

Appendix A

Table A1. Formulas and objectives of indicators [9].

ID	Formula	Objective	ID	Formula	Objective
A1.1	$A1.1 = \frac{r_a}{s_a}$	Maximize	EC2.2	$EC2.2 = \frac{x_e}{GDP}$	Minimize
A1.2	$A1.2 = \frac{r_b}{s_b}$	Maximize	EC2.3	$EC2.3 = \frac{c_e}{e_u}$	Minimize
A1.3	$A1.3 = \frac{r_c}{s_c}$	Maximize	EC2.4	$EC2.4 = \frac{x_e - x_{e-1}}{GDP}$	Minimize
A2.1	$A2.1 = -\sum_i (p_i \ln p_i)$	Maximize	EC3.1	$EC3.1 = \frac{e_c}{GDP}$	Minimize
A2.2	$A2.2 = -\sum_i (q_i \ln q_i)$	Maximize	EC3.2	$EC3.2 = \frac{e_{c,1}}{GDP_1}$	Minimize
A3.1	$A3.1 = \frac{e_z}{e_y}$	Minimize	EC3.3	$EC3.3 = \frac{e_{c,2}}{GDP_2}$	Minimize
A3.2	$A3.2 = -\sum_k (r_k \ln r_k)$	Maximize	EC3.4	$EC3.4 = \frac{e_{c,3}}{GDP_3}$	Minimize
A3.3	$A3.3 = -\sum_i (c_{3,i} p_i \ln p_i)$	Maximize	EC3.5	$EC3.5 = \frac{e_{c,4}}{pl}$	Minimize
A4.1	$A4.1 = \frac{e_{gen,p,s}}{e_{gen}}$	Maximize	EC3.6	$EC3.6 = \frac{e_{c,5}}{vh}$	Minimize

Table A1. *Cont.*

ID	Formula	Objective	ID	Formula	Objective
A4.2	$A4.2 = \frac{e_{gen,p,w}}{e_{gen}}$	Maximize	EC3.7	$EC3.6 = \frac{e_{c,o}}{GDP_o}$	Minimize
I1.1	$I1.1 = \frac{P}{D_{peak}}$	Maximize	EN1.1	$EN1.1 = \frac{e_r}{e_p}$	Maximize
I1.2	$I1.2 = \frac{P_{trans}}{D_{peak}}$	Maximize	EN1.2	$EN1.2 = \frac{P_r}{P}$	Maximize
I1.3	$I1.3 = \frac{pl_e}{pl}$	Maximize	EN2.1	$EN2.1 = \frac{GHG}{pl}$	Minimize
I2.1	$I2.1 = \frac{e_{gen,f}}{e_{gen,f,max}}$	Maximize	EN2.2	$EN2.2 = \frac{GHG}{GDP}$	Minimize
I2.2	$I2.2 = \frac{e_{gen,r}}{e_{gen,r,max}}$	Maximize	G1.1	Direct value	Maximize
I2.3	$I2.3 = \frac{e_l}{e_c}$	Maximize	G1.2	Direct value	Maximize
I3.1	$I3.1 = \frac{S_{pump}}{P}$	Maximize	G1.3	Direct value	Maximize
I3.2	$I3.2 = \frac{PtX}{P}$	Maximize	G2.1	Direct value	Maximize
I3.3	$I3.3 = \frac{P_{gas}}{P}$	Maximize	G2.2	Direct value	Maximize
I3.4	$I3.4 = \frac{P_{dis}}{P}$	Maximize	R1.1	Direct value	Maximize
I3.5	$I3.5 = \frac{L_{int}}{P}$	Maximize	R1.2	Direct value	Maximize
EC1.1	$EC1.1 = \frac{e_c}{TPES}$	Maximize	R2.1	Direct value	Maximize
EC2.1	$EC2.1 = \frac{x_e}{pl}$	Minimize			

EC3.5 consists of a proxy measure; household energy intensity is considered to be domestic electrical consumption per capita.

Table A2. PSIx variables [9].

Variable	Description	Units	Variable	Description	Units
r_a	Crude oil reserves	b	e_l	Electricity supplied to the power lines	kWh
s_a	Crude oil production	b	e_c	Electricity consumption	kWh
r_b	Natural gas reserves	cu m	PtX	Power-to-X installed capacity	MW
s_b	Natural gas production	cu m	P_{gas}	Installed capacity of gas-fired power plants	MW
r_c	Coal reserves	ton	P_{dist}	Installed capacity of distributed generation facilities	MW
s_c	Coal production	ton	L_{int}	International interconnections	MW
p_i	Share of energy source i in the total electricity generation matrix	-	$TPES$	Total primary energy supply	MWh
q_i	Share of energy source i in the total installed capacity matrix	-	x_e	Electrical energy expenditures	USD
e_z	Net imported electricity	kWh	GDP	Gross domestic product	USD
e_y	Net consumed electricity	kWh	$e_{c,1}$	Electricity consumption by industrial activities	kWh
r_k	Share of electrical energy imported from k region	%	GDP_1	Gross domestic product of industrial activities	USD
c_3	Correction factor for pi, political stability	-	$e_{c,2}$	Electricity consumption by agricultural activities	kWh
e_{gen}	Total electricity generation	kWh	GDP_2	Gross domestic consumption of agricultural activities	USD
$e_{gen,p,s}$	Potential for power generation from solar sources	MW	$e_{c,3}$	Electricity consumption by service/commercial activities	kWh
$e_{gen,p,w}$	Potential for power generation from wind sources	MW	GDP_3	Gross domestic product of service/commercial activities	USD
P	Power generation capacity	MW	$e_{c,4}$	Household electricity consumption	kWh
D_{peak}	Peak demand	MW	$e_{c,5}$	Electricity consumption by transport	kWh
pl	Total population	people	vh	Number of vehicles	-
pl_e	Population with access to electricity	people	$e_{c,o}$	Electricity consumption by other activities	kWh

Table A2. *Cont.*

Variable	Description	Units	Variable	Description	Units
$e_{gen,f}$	Produced electricity from fossil-fuel-based installations	kWh	GDP_o	Gross domestic product of other activities	USD
$e_{gen,f,max}$	Maximum possible produced electricity from fossil-fuel-based installations	kWh	c_e	Cost of electricity	USD/kWh
$e_{gen,r}$	Produced electricity from renewable energy installations	kWh	e_u	Electrical energy unit	kWh
$e_{gen,r,max}$	Maximum possible produced electricity from renewable energy installations	kWh	e_r	Electricity produced by renewable sources	kWh
S_{pump}	Pumped-storage capacity	MW	e_p	Electricity production	kWh
$e_{gen,max}$	Maximum generation energy	kWh	P_r	Installed capacity of renewable energy facilities	MW
P_{trans}	Transformers power	MW	GHG	Greenhouse gases emissions	ton

References

1. World Energy Council. *World Energy Trilemma Index*; World Energy Council: London, UK, 2018.
2. IRENA. *Power System Flexibility for the Energy Transition, Part 1: Overview for Policy Makers*; International Renewable Energy Agency: Abu Dhabi, United Arab Emirates, 2018.
3. Hoggett, R.; Bolton, R.; Candelise, C.; Kern, F.; Mitchell, C.; Yan, J. Supply chains and energy security in a low carbon transition. *Appl. Energy* **2014**, *123*, 292–295. [CrossRef]
4. Kruyt, B.; Van Vuuren, D.; De Vries, H.; Groenenberg, H. Indicators for energy security. *Energy Policy* **2009**, *37*, 2166–2181. [CrossRef]
5. Cox, E. Opening the black box of energy security: A study of conceptions of electricity security in the United Kingdom. *Energy Res. Soc. Sci.* **2016**, *21*, 1–11. [CrossRef]
6. Fuentes, S.; Villafafila-Robles, R.; Olivella-Rosell, P.; Rull-Duran, J.; Galceran-Arellano, S. Transition to a greener Power Sector: Four different scopes on energy security. *Renew. Energy Focus* **2020**, *33*, 23–36. [CrossRef]
7. Chester, L. Conceptualising energy security and making explicit its polysemic nature. *Energy Policy* **2010**, *38*, 887–895. [CrossRef]
8. Cherchye, L.; Moesen, W.; Rogge, N.; Van Puyenbroeck, T. Constructing composite indicators with imprecise data: A proposal. *Expert Syst. Appl.* **2011**, *38*, 10940–10949. [CrossRef]
9. Fuentes, S.; Villafafila-Robles, R.; Lerner, E. Composed index for the evaluation of the energy security of power systems: Application to the case of argentina. *Energies* **2020**, *13*, 3998. [CrossRef]
10. Sovacool, B.K.; Mukherjee, I.; Drupady, I.M.; D'Agostino, A.L. Evaluating energy security performance from 1990 to 2010 for eighteen countries. *Energy* **2011**, *36*, 5846–5853. [CrossRef]
11. Chang, Y.; Yong, J. Differing perspectives of major oil firms on future energy developments: An illustrative framework. *Energy Policy* **2007**, *35*, 5466–5480. [CrossRef]
12. Tongsopit, S.; Kittner, N.; Chang, Y.; Aksornkij, A.; Wangjiraniran, W. Energy security in ASEAN: A quantitative approach for sustainable energy policy. *Energy Policy* **2016**, *90*, 60–72. [CrossRef]
13. Holley, C.; Lecavalier, E. Energy governance, energy security and environmental sustainability: A case study from Hong Kong. *Energy Policy* **2017**, *108*, 379–389. [CrossRef]
14. Global Energy Institute. *Index of U.S. Energy Security Risk*; Global Energy Institute: Washington, DC, USA, 2017.
15. Ang, B.; Choong, W.; Ng, T. Energy security: Definitions, dimensions and indexes. *Renew. Sustain. Energy Rev.* **2015**, *42*, 1077–1093. [CrossRef]
16. International Energy Agency; International Renewable Energy Agency. *Perspectives for the Energy Transition: Investment Needs for a Low Carbon Energy System*; International Energy Agency: Paris, France, 2017.
17. Institute for the 21st Century Energy. *International Index of Energy Security Risk*; Institute for the 21st Century Energy: Washington, DC, USA, 2016.
18. OLADE. Sistema de Información Energética de Latinoamérica y el Caribe. Available online: http://sielac.olade.org/ (accessed on 1 March 2020).
19. Little, R.J.A.; Rubin, D.B. *Statistical Analysis with Missing Data*; John Wiley & Sons, Inc.: Hoboken, NJ, USA, 2002; ISBN 9781119013563.
20. Rubin, D.B. Inference and missing data. *Biometrika* **1976**, *63*, 581–592. [CrossRef]
21. Joenssen, D.W.; Bankhofer, U. Hot Deck Methods for Imputing Missing Data: The Effects of Limiting Donor Usage. In *Proceedings of the Lecture Notes in Computer Science (Including Subseries Lecture Notes in Artificial Intelligence and Lecture Notes in Bioinformatics)*; LNAI: Berlin, Germany, 2012; Volume 7376, pp. 63–75.
22. Decancq, K.; Lugo, M.A. Weights in multidimensional indices of wellbeing: An overview. *Econ. Rev.* **2013**, *32*, 7–34. [CrossRef]

23. Hair, J.F.; Anderson, R.E.; Tatham, R.L.; Black, W.C. *Análisis Multivariante*, 5th ed.; Otero, A., Ed.; Prentice Hall Iberia: Madrid, Spain, 1999; ISBN 9788483220351.
24. OECD. *Handbook on Constructing Composite Indicators—Methodology and User Guide*; OECD: Paris, France, 2008.
25. Jolliffe, I.T. *Principal Component Analysis*, 2nd ed.; Springer: New York, NY, USA, 2002; ISBN 0387954422.
26. International Atomic Energy Agency. *Energy Indicators for Sustainable Development: Guidelines and Methodologies*; International Atomic Energy Agency: Vienna, Austria, 2005.
27. Gan, X.; Fernandez, I.C.; Guo, J.; Wilson, M.; Zhao, Y.; Zhou, B.; Wu, J. When to use what: Methods for weighting and aggregating sustainability indicators. *Ecol. Indic.* **2017**, *81*, 491–502. [CrossRef]
28. Arrow, K.J. *Social Choice and Individual Values*, 2nd ed.; Wiley: New York, NY, USA, 1963.
29. Podinovskii, V.V. *Criteria Importance Theory*; Elsevier Science: Moscow, Russia, 1994; Volume 27.

energies

MDPI

Article

The Interdependencies between Economic Growth, Energy Consumption and Pollution in Europe

Ane-Mari Androniceanu [1,*], Raluca Dana Căplescu [2,*], Manuela Tvaronavičienė [3,4] and Cosmin Dobrin [5]

1 Management Doctoral School, The Bucharest University of Economic Studies, 6 Piata Romana, 010374 Bucharest, Romania
2 Department of Statistics and Econometrics, The Bucharest University of Economic Studies, 6 Piata Romana, 010374 Bucharest, Romania
3 Department of Business Technologies and Entrepreneurship Vilnius Gediminas Technical University, eLT-10223 Vilnius, Lithuania; manuela.tvaronaviciene@vgtu.lt
4 Institute of Humanities and Social Sciences, Daugavpils University, Parades Str. 1-421, LV-5401 Daugavpils, Latvia
5 Management Department, The Bucharest University of Economic Studies, 6 Piata Romana, 010374 Bucharest, Romania; cosmin.dobrin@man.ase.ro
* Correspondence: ane.androniceanu.drd@gmail.com (A.-M.A.); raluca.caplescu@csie.ase.ro (R.D.C.)

Abstract: The strong interdependency between economic growth and conventional energy consumption have led to significant environmental impact, especially with respect to greenhouse gas emissions. Conventional energy-intensive industries release increasing quantities every year, which has prompted global leaders to consider new approaches based on sustainable consumption. The main purpose of this research is to propose a new energy index that accounts for the complexity and interdependences between the research variables. The methodology is based on Principal Component Analysis (PCA) and combines the key components determined into a score that allows for both temporal and cross-country comparisons. All data analyses were performed using IBM SPSS Statistics 25™. The main findings show that most countries improved their economic performance since 2014, but the speed of the improvement varies a lot from one country to another. The final score determined reflects the complex changes taking place in each country and the efficiency of the governmental measures for sustainable economic growth based on low energy consumption and low environmental pollution.

Keywords: economic growth; energy efficiency; pollution; renewable energy

Citation: Androniceanu, A.-M.; Căplescu, R.D.; Tvaronavičienė, M.; Dobrin, C. The Interdependencies between Economic Growth, Energy Consumption and Pollution in Europe. *Energies* **2021**, *14*, 2577. https://doi.org/10.3390/en14092577

Academic Editor: Miguel-Angel Tarancon

Received: 29 March 2021
Accepted: 26 April 2021
Published: 30 April 2021

1. Introduction

Economic growth and modern life are inconceivable without electricity, and most of the last century's discoveries would not have been possible without electricity. The European Union (EU) aims to give up using coal entirely by 2050 but will need significant help from European banks, which still finance 26% of all coal-fired power plants in the world [1]. Many Western European countries (including Italy and Spain) target total coal abandonment by 2030, while Germany (where coal still supplies 40% of energy needs) plans to reach this target by 2038 [2]. However, there still is a long road to full transition, with only 38 of Europe's 287 active coal-fired power plants (EU-27, plus the Balkans and Turkey) being officially planned to shut down in the foreseeable future. This represents a capacity reduction of only 18,162 megawatts out of a total of 179,157 MW [3].

Eastern and Central European countries rely largely on coal in their electricity production and fear that an unconsolidated transition to other forms of energy production could have a negative impact on their economic growth. Visegrad countries have been experiencing challenging situations, as the economies of Czechia, Hungary, Poland and Slovakia are more dependent on coal than Western European economies [4,5]. Poland plans to build three new coal-fired power plants, representing an increase in capacity of about 5000 megawatts—by far the most significant increase in EU countries [6]. Hungary, Romania and Bulgaria are also looking into increasing coal-based energy output rather than reducing it, albeit to a lesser extent than the four countries previously mentioned [7]. The effects on the environment and population health are major [8]. Emissions from coal-fired power plants in Europe contribute significantly to the share of diseases caused by environmental pollution. The latest published data show that the impact in the European Union amounts to over 18 thousand premature deaths, approximately 8.6 thousand new cases of chronic bronchitis and over 4 million working days lost annually [9].

The economic costs of the impact of coal-fired power plants on energy in Europe are estimated at around €43.1 billion a year. According to The European Environmental Agency, in 2018, approximately 379 thousand premature deaths were attributable to air pollution in the 27 EU Member States and the United Kingdom [10]. These costs are mainly associated with respiratory and cardiovascular diseases, the most important groups of chronic diseases in Europe. Together, coal-fired power plants in Poland, Romania and Germany are responsible for more than half of the health effects. Other important effects are attributed to coal burning in Bulgaria, Czechia, France, Greece, Serbia, Turkey and the United Kingdom [11]. It is well known that the use of coal for energy production is one of the major obstacles to reducing emissions. Currently, energy consumption is constantly on the rise, and stagnation, let alone regression, is hardly likely in the foreseeable future. Sustainable development concepts should be integrated in targets for economic development, energy productivity and monitoring of energy consumption [12,13]. Increasing the wealth of a state that does not account for more efficient use of energy in order to protect and conserve natural resources and the environment is neither sustainable nor conceivable anymore. One of the fastest and most effective ways to boost organizational performance improvement with respect to social and environmental protection [14–17] is to act upon energy efficiency [18,19].

For a long time, it was considered that there is an intrinsic link between economic growth, energy production and consumption and pollution. The main opportunities for energy savings in the future will come from the optimal selection of production processes [20] and a reduction or even removal of wastes from the system [21]. That is, economic growth can only be achieved by assuming a higher consumption and production of energy and implicitly greater pollution of the planet. During the last decade, decoupling between energy production and consumption, pollution by greenhouse gas emissions and economic growth was observed globally and attributed to increases in energy productivity [22]. Very recent analyses conducted by the International Energy Agency (AIE) show that CO_2 emissions have stagnated globally for the second year in a row, while the global economy has grown by more than 3% [23]. Preliminary AIE data suggest that electricity generated from renewable energy played a crucial role, accounting for about 90% of total new energy generated in 2015–2019. This new decoupling trend is found in 21 states that have managed to reduce greenhouse gas emissions while increasing gross domestic product. Of those countries, 16 are EU Member States [24].

To achieve this independence between economic growth and negative environmental impact, various countries have several measures, ranging from carbon taxation to increased investment in renewable energy sources and technologies, and even shifts from emission-intensive industry to more environmentally friendly approaches [25]. In Europe, the transition to a low-carbon economy would require an additional investment of €270 billion or 1.5% of the EU's annual GDP over the next four decades [26]. A simultaneous analysis of GDP per capita, energy productivity, energy consumption and pollution reflects the decoupling between economic growth and energy consumption.

In this context, the present study aims at contributing to answering the question regarding the interdependences between economic growth, energy consumption and pollution in the EU countries by proposing a new index to measure progress towards more sustainable economic growth. To this aim, we chose a series of variables that focus on energy productivity, private and industrial energy consumption, the share of renewable energy in total energy consumption and population exposure to pollutants that pose health risks [27–30]. The results of this research are relevant both theoretically, because a new indicator that reflects the interdependencies between the selected variables was designed, and from a practical point of view, because it can contribute to the development of viable and sustainable economic development strategies in European countries.

2. Literature Review

Economic growth results from the interaction of production factors, namely the productive activities of private economic agents. Various models for economic growth, its determinants and its measures were identified in the literature [31]. As the expression of an economy's capacity to produce goods and services [32], economic growth is measured based on the increase in real Gross Domestic Product (GDP) [33] and is dependent on production dynamics, inflation and unemployment as main drivers of economic cycles [34]. These alternative periods are called booms and recessions, respectively. The 21st century sees a more rapid succession between periods of growth and recessions, the most notable being the global financial crisis [35], the debt crisis and, more recently, the COVID-19 crisis [36]. An economic cycle represents the fluctuation of a country's economic activity, characterized by an increase in aggregate economic indicators followed by a decrease [37].

As a continuous process, economic growth has greatly benefited from the last three industrial revolutions [38], which has led to further and more rapid growth, producing a virtuous economic circle [39]. Several studies have found that the quality of the business environment [40,41] and the services sector [42,43] play an important role in the economic growth of a state. Besides the focus on the productive capacity of an economy, the concept of economic growth encompasses citizens' quality of life, particularly for those actively contributing to it [44].

Recent research [45] has shown that the increase in energy demand is decoupling from economic growth, as a result of the reduction in energy intensity required for the same unit of GDP, with doubling of the economic indicator being accompanied by only a 14% increase in energy consumption by 2050 according to the Global Energy Perspective report published in 2019 [46]. The main driver of these trends is the decrease in energy intensity of the economic processes, which compensates for the increase in population consumption triggered by higher incomes. According to the same report, more efficiently used energy contributes to a slowdown in the growth of energy demand.

It is estimated that the importance of renewable energy resources will increase globally enough to cover more than half of the electricity generation capacity by 2035 [47] and by 2050, together with nuclear production, will account for 34% of energy produced. Based on these projections, the methodology we propose also includes the share of renewable energy in gross final energy consumption.

Although ambient air quality in Europe has improved in recent years, air pollution continues to pose a major threat to public health. The European Environment Agency [48] (EEA) estimates that 80–90% of Europe's urban population is currently exposed to higher concentrations of suspended dust and ozone than the values recommended by the World Health Organization [49]. The process of producing electricity begins inside the power plant, where the conversion of primary energy into electricity takes place by burning fossil fuels or renewable resources, coal, natural gas and hydroelectric or wind energy and generating a water vapor, used to operate a turbine connected to the generator, which drives the generator (alternating current).

Biomass energy (bioenergy) is stored chemical energy and includes any solid, liquid or gaseous fuel or any electricity or useful chemical, derived from organic matter, either directly from plants or indirectly from industrial waste derived from plants, commercial and urban waste or agricultural and forestry residues. During the conversion processes, such as burning, biomass releases energy, usually in the form of heat, and carbon is re-oxidized to CO_2 to replace that consumed while the plant has grown (using biomass for energy is a reverse process of photosynthesis).

Coal-based energy production further contributes to the already poor air quality in Europe caused by the transport sector, industrial processes, residential heating systems and agriculture. Coal power plants emit significant amounts of suspended dust, sulfur dioxide and nitrogen oxides—the latter indirectly contributing to ozone depletion [50]. Of these, the most worrying for health are particulate matter (PM2.5) and ozone. Since pollutants can travel long distances, including across borders, the entire European population is affected by air pollution caused by the use of coal. Following several decades of reduction in coal use for energy production, the trend is once again on the rise. Coal is still an important source of energy in Europe, covering about a quarter of electricity production. About 50 new coal-fired power plants are planned. However, the continued use of coal also has a price that decision-makers are very unaware of the unpaid health bill. This health bill is paid by citizens, national health insurance budgets and the economy in general, due to productivity losses.

There is a significant amount of evidence on how these air pollutants affect the lungs and heart; the literature on the topic traces their impact to chronic respiratory diseases, such as chronic bronchitis, emphysema and lung cancer, and cardiovascular diseases such as myocardial infarction, congestive heart failure, ischemic heart disease and cardiac arrhythmias. Acute effects include respiratory symptoms, such as chest pain and cough, as well as violent asthma attacks. Children, the elderly and other patients are more susceptible to these effects. Recent studies show that air pollution can lead to low birth weight and premature birth due to exposure during pregnancy. Of particular concern is the high emissions of mercury from coal-fired power plants, because mercury can affect children's cognitive development and cause irreversible damage to the vital organs of the child. Coal-fired power plants are the most important source of mercury pollution in Europe, and the EU is starting to focus on technical options to reduce these emissions under a new United Nations treaty [51].

Although coal-fired power plants are responsible for only a small part of the total ambient air pollution, they are the most important source of industrial air pollutants. A large coal-fired power plant emits several thousand tons of hazardous air pollutants annually and has an average lifespan of at least 40 years. The construction of new coal-fired power plants would mean that hazardous emissions and their effects on health would be maintained for many years. It would also cancel out the short-term reduction in air pollution in other sectors. In order to highlight the negative impact of pollution on population health, the following two specific variables were integrated into our dataset: pollution, grime or other environmental problems and exposure to air pollution by particulate matter.

Among fossil fuel sources, the only one that will increase by 2035 will be the share of natural gas, which will also cap after this time horizon. The additional natural gas-fired electricity generation capacity will amount to 675 GW by 2035, which three times the installed capacity of Organization for Economic Cooperation and Development (OECD) member countries in Europe. Demand for oil will peak at 108 million barrels per day by 2030, after which it will "drop substantially" [52].

The chemical industry will account for more than half of the increase in oil demand over the next 15 years, after which time the contribution of this sector will decrease as a result of the reduction in demand for plastics and growing recycling efforts. The strongest decline in oil demand will be in electricity generation and transportation. Electric vehicle sales will exceed the 100 million mark by 2035 and are expected to become a cheaper alternative in five to 10 years [53]. It is estimated that by 2022, the costs of autonomous energy generation and storage will be similar to the cost of purchasing energy from a supplier, according to a new study by Ernst and Young [54]. The study also shows that in all markets, by 2025, electric vehicles (EVs) will reach parity with traditional vehicles with internal combustion in terms of costs and performance.

The level of electricity consumption from conventional sources will be greatly impacted by changing consumer preferences for renewable energy. In turn, this preference would be facilitated by making this type of energy available at prices comparable to conventional energy and the development of infrastructure that makes consumption of energy from renewable sources convenient [55,56]. Based on current trends in research and development, as well as implementation of newly found technology in the renewable energy sector, it is estimated that in Europe and Oceania, renewable energy will account for about 50% of energy demand by 2050 [57].

Many European countries have already begun to change their energy-based business models [58] in response to legislative and regulatory pressures aiming at increased uptake of renewable resources and ambitious carbon footprint reduction targets. Forecasts show that in the next decade, the revenues of the traditional utility companies will decrease significantly [59–63]. This change is determined on the one hand by changes in business models [64] that integrate modern information technologies and opt for renewable energy production and on the other hand by private investments in household renewable energy production with the view to optimize energy costs for the family [65].

This research highlights the above changes in state economies and among the population during the period 2014–2019 and shows that, while the interdependencies between economic growth, energy production and consumption, and pollution in EU countries are quite strong, the economic development model [66] is seeing significant changes compared to the previous one [67,68]. Sustainable economic growth has become part of the agenda for most states worldwide and, while current data do not suggest this at scale, important shifts are expected in the not-so-distant future [69,70]. Based on the indicator we propose, the EU states are already transitioning towards a more sustainable and environmentally friendly economic model, so reaching that goal is mostly a matter of time and consistency from here onward [71,72].

3. Data and Methods

This section presents the data and pre-processing steps, together with the methodology for developing an energy index for Europe. All analysis was performed using IBM SPSS Statistics 25™(IBM SPSS Statistics for Windows, Version 25.0. IBM Corp; Armonk, NY, USA).

3.1. Data Source

All data were extracted from Eurostat database and refer to years 2014, 2018 and 2019 for 27 European countries, namely EU-28 (before 2020) without Malta. The rationale behind the choice of countries was based on data availability, and Malta was eliminated because it had missing information for most variables of interest. Given that the purpose of the present study is to propose a new index together with the methodology for computing it, we only chose three years for which we computed the new index and analyzed the results. The choice of the three years was made based on the following criteria:

- 2014 was chosen in relation to the 2030 Agenda for Sustainable Development adopted by the UN General Assembly in 2015, assuming these data were the most recent available at the time of preparation of the document; therefore, it seemed reasonable to use it as starting point;
- Data for 2019 comprise the most recent available information for all variables of interest;
- The year 2018 was selected for validation purposes. While 2014 data may also serve as validation data, five years is a long period in the current fluctuating context so the proposed methodology was also tested on the closest year to the development sample for which data were available.

As there were no major disruptions during any of the chosen years, they were considered stable enough for a comparative analysis and validation of the proposed methodology.

The list of variables chosen, together with their descriptions and sources, is presented in Table 1 in Appendix A.

3.2. Data Pre-Processing

The raw data was mostly clean, though there were some missing values to be dealt with for 2014 and 2019. For the latter year, missing information about energy taxes for Cyprus and Latvia and the percentage of households exposed to pollution, grime and other environmental problems were replaced by the values in the previous year (2018).

The 2014 data had a single missing value, namely Exposure to air pollution for particulate matter less than 2.5 µm for Greece. After analyzing the year-to-year evolution of the available information both for Greece and for the other countries, the missing information was replaced with the mean of the values for 2013 and 2015.

Given the largely different countries both in terms of population size and economy, the following variables were created during the pre-processing stage in order to ensure cross-country comparability:

- GDP per capita and Energy taxes per capita (both expressed in million Euros) were derived by dividing GDP and energy taxes, respectively, by population size and multiplying the result by 1 million to convert it to Euros;
- Industry energy consumption and Services energy consumption (expressed in thousand tons of oil equivalent) were divided by GDP and multiplied by 1000 in order to express them as tons of oil equivalent consumed per Euro produced;
- Household energy consumption per capita (expressed in kg of oil equivalent per capita) was also transformed to ton consumed per Euro produced by dividing the original variable by 1000, multiplying it to the population and dividing the result by GDP, so that it became comparable with energy consumption from the economic sector.

Following these transformations, all initial variables were discarded and only resulting variables were used further.

3.3. Index Methodology

The proposed methodology is based on identifying underlying components using Principal Component Analysis (PCA) and combining them into a score that is then scaled so that it allows for both temporal and cross-country comparisons. Results were validated by applying the same methodology to two other years. The remainder of this section will be dedicated to describing in detail the methodology proposed.

The reason why we chose PCA instead of more widely used methods (like data envelopment analysis or decomposition analysis) is that it treats all variables as input and allows accounting for interdependencies between economic and environmental variables, as opposed to the frequently used methods that only account for the relationship between each input variable and the outcome.

The first step is to check for data normality, as PCA results are influenced by data distribution. For variables that were not normally distributed according to the Shapiro–Wilk test [73]. ln-transformations were applied to normalize them, followed by a new test.

The null hypothesis of the Shapiro–Wilk test is that a sample comes from a normally distributed population and the statistic is computed as:

$$W = \frac{\left(\sum_{i=1}^{n} a_i \cdot x_{(i)}\right)^2}{\sum_{i=1}^{n} (x_i - \overline{x})^2} \quad (1)$$

where:

- $x_{(i)}$—ith smallest number in the sample
- x_i—the ith element of the sample
- x_i—the sample mean
- a_i is given by $(a_1, \ldots, a_n) = \frac{m^T \cdot V^{-1}}{C}$, with:
  - $C = ||V^{-1} \cdot m||$

- $m = (m_1, \ldots, m_n)^T$—expected values of the order statistics of independent and identically distributed random variables sampled from the standard normal distribution
- V—covariance matrix of the normal order statistics

The second step was to check for and analyze the impact of outliers. Natural outliers explained by a country's population or economy size should have been eliminated during the pre-processing stage, and the impact of those causing data to be non-normally distributed was diminished by the transformation applied to normalize the variables. To ensure a balance between the need for data quality and processing complexity, the best approach for any remaining outliers was chosen by comparing the impact of no treatment with the elimination of the observation(s) with outliers and ln-transformation of the variable.

The third step was to standardize the variables to prevent the disproportionate contribution of a variable measured on a scale that is several orders of magnitude above others. This step was performed by using the *z-score* as presented below, and it effectively converted all variables to the same measurement unit, namely number of standard deviations from the mean:

$$z - score = \frac{x_i - \overline{x}}{\sigma} \quad (2)$$

where x_i is the ith element, $\overline{x}$ is the mean and σ is the standard deviation of the sample.

PCA [74] can be used as a dimensionality reduction technique by uncovering the underlying factors, called components. To be useful, PCA must be applied to data that contain clusters of correlated variables, so once the data were deemed of satisfactory quality, the next step was to compute the correlation matrix of the variables. Since data pre-processing ensured all variables were normally distributed, the Pearson correlation coefficient was computed for a sample as follows:

$$r_{xy} = \frac{\sum_{i=1}^{n}(x_i - \overline{x})(y_i - \overline{y})}{\sqrt{\sum_{i=1}^{n}(x_i - \overline{x})^2 \cdot \sum_{i=1}^{n}(y_i - \overline{y})^2}} \quad (3)$$

where (x, y) is the pair of variables for which the correlation is computed. Each correlation is then tested for significance.

Given that Pearson's correlation coefficient is only useful for identifying linear correlations, a scatterplot matrix is also presented along with the results in order to detect potential non-linear relationships in the data.

The expectation is that PCA will confirm the results indicated by the correlation analysis and that the clusters of variables that were inter-correlated in the previous analysis will also be strongly correlated to the same component. Components generated by PCA are orthogonal (independent) and can be used further to compute the final score for each country as follows:

$$final\ score = \sum_{i=1}^{k} c_i \cdot f_i \quad (4)$$

where k is the number of components retained based on the PCA and correlation analysis results, f_i are the retained components and

$$c_i = \begin{cases} -1, & if\ higher\ factor\ scores\ lead\ to\ desired\ outcome \\ +1, & if\ lower\ factor\ scores\ lead\ to\ desired\ outcome \end{cases}$$

Since the final score is not easily interpretable as such, a rating scale was created using the following logic (the result is rounded to the nearest integer):

$$rating = \frac{best\ case\ scenario - score_i}{best\ case\ scenario - worst\ case\ scenario} \cdot 100 \quad (5)$$

The idea is to create best and worst scores for each component and combine them in a final score the same way as for the actual final scores. These scenarios were then used for scaling the scores so that the closer the score to 100, the closer the country is to the best-case scenario. Conversely, the closer the score is to zero, the closer the country is to the worst-case scenario. The resulting rating can then be used to compare countries or analyze a country's progress over time.

The validation of the results was done for two years, namely for 2014 and 2018. To test the new methodology, the data for each year were passed through the same transformation pipeline as the 2019 one, namely:

- Logarithmation of the same variables as for 2019;
- Identification and treatment of outliers;
- Standardization using the mean and standard deviation for 2019 to prevent data leakage;
- Determination of components using the 2019 coefficients;
- Calculation of the final score for each country.

In order to test the time consistency of the estimated scores, a correlation matrix was computed. The expectation was that the correlation would be high and monotonously decreasing, so that for years further apart, the relationship would be weaker than for closer years. Nevertheless, all correlation coefficients are expected to be significant and high (above 0.95).

For creating the best- and worst-case scenarios, the data for 2014 were chosen as they were likely the most recent available before the adoption of the 2030 Agenda for Sustainable Development, which includes Sustainable Development Goal 7, focused on energy. Other scenarios can be easily incorporated into the rating scale as well.

4. Research Results and Discussion

The main goal of the research is to design an energy index based on the correlation between economic growth, energy consumption and environmental pollution and to apply it for the states of the European Union.

The final variables resulted from the pre-processing stage, and their descriptive statistics for each year are presented in Table 1. For this research, variables were selected according to the criterion of relevance and importance for each of the three components analyzed: economic growth; energy consumption and pollution.

Table 1. Descriptive statistics of the final variables selected for analysis.

Variables	Min	Max	Mean	Std. Dev.	Skewness	Kurtosis
		2014				
Energy productivity	2.23	14.00	6.97	3.01	0.72	0.17
Renewable energy	4.47	51.82	19.91	11.66	0.93	0.65
Exposure to particulates <2.5 μm	7.40	26.10	14.89	4.84	0.71	0.34
Exposure to particulates <10 μm	13.50	41.20	23.11	6.52	0.85	1.08
Pollution, grime, other	4.50	23.20	13.16	4.78	0.27	−0.13
GDP per capita	5919	90,643	26,569	18,215	1.71	4.55
Energy taxes per capita	146.01	1645.56	518.60	324.94	1.73	4.52
Industry final energy consumption	7.98	61.03	26.20	13.12	0.80	0.42
Services final energy consumption	6.72	25.78	13.02	5.47	0.87	−0.25
Household final energy consumption	9.31	52.63	27.74	14.91	0.66	−1.28
		2018				
Energy productivity	2.41	18.58	7.54	3.63	1.35	2.34
Renewable energy	7.34	54.65	21.59	11.71	1.11	0.93

Table 1. *Cont.*

Variables	Min	Max	Mean	Std. Dev.	Skewness	Kurtosis
Exposure to particulates <2.5 μm	6.20	24.30	13.80	4.77	0.25	−0.55
Exposure to particulates <10 μm	11.50	33.80	22.27	6.30	0.21	−0.56
Pollution, grime, other	6.30	24.80	12.79	4.53	0.66	0.43
GDP per capita	7959	99,755	30,894	20,375	1.69	3.84
Energy taxes per capita	191.62	1562.81	571.66	293.16	1.48	3.77
Industry final energy consumption	7.00	48.66	22.87	11.50	0.70	−0.07
Services final energy consumption	5.27	21.94	11.90	4.30	0.69	−0.11
Household final energy consumption	8.24	44.41	23.95	11.55	0.54	−1.21
		2019				
Energy productivity	2.52	19.64	7.80	3.76	1.54	2.96
Renewable energy	7.05	56.39	22.52	11.94	1.12	1.06
Exposure to particulates <2.5 μm	4.80	19.60	12.06	3.74	0.00	0.04
Exposure to particulates <10 μm	10.20	30.90	20.23	5.59	0.11	−0.51
Pollution, grime, other	5.90	25.20	12.82	4.40	0.73	1.10
GDP per capita	8748	103,465	32,115	21,077	1.74	4.00
Energy taxes per capita	227.10	1654.18	587.46	295.82	1.77	5.41
Industry final energy consumption	6.44	45.34	21.39	10.53	0.64	−0.10
Services final energy consumption	4.95	20.71	11.19	3.88	0.68	0.05
Household final energy consumption	7.19	41.34	22.63	10.35	0.39	−1.20

Energy productivity—Eur/kg of oil equivalent; Renewable energy—% gross final energy consumption; Exposure to particulates <2.5 μm—μg/m^3; Exposure to particulates <10 μm—μg/m^3; Pollution, grime, other—% of households exposed; GDP per capita—Eur; Energy taxes per capita—Eur; Industry final energy consumption—ton/Eur; Services final energy consumption—ton/Eur; Household final energy consumption—ton/Eur. Source: authors' computation.

The variables in Table 1 were selected on the basis of studies in the literature and their importance and relevance in the current economic and social context. Over the last decade, major changes have taken place in energy, economic and environmental policies at the EU and Member State level. Thus, the EU states aim to have economic growth in the future, both in industry and in services, but with low consumption of traditional energy, more renewable energy and less pollution. From this perspective, the variables were selected. They are important for achieving the research objective, namely, the elaboration of an energy index. This index reflects the interdependencies between the selected variables and facilitates the comparison of the EU countries that have different energy, economic and environmental policies during the analyzed period. Through the proposed variables and the new index, the EU states are compared and several particularities were discovered. The main research variables and their relevance for this research are explained below.

Energy productivity was selected for measuring the economic benefit received from each unit of energy used both for economic growth and householders. Renewable energy has been integrated into research for two reasons: first, because it is a priority for the EU countries in their business model based on green energy as a main source of energy consumers, both from the industry and services and for households, and second, because renewable energy significantly reduces environmental pollution.

Industry final energy consumption, services final energy consumption and household final energy consumption were selected because they are the main energy consumers with a direct impact on economic growth and pollution in Europe. This issue is analyzed with the following three variables: Exposure to particulates <2.5 μm, Exposure to particulates <10 μm and Pollution, grime, other. Through these selected variables, a comparative analysis of the level of pollution in each country was developed. These reflect the degree of pollution generated by the main energy consumers in the EU countries during the analyzed period.

The novelty in this research consists in its focus on the interdependencies between these selected variables, which are compared and analyzed using the PCA and the new energy index, created to measure the influences of selected variables in different EU countries during the analyzed period.

The Shapiro–Wilk normality test (Table 2) indicated that five variables were not normally distributed, so a ln-transformation was applied, and the result has been re-tested. The second test confirmed the normality of the distribution.

Table 2. Results of the Shapiro–Wilk test.

Initial Variables	Statistic	df	Sig.
Energy productivity	**0.872**	**27**	**0.003**
Renewable energy	**0.912**	**27**	**0.025**
Exposure to particulates<2.5 μm	0.969	27	0.587
Exposure to particulates<10 μm	0.971	27	0.64
Pollution, grime, other	0.958	27	0.338
GDP per capita	**0.838**	**27**	**0.001**
Energy taxes per capita	**0.853**	**27**	**0.001**
Industry final energy consumption	0.948	27	0.196
Services final energy consumption	0.956	27	0.302
Household final energy consumption	**0.919**	**27**	**0.037**
Transformed variables			
Energy productivity (ln)	0.978	27	0.809
Renewable energy (ln)	0.989	27	0.987
GDP per capita (ln)	0.977	27	0.793
Energy taxes per capita (ln)	0.967	27	0.516
Household final energy consumption (ln)	0.944	27	0.152

Variables in bold are not normally distributed (Sig. < 0.05). Source: authors' computation.

As expected, outliers are not present in most of the normally distributed variables. Since the methodology is developed on 2019 data, unusual values were checked only for that year's data.

The variable regarding the percentage of households exposed to pollution, grime and other environmental problems registered an outlier for Germany. The following three approaches were chosen for testing and for dealing with the presence of this extreme value, namely:

- Removing Germany from the sample: despite having a reduced number of observations in the sample, elimination of Germany because of the outlier might be a valid choice if the resulting reduction in variability compensates for the smaller sample;
- Applying ln-transformation on the variable: this approach results in less variability in the data as the very large values are reduced more than the small ones, but being the most complex of the three, assessing its impact and contribution in is recommended subsequent analysis;
- Not changing anything: this approach is appropriate in a small sample if the treatment of the outlier proves to be too resource-expensive compared to the benefit obtained or if it causes results to worsen due to diminishing the sample.

Subsequent analysis will be performed with all three versions of the variable until the impact can be assessed and the best approach chosen.

All normally distributed variables were standardized using the means and standard deviations for 2019 (Table 3), to prevent information leakage from the other years. Performing this step is the best practice for any analysis and compulsory for datasets where variables are measured on very different scales.

Table 3. Standardization values: means and standard deviations for transformed variables, 2019.

Variables	Mean	Std. Deviation
Energy productivity (ln)	1.96	0.44
Renewable energy (ln)	2.99	0.52
GDP per capita (ln)	10.20	0.60
Energy taxes per capita (ln)	6.27	0.47
Household final energy consumption (ln)	3.01	0.49
Pollution, grime, other (ln)	2.49	0.35
Pollution, grime, other	12.82	4.40
Pollution, grime, other (w/mis)	12.34	3.70
Exposure to particulates <2.5 μm	12.06	3.74
Exposure to particulates <10 μm	20.23	5.59
Industry final energy consumption	21.39	10.53
Services final energy consumption	11.19	3.88

Source: authors' computation.

Following the standardization step, all variables are normally distributed and measured on the same scale, namely in standard deviations around their respective means. Next, the Pearson correlation matrix was computed and tested (Table 4).

Table 4. Correlation matrix—Pearson coefficient.

	EP	RE	GDP	ET	HH	PLN	P	PM *	2.5	10	IND	SRV
EP												
RE	−0.167	1	−0.223	−0.174	0.356	−0.276	−0.231	−0.245	−0.282	−0.238	0.291	
GDP	**0.818**	−0.223	1	0.883	−0.86	−0.129	−0.092	−0.212	−0.647	−0.666	−0.581	
ET	**0.643**	−0.174	**0.883**	1	−0.709	−0.013	0.005	−0.035	−0.571	−0.531	−0.529	
HH	**−0.838**	0.356	**−0.86**	**−0.709**	1	0.075	0.056	0.122	0.462	0.445	0.668	
PLN	−0.195	−0.276	−0.129	−0.013	0.075	1	0.975	0.985	0.199	0.146	0.031	
P	−0.131	−0.231	−0.092	0.005	0.056	**0.975**	1	1	0.163	0.126	-0.022	
PM *	−0.258	−0.245	−0.212	−0.035	0.122	**0.985**	**1**	1	0.239	0.256	0.041	
2.5	**−0.427**	−0.282	**−0.647**	**−0.571**	**0.462**	0.199	0.163	0.239	1	0.896	0.313	
10	**−0.445**	−0.238	**−0.666**	**−0.531**	**0.445**	0.146	0.126	0.256	**0.896**	1	0.211	
IND	**−0.784**	0.291	**−0.581**	**−0.529**	**0.668**	0.031	−0.022	0.041	0.313	0.211	1	
SRV	**−0.875**	0.204	**−0.752**	**−0.534**	**0.783**	0.065	0.029	0.154	0.371	**0.457**	**0.622**	

* computed based on data for 26 countries; bold—significant at 0.05 level. EP—Energy productivity; RE—Renewable energy; GDP—GDP per capita; ET—Energy taxes per capita; HH—Household final energy consumption; PLN—Pollution, grime, other (ln); P—Pollution, grime, other; PM—Pollution, grime, other (with missing); 2.5—Exposure to particulates <2.5 μm; 10—Exposure to particulates <10 μm; IND—Industry final energy consumption; SRV—Services final energy consumption. Source: authors' computation.

For ease of interpretation, the coef ficients were color-coded such that an intense color indicates a strong correlation, blue color shows direct correlations and red color highlights inverse relationships between variables. The matrix shows that there are clusters of inter-correlated variables, which means the dataset is suitable for PCA.

Since Pearson's correlation coefficient only shows linear correlations, a scatterplot matrix was plotted to assess the linearity of the variable pairs. The closer the shape of the dots in each square is to a line, the stronger the linear the relationship is. The resulting plot confirms that where the correlation coefficient is high, the shape formed by the dots resembles a line, thus validating the conclusion based on the correlation matrix. Figure 1 contains bivariate analysis that involves measuring the degree of association of the variables considered in terms of direction, intensity and statistical significance. This figure shows how the researched variables are associated in terms of direction, intensity and statistical significance. The bivariate correlation presented refers to the analysis of the correlations between the twelve variables considered, designated as X and Y, mainly for determining the empirical relationship they have.

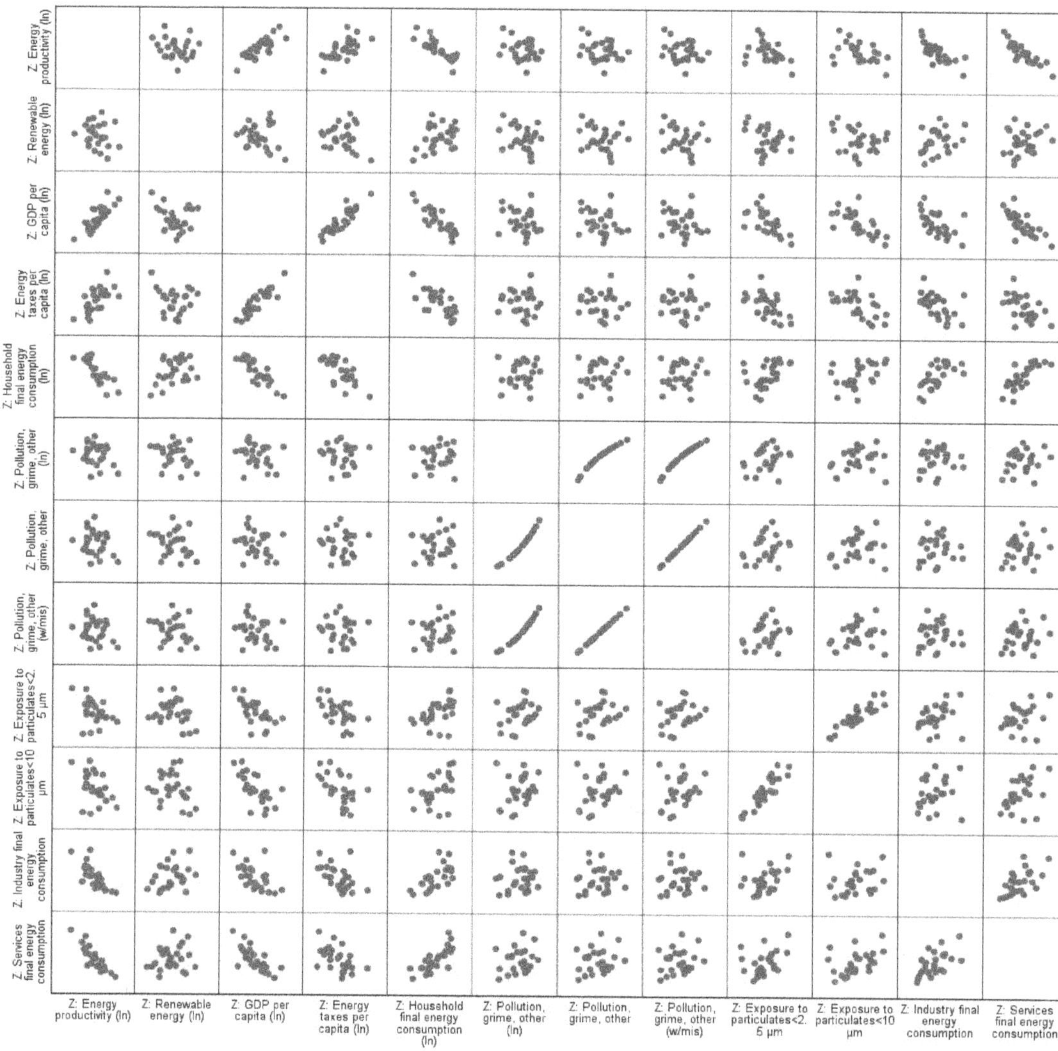

Figure 1. Bivariate relationship between variables in the dataset. Source: authors' computation.

Through bivariate correlation, the association and causality between them were tested. This analysis was used to see if the variables were related to each other and to measure how the investigated variables change together at the same time. The purpose of the bivariate analysis was to examine several relationships between several variables simultaneously. The bivariate correlation helped to understand the correlations between the researched variables.

Outlier impact was assessed during the PCA transformation stage. To this end, seven approaches were tested, and their were results evaluated based on sample adequacy, percentage of variance explained and number of components. The differences between the various approaches are given by the type of outlier treatment and variables included in the input data (Table 5).

Table 5. Approaches tested for PCA transformation.

Approach	Sample Adequacy	Variance Explained	Number of Components	Pollution, Grime, Other	Observations
1	0.733	72.506	2	original	
2	0.733	75.162	2	original	without Renewable energy
3	0.728	73.036	2	without outlier	
4	0.767	75.683	2	without outlier	without Renewable energy
5	0.729	82.793	3	ln-transformed	
6	0.764	75.178	2	ln-transformed	without Renewable energy
7	0.789	83.897	2	none	without Renewable energy

Source: authors' computation.

The fifth approach was selected as being the best one because it performed best among those including all variables of interest in the analysis, and the resulting components make economic sense (Table 6). The last approach was also considered, due to the high percentage of explained variance, but it was discarded in the end because the renewable energy variable was correlated with the resulting energy efficiency component (r = 0.404, $p = 0.037$), but not strongly enough to bring a significant contribution to it.

Table 6. Correlations between components and input variables based on PCA results.

Variables	Final Choice			Second Best Choice	
	Energy Efficiency	Pollution	Renewable Energy and Environmental Impact	Energy Efficiency	Pollution
Z: Energy productivity (ln)	−0.917	-	-	−0.922	-
Z: Household final energy consumption (ln)	0.892	-	-	0.861	-
Z: Services final energy consumption	0.841	-	-	0.852	-
Z: Industry final energy consumption	0.829	-	-	0.848	-
Z: GDP per capita (ln)	−0.799	-	-	−0.747	-
Z: Energy taxes per capita (ln)	−0.658	-	-	−0.627	-
Z: Exposure to particulates <10 μm	-	0.910	-	-	0.941
Z: Exposure to particulates <2.5 μm	-	0.905	-	-	0.931
Z: Pollution, grime, other (ln)	-	-	0.922	-	-
Z: Renewable energy (ln)	-	-	−0.537	-	-

Source: authors' computation.

It can be seen that including the two variables in the last component changes the correlation coefficients to some extent, but both the strength and the direction remain the same.

Three variables have inverse correlation with the energy efficiency component; thus, improvements in the respective areas will lead to lower values for the energy efficiency score. What is more, all three are the ln-transformed versions, which means increases in the original values, will lead to exponential decreases in the component score. Conversely, energy consumption by type of consumer is directly correlated to the component, thus lower consumption also leads to lower values on the factor, and exponentially so for household consumption. This suggests that:

- The smaller the value for this component, the better the performance of the respective country;
- The biggest impact can be achieved by increasing energy productivity and reducing household energy consumption;
- While the focus on energy consumption of industry and services may be beneficial due to the scale effect, the inertia is high in both cases, so measures taken in this direction, while having a non-negligible impact, are less effective than the for the first two components;
- The medium correlation between the component and energy taxes suggests that regulatory measures in this regard are bound to be less effective.

For the pollution component, the relationship it has with the variables defining it is straightforward, the less the population is exposed to fine particle matter, the lower the component value.

The last component shows that reducing pollution would have a great environmental impact, and increasing the share of renewable energy in gross final energy consumption can contribute to this. Both variables have exponential impact. Similar to the previous component, in the case of this component lower values are indicative of better performance [75]. The estimated coefficient matrix (Table 7) was used to compute the values for each country and year. Given that all components have the same relationship with the final score, the latter was obtained by adding the three components.

Table 7. Component Score Coefficient Matrix.

Variables	Energy Efficiency	Pollution	Renewable Energy and Environmental Impact
Z: Energy productivity (ln)	−0.246	0.098	−0.183
Z: Household final energy consumption (ln)	0.209	−0.029	−0.027
Z: Services final energy consumption	0.213	−0.062	0.074
Z: Industry final energy consumption	0.246	−0.16	0.074
Z: GDP per capita	−0.127	−0.135	0.069
Z: Energy taxes per capita	−0.074	−0.182	0.191
Z: Exposure to particulates <10 μm	−0.106	0.433	−0.074
Z: Exposure to particulates <2.5 μm	−0.099	0.42	−0.026
Z: Pollution, grime, other (ln)	0.116	−0.177	0.814
Z: Renewable energy (ln)	0.191	−0.249	−0.346

Source: authors' computation.

The best- and worst-case scenarios for 2014 were selected for creating the rating scale. These scenarios are based on the smallest and largest values for each component. According to the results presented so far, smaller values represent a better situation, so the minimum values for each component were added together to obtain the best-case scenario. Similarly, the maximum values were summed up to obtain the worst-case scenario. The results are presented in Table 8 below.

Table 8. Original values for countries contributing to the best- and worst-case scenarios of the rating scale.

Variables	Best Case		Worst Case		Component
	Value	Country	Value	Country	
Energy productivity	10.506	Luxembourg	2.226	Bulgaria	Energy efficiency
Household final energy consumption	9.31	Luxembourg	50.68	Bulgaria	
Services final energy consumption	7.24	Luxembourg	23.13	Bulgaria	
Industry final energy consumption	13.19	Luxembourg	61.03	Bulgaria	
GDP per capita	90,643	Luxembourg	5919	Bulgaria	
Energy taxes per capita	1645.56	Luxembourg	61.03	Bulgaria	
Exposure to particulates <10 µm	13.7	Finland	35.1	Poland	Pollution
Exposure to particulates <2.5 µm	8.4	Finland	26.1	Poland	
Pollution, grime, other	5.7	Croatia	15.4	Luxembourg	Renewable energy and environmental impact
Renewable energy	27.8	Croatia	4.5	Luxembourg	

Source: authors' computation.

The final scores were rescaled from 0 to 100 such that the worst case is 0 and the best case is 100. The closer the new values are to 100, the smaller the gap between them and the best-case scenario. The resulting rating can then be used to compare countries or analyze a country's progress over time.

For validating the results obtained for 2019 the analysis was replicated for 2018 and 2014 by taking the data through the same steps as for 2019, namely ln-transformation, standardization, estimation of components based on the Score Matrix (Table 7), and scaling using the same best- and worst-case scenarios. As expected, the correlations for each pair of years are high (above 0.95) and monotonously decreasing (the strongest correlation is between 2019 and 2018 and weakest between 2019 and 2014).

To further illustrate the use of the proposed index, a small analysis of the three years is presented. The final scores were computed for each country and year and ordered from largest to smallest based on 2019 values (Table 9). One thing to note is that the majority of the countries improved their performance since 2014. Furthermore, the rankings indicate much variation, with very few countries maintaining their relative position. This suggests that the improvement speed varies greatly from one country to another. While the proposed index cannot keep track of the individual evolutions for each variable, the resulting final score reflects the complex changes taking place and the variety of national legislations.

Interestingly, if compared either to the best-performing country for a particular year or to the best-case scenario (100 points), the results suggest a trend towards homogenization, particularly visible at the bottom of the ranking (large values in Figure 2). What this means is that while distances between best-performing countries remained relatively stable, countries that occupy the last positions in the ranking made significant progress towards the best-case scenario despite remaining in the last positions.

Figure 2 reflects the distance between the best-performing European country (left) and the best-case scenario (right). Thus, an interesting and useful comparison can be made by referring to the country with the best performance and the best-case scenario.

Going into more detail about improvements, the highest-achiever countries were mostly Eastern European, with top 10 being dominated by former communist states (Figure 3). Despite constantly being at the bottom of the ranking, Bulgaria made the biggest progress among considered countries. Starting from a score of 12, the closest any of the countries was to the worst-case scenario in any of the years, the country's performance went up to 29 points in 2019.

Table 9. Final scores and resulting country rankings, sorted by 2019 values.

Countries	Final Score			Country Rank		
	2019	2018	2014	2019	2018	2014
Sweden	81	81	75	1.5	1	3
Ireland	81	79	78	1.5	2	1
Denmark	78	76	76	3	3	2
Finland	68	67	63	4	4	6
Austria	67	65	64	5	5	4.5
Spain	62	60	59	6	6	7
Estonia	61	58	46	7	9	13
United Kingdom	60	59	64	8	7.5	4.5
Luxembourg	58	59	53	9.5	7.5	10
Portugal	58	57	54	9.5	10	9
France	57	56	57	11.5	11	8
Cyprus	57	53	45	11.5	13	15
Croatia	56	50	52	13	14.5	11
Italy	55	54	46	14	12	13
Netherlands	54	50	46	15	14.5	13
Germany	50	47	44	17	17	16
Belgium	50	48	43	17	16	18
Slovakia	50	44	37	17	19	21.5
Slovenia	48	43	43	19	20	18
Czechia	47	40	32	20	22.5	23
Lithuania	45	46	43	21	18	18
Romania	44	39	37	22	24	21.5
Latvia	43	38	31	23	25	24
Greece	42	42	40	24	21	20
Hungary	41	40	29	25	22.5	25
Poland	35	27	27	26	26	26
Bulgaria	29	24	12	27	27	27
Number of countries that, compared to previous year *, scored/ranked:						
Higher	23	21	25	10	11	10
Lower	2	3	1	10	13	14
Same	2	3	1	7	3	3

* values in the 2014 columns reflect comparisons between 2019 and 2014. Source: authors' computation.

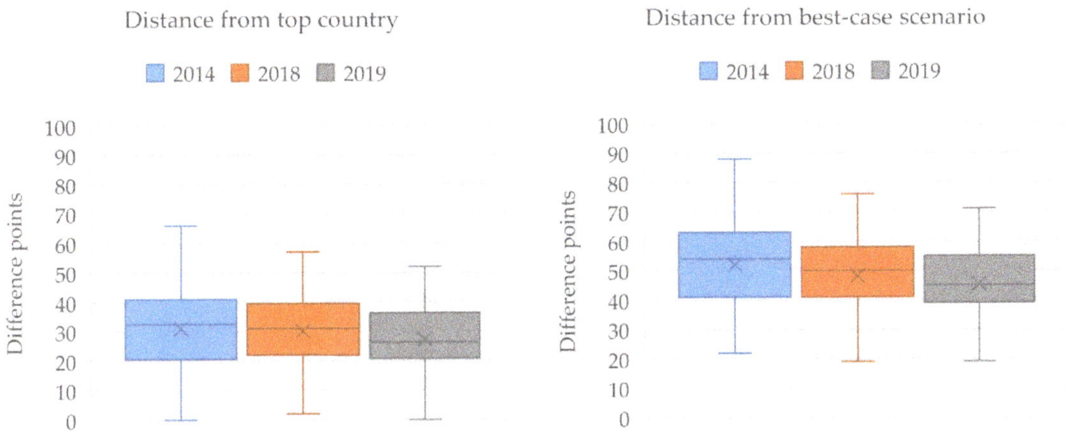

Figure 2. Distance from best-performing country of the year and best-case scenario. Source: authors' computation.

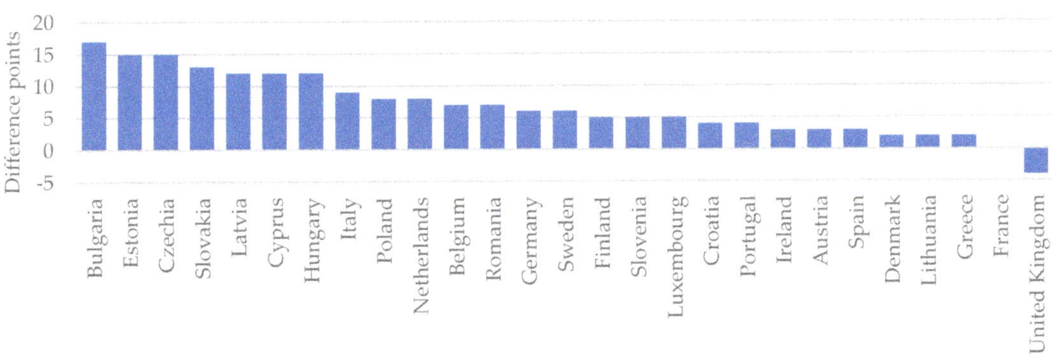

Figure 3. Overall changes in country scores in 2019 compared to 2014. Source: authors' computation.

What is more, it got closer to the second-last country in the sample, Poland, reducing the distance from 15 points to only 6. Another remarkable improvement is registered by Estonia, with a 15-point improvement that also translated into moving six positions ahead in the ranking, from 13th place (tied with Italy and Netherlands) in 2014 to the seventh country in the top. No less notable of an improvement was also made by Czechia, also 15 points, which started as fifth from the bottom and moved up three positions by 2019. At the other end of the spectrum, UK's context seems to have worsened, both in absolute and in relative terms, having dropped from fourth position (tied with Austria) to eighth, with a score decreasing by 4 points in 2019 compared to 2014.

Between the most recent two years analyzed, the top 10 seem to be populated mostly by the same countries, but in different positions (Figure 4). The biggest progress in terms of score was registered by Poland, but this did not translate into a higher ranking position. On the other hand, Czechia, Slovakia and Latvia went up two positions the list in 2018 compared to 2019. Estonia, Hungary and Italy exited the top 10 and were replaced by Croatia, Romania and Slovenia, all of which also moved up two positions in the ranking. Despite the medium-low change in score registered by Estonia, the country also moved up two positions, into seventh place.

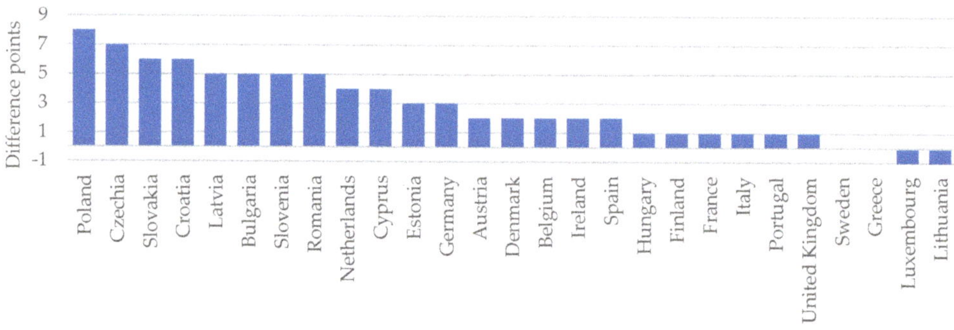

Figure 4. Changes in country score in 2019 compared to 2018. Source: authors' computation.

The two figures above suggest shifts in the evolutions of the countries considered. In the case of some countries, an important part of the progress registered over the entire period (Figure 3) happened during the last year (Figure 4), which is indicative of intensifying the efforts towards a cleaner and more environmentally friendly economy. Among countries at the bottom of the list in 2019 that show significant improvements are Poland, Romania, Latvia and Bulgaria. Notable progress was also made by some of the larger economies of the EU, for example, Germany, Netherlands and Spain. Given the initial status and size of economy, while the progress registered by the latter countries is smaller in absolute terms, the impact on the environment is large enough to matter.

Limitation and Future Research

The instrument proposed is useful and reliably shows interdependencies between economic perspective and environmental impact. To be able to show the usefulness of the index, the year 2014 was chosen as base for determining the best- and worst-case scenarios. The choice made was by highlighting that the information was most likely the latest available when UN's Sustainable Development Goal 7 was adopted together with the 2030 Agenda. This being said, different scaling scenarios can be applied, like, for example, using 2014 data to determine the worst-case scenario and target values for the indicators in the index as the best-case scenarios. This way, the instrument can be used for measuring countries' progress towards the set goals. The drawback to this approach is that the target values needed must be set to each of the 10 variables included in the index calculation methodology, which requires a relatively thorough analysis, especially at multi-state level.

A future study starting from current results would be to analyze COVID-19 and subsequent data to verify if the results still hold true [76,77]. As the period up to 2019 inclusively was relatively stable, data for 2020 are very likely impacted by worldwide lockdowns and economic activity disruption. For this reason, new analysis is recommended on newer data and comparisons with results presented in this research paper.

5. Conclusions

As efforts are made globally to reduce pollution, the proposed methodology can represent a valuable instrument for tracking country progress by taking into account the complexity and interconnections between the impacts economic and private activities have on the environment.

The main contribution of the present study is the energy index proposed for tracking countries' progress in time and in comparison with other countries. Its novelty consists in the fact that it accounts for complex interdependencies between variables rather than analyzing them in pairs of input–outcome variables. The three components identified, energy efficiency, pollution, and renewable energy and environmental impact, while independent from each other, capture the complex relationships between economic growth, energy productivity and consumption, pollution and efforts towards a more sustainable economic growth [78]. The index incorporates the three dimensions targeted by the EC, namely reducing greenhouse gas emissions, increasing share of renewable energy in consumption and improving energy efficiency and combines them into an instrument that is easy to use both for cross-country comparisons, and for time series analysis.

The research was performed in two main directions, namely on a correlative analysis of the variables considered within the three components and on a comparative analysis of them in the states included in the research. During the pre-processing stage, the following variables were created in order to ensure cross-country comparability: GDP per capita and energy taxes per capita, industry energy consumption and services energy consumption, and household energy consumption per capita. These were grouped into three components: energy efficiency, pollution and renewable energy and environmental impact. The components were combined into a score that was then scaled two allow for both temporal and cross-country comparisons. In order to prevent a disproportionate contribution of the states, the research variables were standardized by using the z-score. The Shapiro–Wilk normality test indicated that five variables were not normally distributed. Another result was related to the Pearson correlation matrix. The matrix shows that there are clusters of inter-correlated variables, which means the dataset is suitable. Bivariate relationship between variables in the dataset shows that all existing relationships in the data are linear.

A notable result is that the biggest impact in terms of energy efficiency can be obtained by increasing energy productivity, coupled with lower of more efficient consumption in households. Since industrial change is slower to achieve and legislative pressure does not seem to have significant impact, it follows that incentivizing more responsible household consumption could be the most lucrative direction to begin with. In parallel, investments in renewable energy production, both at the national and at the household level, would lead to faster decrease of pollution since they would allow for smaller demand of traditional energy production.

The results of the research can be used by the EU governments to adapt their economic, energy and environmental policies by developing a renewable energy business model that ensures sustainable economic growth and low environmental pollution in Europe [79].

Author Contributions: Conceptualization: A.-M.A., M.T. and C.D.; Methodology: R.D.C.; Validation: A.-M.A., M.T. and C.D.; Formal analysis: R.D.C.; Investigation: A.-M.A.; Resources: all authors; Data curation: R.D.C.; Writing—original draft preparation: A.-M.A., R.D.C., M.T. and C.D.; writing—review and editing: A.-M.A. and R.D.C.; Visualization: A.-M.A. and R.D.C.; Supervision: M.T. and C.D. All authors have read and agreed to the published version of the manuscript.

Funding: This research received no external funding.

Institutional Review Board Statement: Not applicable.

Informed Consent Statement: Not applicable.

Data Availability Statement: Not applicable.

Acknowledgments: We had the administrative and technical support of the International Centre for Public Management from the Bucharest University of Economic Studies and from Vilnius Gediminas Technical University.

Conflicts of Interest: The authors declare no conflict of interest.

Appendix A

Table 1. Data definitions and sources.

Indicator	Source
Energy productivity	Available online: https://ec.europa.eu/eurostat/databrowser/view/T2020_RD310/default/table (accessed on 15 March 2021)
Final energy consumption in households per capita	Available online: https://ec.europa.eu/eurostat/databrowser/view/SDG_07_20/default/table (accessed on 15 March 2021)
Final energy consumption in industry	Available online: https://ec.europa.eu/eurostat/databrowser/view/TEN00129/default/table (accessed on 15 March 2021)
Final energy consumption in services	Available online: https://ec.europa.eu/eurostat/databrowser/view/TEN00128/default/table (accessed on 15 March 2021)
Share of renewable energy in gross final energy consumption	Available online: https://ec.europa.eu/eurostat/databrowser/view/T2020_RD330/default/table (accessed on 15 March 2021)
GDP	Available online: https://ec.europa.eu/eurostat/databrowser/view/NAMA_10_GDP$DEFAULTVIEW/default/table (accessed on 15 March 2021)
Energy taxes	Available online: https://ec.europa.eu/eurostat/databrowser/product/view/ENV_AC_TAXIND2 (accessed on 15 March 2021)
Exposure to air pollution by particulate matter	Available online: https://ec.europa.eu/eurostat/databrowser/view/SDG_11_50/default/table (accessed on 15 March 2021)
Population on 1 January	Available online: https://ec.europa.eu/eurostat/databrowser/view/DEMO_PJAN$DEFAULTVIEW/default/table (accessed on 15 March 2021)
Pollution, grime or other environmental problems - EU-SILC survey	Available online: Available online: https://ec.europa.eu/eurostat/databrowser/product/page/ILC_MDDW02 (accessed on 15 March 2021)
Number of private households	Available online: https://ec.europa.eu/eurostat/databrowser/view/LFST_HHNHWHTC$DEFAULTVIEW/default/table (accessed on 15 March 2021)

References

1. European Commission. Energy Roadmap. 2050. 2012. Available online: https://ec.europa.eu/energy/sites/ener/files/documents/2012_energy_roadmap_2050_en_0.pdf (accessed on 3 December 2020).
2. European Commission. A sustainable Europe by 2030. 2019. Available online: https://ec.europa.eu/info/publications/reflection-paper-towards-sustainable-europe-2030_en (accessed on 9 December 2020).
3. International Renewable Energy Agency (IRENA). Global Energy Transformation. A Road Map to 2050. 2018. Available online: https://www.irena.org/-/media/Files/IRENA/Agency/Publication/2018/Apr/IRENA_Report_GET_2018.pdf (accessed on 7 December 2020).
4. Pach-Gurgul, A.; Ulbrych, M. Progress of the V4 Countries towards the EU's Energy and Climate Targets in the Context of Energy Security Improvement. *Entrep. Bus. Econ. Rev.* **2019**, *7*, 175–197. [CrossRef]
5. Meyer, D.F.; Masehla, T.M.; Kot, S. The relationship between economic growth and economic development: A regional assessment in South Africa. *J. Adv. Res. Law Econ.* **2017**, *8*, 1377–1385.
6. Kurowska-Pysz, J.; Szczepańska-Woszczyna, K.; Štverková, H.; Kašík, J. The catalysts of cross-border cooperation development in euroregions. *Pol. J. Manag. Stud.* **2018**, *18*, 180–193. [CrossRef]
7. European Council. The 2030 Climate and Energy Framework. 2014. Available online: https://www.consilium.europa.eu/ro/policies/climate-change/2030-climate-and-energy-framework/ (accessed on 8 December 2020).
8. Androniceanu, A. Hospital management based on the relationship between doctors and patients. *Adm. Manag. Public* **2017**, *29*, 41–53.
9. Grondys, K.; Androniceanu, A.; Dacko-Pikiewicz, Z. Energy Management in the Operation of Enterprises in the Light of the Applicable Provisions of the Energy Efficiency Directive (2012/27/EU). *Energies* **2020**, *13*, 4338. [CrossRef]

10. European Environmental Agency. Energy Efficiency. 2018. Available online: https://www.eea.europa.eu/airs/2018/resource-efficiency-and-low-carbon-economy/energy-efficiency (accessed on 14 December 2020).
11. Statista.Leading Countries in Primary Energy Consumption Worldwide in 2019. 2019. Available online: https://www.statista.com/statistics/263455/primary-energy-consumption-of-selected-countries/ (accessed on 15 December 2020).
12. Networked and Integrated Urban Technologies in Sustainable Smart Energy Systems. *Geopolit. Hist. Int. Relations* **2020**, *12*, 45. [CrossRef]
13. Androniceanu, A.; Tvaronavičienė, M. Developing a holistic system for social assistance services based on effective and sustainable partnerships. *Adm. Manag. Public* **2019**, *1*, 103–118. [CrossRef]
14. Tvaronavičienė, M.; Prakapienė, D.; Garškaitė-Milvydienė, K.; Prakapas, R.; Nawrot, Ł. Energy Efficiency in the Long-Run in the Selected European Countries. *Econ. Sociol.* **2018**, *11*, 245–254. [CrossRef] [PubMed]
15. Brookes, L. Energy efficiency fallacies revisited. *Energy Policy* **2000**, *28*, 355–366. [CrossRef]
16. Worrell, E.; A Laitner, J.; Ruth, M.; Finman, H. Productivity benefits of industrial energy efficiency measures. *Energy* **2003**, *28*, 1081–1098. [CrossRef]
17. Rosenberg, N. How Far Can the World Get on Energy Efficiency Alone. In *Energy Efficient Technologies: Past, Present and Future Perspectives*; Oak Ridge National Laboratory: Oak Ridge, TN, USA, 1989.
18. Igaliyeva, L.; Niyazbekova, S.; Serikova, M.; Kenzhegaliyeva, Z.; Mussirov, G.; Zueva, A.; Tyurina, Y.G.; Maisigova, L.A. Towards environmental security via energy efficiency: A case study. *Entrep. Sustain. Issues* **2020**, *7*, 3488–3499. [CrossRef]
19. Androniceanu, A. Social Responsibility, an Essential Strategic Option for a Sustainable Development in the Field of Bio-Economy. *Amfiteatru Econ.* **2019**, *21*, 503–519. [CrossRef]
20. Gajdzik, B.; Sroka, W. Resource Intensity vs. Investment in Production Installations—The Case of the Steel Industry in Poland. *Energies* **2021**, *14*, 443. [CrossRef]
21. Shpak, N.; Kuzmin, O.; Melnyk, O.; Ruda, M.; Sroka, W. Implementation of a Circular Economy in Ukraine: The Context of European Integration. *Resources* **2020**, *9*, 96. [CrossRef]
22. The World Bank. Electric Power Consumption (kWh Per Capita). 2014. Available online: https://data.worldbank.org/indicator/EG.USE.ELEC.KH.PC (accessed on 15 December 2020).
23. IEA. *Global Energy Review 2020*; IEA: Paris, France, 2020. Available online: https://www.iea.org/reports/global-energy-review-2020 (accessed on 7 January 2021).
24. International Energy Agency. Energy Efficiency. 2020. Available online: https://www.iea.org/topics/energy-efficiency (accessed on 6 April 2021).
25. Mura, L.; Buleca, J. Evaluation of Financing Possibilities of Small and Medium Industrial Enterprises. *Procedia Econ. Financ.* **2012**, *3*, 217–222. [CrossRef]
26. European Commission. Low Carbon-Economy. 2014. Available online: https://ec.europa.eu/regional_policy/en/policy/themes/low-carbon-economy/ (accessed on 6 April 2021).
27. Toman, M.T.; Jemelkova, B. Energy and Economic Development: An Assessment of the State of Knowledge. *Energy J.* **2003**, *24*, 93–112. [CrossRef]
28. Jackson, T. *Prosperity without Growth?—The Transition to a Sustainable Economy*; Sustainable Development Commission: London, UK, 2009.
29. Bélaïd, F.; Youssef, M. Environmental degradation, renewable and non-renewable electricity consumption, and economic growth: Assessing the evidence from Algeria. *Energy Policy* **2017**, *102*, 277–287. [CrossRef]
30. Cleveland, C.J.; Kaufmann, R.K.; I Stern, D. Aggregation and the role of energy in the economy. *Ecol. Econ.* **2000**, *32*, 301–317. [CrossRef]
31. Chlebisz, A.; Mierzejewski, M. Determinants of GDP growth in Scandinavian countries with special reference to scientific progress. *Int. Entrep. Rev.* **2020**, *6*, 21–35. [CrossRef]
32. Mura, L.; Daňová, M.; Vavrek, R.; Dúbravská, M. Economic freedom—classification of its level and impact on the economic security. *AD ALTA J. Interdiscip. Res.* **2017**, *7*, 154–157.
33. Srovnalíková, P.; Semionovaitė, E.; Baranskaitė, E.; Labanauskaitė, D. Evaluation of the Impact of Sharing Economy on Hotel Business. *J. Tour. Serv.* **2020**, *11*, 150–169. [CrossRef]
34. Caballero-Morales, S.-O.; Guridi, J.D.J.C.; Alvarez-Tamayo, R.I.; Cuautle-Gutiérrez, L. Education 4.0 to Support Entrepeneurship, Social Development and Education in Emerging Economies. *Int. J. Entrep. Knowl.* **2020**, *8*, 89–100. [CrossRef]
35. Pawęta, B. Impact of the Global Financial Crisis on the Business Cycle in the Visegrad Group. *Entrep. Bus. Econ. Rev.* **2018**, *6*, 43–58. [CrossRef]
36. Androniceanu, A. Major structural changes in the EU policies due to the problems and risks caused by COVID-19. *Adm. Manag. Public* **2020**, *34*, 137–149. [CrossRef]
37. Mazzoni, F. Circular economy and eco-innovation in Italian industrial clusters. Best practices from Prato textile cluster. *Insights Reg. Dev.* **2020**, *2*, 661–676. [CrossRef]
38. Androniceanu, A.-M.; Kinnunen, J.; Georgescu, I.; Androniceanu, A. A Multidimensional Approach to Competitiveness, Innovation and Well-Being in the EU Using Canonical Correlation Analysis. *J. Compet.* **2020**, *12*, 5–21. [CrossRef]
39. Lampropoulos, G.; Siakas, K.; Anastasiadis, T. Internet of Things in the Context of Industry 4.0: An Overview. *Int. J. Entrep. Knowl.* **2019**, *7*, 4–19. [CrossRef]

40. Čepel, M. Social and Cultural Factors and Their Impact on the Quality of Business Environment in the SME Segment. *Int. J. Entrep. Knowl.* **2019**, *7*, 65–73. [CrossRef]
41. Virglerova, Z.; Conte, F.; Amoah, J.; Massaro, M.R. The Perception of Legal Risk and Its Impact on the Business of SMES. *Int. J. Entrep. Knowl.* **2020**, *8*, 1–13. [CrossRef]
42. Dvorský, J.; Petráková, Z.; Khan, K.A.; Formánek, I.; Mikoláš, Z. Selected aspects of strategic management in the service sector. *J. Tour. Serv.* **2020**, *11*, 109–123. [CrossRef]
43. Belas, J.; Amoah, J.; Petráková, Z.; Kliuchnikava, Y.; Bilan, Y. Selected Factors of SMEs Management in the Service Sector. *J. Tour. Serv.* **2020**, *11*, 129–146. [CrossRef]
44. Gavurova, B.; Ivankova, V.; Rigelsky, M.; Přívarová, M. Relations Between Tourism Spending and Global Competitiveness—an Empirical Study in Developed OECD Countries. *J. Tour. Serv.* **2020**, *11*, 38–54. [CrossRef]
45. Haseeb, M.; Hussain, H.I.; Kot, S.; Androniceanu, A.; Jermsittiparsert, K. Role of Social and Technological Challenges in Achieving a Sustainable Competitive Advantage and Sustainable Business Performance. *Sustainability* **2019**, *11*, 3811. [CrossRef]
46. Kinsey, M.C. Global Energy Perspective-2019. 2019. Available online: https://www.mckinsey.com/~{}/media/McKinsey/Industries/Oil%20and%20Gas/Our%20Insights/Global%20Energy%20Perspective%202019/McKinsey-Energy-Insights-Global-Energy-Perspective-2019_Reference-Case-Summary.ashx (accessed on 6 April 2021).
47. United Nations. Transforming Our World: The 2030 Agenda for Sustainable Development. 2015. Available online: https://sdgs.un.org/2030agenda (accessed on 9 December 2020).
48. European Environment Agency. Air Pollution: How It Affects Our Health. 2020. Available online: https://www.eea.europa.eu/themes/air/health-impacts-of-air-pollution (accessed on 3 February 2021).
49. WHO Health Organization. Ambient (Outdoor) Air Pollution. Available online: https://www.who.int/news-room/fact-sheets/detail/ambient-(outdoor)-air-quality-and-health (accessed on 3 March 2020).
50. The European Environment Agency. Urban Air Quality. 2019. Available online: https://www.eea.europa.eu/themes/air/urban-air-quality (accessed on 9 December 2020).
51. International Energy Agency. 2030 Climate and Energy Framework. 2017. Available online: https://www.iea.org/policies/1494-2030-climate-and-energy-framework (accessed on 8 December 2020).
52. European Commission. A Clean Planet for All. A European Strategic Long-Term Vision for A Prosperous, Modern, Competitive and Climate Neutral Economy. 2018. Available online: https://eur-lex.europa.eu/legal-content/EN/TXT/?uri=CELEX:52018DC0773 (accessed on 3 December 2020).
53. Hnatyshyn, M. Decomposition analysis of the impact of economic growth on ammonia and nitrogen oxides emissions in the European Union. *J. Int. Stud.* **2018**, *11*, 201–209. [CrossRef] [PubMed]
54. Erst and Young. If Green Energy is the Future, How Can Technology Lead the Way? 2020. Available online: https://assets.ey.com/content/dam/ey-sites/ey-com/en_gl/topics/power-and-utilities/ey-recai-56-edition.pdf (accessed on 19 February 2021).
55. Sensor-based Big Data Applications and Computationally Networked Urbanism in Smart Energy Management Systems. *Geopolit. Hist. Int. Relat.* **2020**, *12*, 52. [CrossRef]
56. Abdollahi, H.; Ebrahimi, S.B. A new hybrid model for forecasting Brent crude oil price. *Energy* **2020**, *200*, 117520. [CrossRef]
57. Chehabeddine, M.; Tvaronavičienė, M. Securing regional development. *Insights Reg. Dev.* **2020**, *2*, 430–442. [CrossRef]
58. Androniceanu, A. The three-dimensional approach of Total Quality Management, an essential strategic option for business excellence. *Amfiteatru Econ.* **2019**, *19*, 61–78.
59. Howarth, R.; Norgaard, R.B.; Hall, C.A.S.; Cleveland, C.J.; Kaufmann, R. Energy and Resource Quality: The Ecology of the Economic Process. *Land Econ.* **1988**, *64*, 311. [CrossRef]
60. Connected and Autonomous Vehicle Mobility: Socially Disruptive Technologies, Networked Transport Systems, and Big Data Algorithmic Analytics. *Contemp. Read. Law Soc. Justice* **2020**, *12*, 61. [CrossRef]
61. Berndt, E.R. Energy use, technical progress and productivity growth: A survey of economic issues. *J. Prod. Anal.* **1990**, *2*, 67–83. [CrossRef]
62. Kummel, R.; Lindenberger, D.; Eichhorn, W. The productive power of energy and economic evolution. *Indian J. Appl. Econ.* **2000**, *8*, 231–262.
63. Sanstad, A.H.; Roy, J.; Sathaye, J.A. Estimating energy-augmenting technological change in developing country industries. *Energy Econ.* **2006**, *28*, 720–729. [CrossRef]
64. Richmond, A.K.; Kaufmann, R.K. Is there a turning point in the relationship between income and energy use and/or carbon emissions? *Ecol. Econ.* **2006**, *56*, 176–189. [CrossRef]
65. Androniceanu, A.-M.; Georgescu, I.; Tvaronavičienė, M.; Androniceanu, A. Canonical Correlation Analysis and a New Composite Index on Digitalization and Labor Force in the Context of the Industrial Revolution 4.0. *Sustainability* **2020**, *12*, 6812. [CrossRef]
66. Androniceanu, A.; Popescu, C.R. An inclusive model for an effective development of the renewable energies public sector. *Adm. Manag. Public* **2017**, *28*, 81–96.
67. Moutinho, V.; Varum, C.; Madaleno, M. How economic growth affects emissions? An investigation of the environmental Kuznets curve in Portuguese and Spanish economic activity sectors. *Energy Policy* **2017**, *106*, 326–344. [CrossRef]
68. Haseeb, M.; Kot, S.; Hussain, H.I.; Kamarudin, F. The natural resources curse-economic growth hypotheses: Quantile–on–Quantile evidence from top Asian economies. *J. Clean. Prod.* **2021**, *279*, 123596. [CrossRef]

69. Meyer, D.F. Predictors of Good Governance and Effective Government Management: The Case of Poland. *Pol. J. Manag. Stud.* **2018**, *18*, 206–217. [CrossRef]
70. Vlasov, A.I.; Shakhnov, V.A.; Filin, S.S.; Krivoshein, A.I. Sustainable energy systems in the digital economy: Concept of smart machines. *Entrep. Sustain. Issues* **2019**, *6*, 1975–1986. [CrossRef]
71. Hall, C.A.S.; Powers, R.; Schoenberg, W. Peak Oil, EROI, Investments and the Economy in an Uncertain Future. In *Renewable Energy Systems: Environmental and Energetic Issues*; Pimentel, D., Ed.; Elsevier: London, UK, 2008; pp. 113–136.
72. Barker, T.; Dagoumas, A.; Rubin, J. The macroeconomic rebound effect and the world economy. *Energy Effic.* **2009**, *2*, 411–427. [CrossRef]
73. Shapiro, S.S.; Wilk, M.B. An Analysis of Variance Test for Normality (Complete Samples). *Biometrika* **1965**, *52*, 591. [CrossRef]
74. Pearson, K. On lines and planes of closest fit to systems of points in space. *Lond. Edinb. Dublin Philos. Mag. J. Sci.* **1901**, *2*, 559–572. [CrossRef]
75. Androniceanu, A.; Kinnunen, J.; Georgescu, I. E-Government clusters in the EU based on the Gaussian Mixture Models. *Adm. Manag. Public* **2020**, *1*, 6–20. [CrossRef]
76. Škare, M.; Radolović, S.; Tvaronavičienė, M. Managing the institutional environment and impact on the competitiveness of transitional economies. *Entrep. Sustain. Issues* **2021**, *8*, 504–526. [CrossRef]
77. Dobrowolski, Z. After COVID-19. Reorientation of crisis management in crisis. *Entrep. Sustain. Issues* **2020**, *8*, 799–810. [CrossRef]
78. Kosach, I.; Duka, A.; Starchenko, G.; Myhaylovska, O.; Zhavoronok, A. Socio-economic viability of public management in the context of European integration processes. *Adm. Manag. Public* **2020**, *1*, 139–152. [CrossRef]
79. Haller, A. From classical and neoclassical economic growth to degrowth in Europe. Challenges for public administration. *Adm. Manag. Public* **2020**, *34*, 150–170. [CrossRef]

energies

MDPI

Article

Improving the Development Technology of an Oil and Gas Company Using the Minimax Optimality Criterion

Alex Borodin [1,*], Manuela Tvaronavičienė [2,3], Irina Vygodchikova [4], Andrey Kulikov [5], Marina Skuratova [1] and Natalia Shchegolevatykh [6]

1 Department of Finance, Plekhanov Russian University of Economics, 117997 Moscow, Russia; Skuratova.MA@rea.ru
2 Vilnius Gediminas Technical University (Vilnius Tech), LT-10223 Vilnius, Lithuania; manuela.tvaronaviciene@vilniustech.lt
3 Institute of Humanities and Social Sciences, Daugavpils University, Parades Str. 1-421, LV-5401 Daugavpils, Latvia
4 Department of Differential Equations & Mathematic Economics, National Research Saratov State University Named after N. G. Chernyshevsky, 410012 Saratov, Russia; irinavigod@yandex.ru
5 Department of Organization of Medical Provision and Pharmacoeconomics, I.M. Sechenov First Moscow State Medical University (Sechenov University), 119991 Moscow, Russia; 7677041@mail.ru
6 Department of Finance, SUEK JSC, 115054 Moscow, Russia; shchegolevatykh@sibgenco.ru
* Correspondence: aib-2004@yandex.ru

Abstract: The article deals with the problem of adaptation of the Russian oil and gas company (Novatek, Russia) to the rapidly changing external environment, the avalanche of data from competitors, and the need to filter important information for business development and the prosperity of the industry as a whole. The approach is based on the system of integrated software monitoring of key business processes at the enterprise developed by the authors—from the formation of the idea of a new product to its implementation to paying customers. The scientific novelty lies in the use of an optimization model that allows for minimizing the maximum losses of the investor at all levels of decision-making, from the distribution of capital between companies, to the optimization of internal reserves to increase the competitiveness of the company. The toolkit is a minimax model that allows you to redistribute the shares of investor influence at the portfolio level, and then within the business processes of each company selected by investors, in order to achieve the optimal solution in accordance with the selected estimated indicators. Application of the well-known portfolio investment models of Markowitz, Tobin, Sharp, etc. is not possible due to the lack of necessary data on the basis of which the probabilistic parameters involved in the model are estimated. Even if we get them, it is necessary to take into account the level of correlation influence of the technological process in the composition of each subsystem, which is unacceptable for the data used, as it leads to a strong increase in errors. Using minimax and a systematic approach allows you to minimize such errors by choosing a balanced concentration of distributed assets for both the investor and the buyer. To this end, a three-way analysis of the company's development was carried out and a technology for comprehensive improvement of the company's activities was developed in the following areas: the company's rating in the industry, financial condition, and interaction with counterparties using merchandising technologies. Tools for optimal image zoning at the Novatek site using the minimax approximation criterion have been developed. The technology provides a procedure for creating a comfortable mode of image perception based on high-tech visualization of merchandising, zoning of the screen area, and a mathematical approach that allows you to develop a calculation algorithm.

Keywords: oil and gas industry; investments; rating; competitiveness; financial analysis; risk; profitability; merchandising technologies

Citation: Borodin, A.; Tvaronavičienė, M.; Vygodcyikova, I.; Kulikov, A.; Skuratova, M.; Shchegolevatykh, N. Improving the Development Technology of an Oil and Gas Company Using the Minimax Optimality Criterion. *Energies* **2021**, *14*, 3177. https://doi.org/10.3390/en14113177

Academic Editors: Beata Zofia Filipiak and Yuriy Bilan

Received: 7 April 2021
Accepted: 26 May 2021
Published: 28 May 2021

1. Introduction

The distribution of investments in the oil and gas sector is particularly relevant. This is due to the huge capital involved in this industry. A competent resource-saving strategy can be achieved only by applying a comprehensive analysis of internal and external business development prospects. The least researched and most popular problems lie in the application of optimization models in visual marketing solutions based on online technologies (attracting customers for each company of interest), as well as portfolio structuring. This problem is an important and urgent task of investment analysis [1–3]. However, for the oil and gas sector, the initial indicators are not enough, and a serious modification is needed based on important criteria of fundamental analytics. Turning to the history of the issue of investment analysis and modeling, it should be noted that the first model for optimizing the capital distributed among the investor's assets was proposed by [4]. Currently, many approaches based on the solution of the Markowitz problem and its development allow us to achieve an optimal share structure of investments distributed among several investment entities. However, despite the high accuracy of forecasting by statistical indicators, in practice, models of this kind are little used, since their application requires a wide array of constantly updated statistical information about the dynamics of profitability, which is necessary for the construction (evaluation of real coefficients) of the covariance matrix [5,6]. In addition, the work is based on the specifics of key models for analyzing the external environment of an industrial enterprise, such as strategic matrix models or the business models of A. Osterwalder and I. Pinier [7].

With modern high-tech business projects and high-speed decision-making systems, the use of traditional investment portfolio optimization models is associated with difficulties associated with obtaining quantitative risk estimates used in the model [8]. In particular, it is not possible to form the covariance matrix of returns on capital invested in business projects required for the Markowitz model, since the development of each high-tech project in an enterprise requires some time and a set of resources mobilized at this enterprise. Therefore, such a process is almost impossible to perform in real time.

In view of the above, the current direction of research is the development of a procedure for the shared distribution of investments between high-tech projects, based on the revision of the set of quantitative indicators of the model and the use of new mathematical methods and models. Such a construction is advisable to perform in a hierarchical mode, and the minimax model makes it possible to expand the range of indicators used and consider the requirements of the investor for various criteria of financial analysis.

The difficulties faced by industrial entities include significant changes in the business environment and the influence of its factors on the results of their activities. Today, the business environment of oil and gas companies is subject to a high degree of volatility (risk, volatility). With the advent of new technologies of the digital economy, the company's website has become the most important area of activity. Activity, therefore, the optimization of merchandising technology ("optimal display of goods on the showcase") has moved into the sphere of optimizing the company's website in terms of improving the technological profile of virtual ideas about the specifics of the company's work [9].

One of the objectives of the article is to develop economic and mathematical tools to justify the modernization of an oil and gas company from a technological point of view.

Research hypothesis: A comprehensive toolkit for improving market position, increasing profitability, and making Novatek more attractive to investment: the main mechanism in a short period is the visual presentation of their services based on high-tech digital technologies, in the average (from one month to two years) period, the revision of the capital structure in favor of increasing equity capital, and reducing the borrowed capital to the maximum allowable limits (without reducing the total amount of borrowed and equity capital) has a stronger effect.

In a long (more than two years) period, one should be guided by the integral rating of Novatek among other industries.

In this context, it is obvious that Russian industrial entities must engage in thorough research, both practical and theoretical.

The issue of timely response to changes is especially acute. At the same time, it is necessary to pay attention to the fact that the concept of adaptation and the system that implements it in the practice is new and little studied, since their experience in the market is not long. In this context, it is obvious that the issues of adaptation of domestic industrial enterprises to changes in the business environment require in-depth research both in theoretical and practical terms.

In real-world practice, the adaptation of industrial entities to changes in the external environment involves the introduction of certain changes in the internal systems and processes of business entities in order to ensure that their functioning corresponds to the state of the business environment and, ultimately, to achieve efficiency in general. This is realized by accelerating the release of new, high-quality, competitive products, replacing physically and morally obsolete fixed assets, taking into account the latest achievements of science and technology, and introducing energy and resource-saving technologies into the production process [10–12].

A mature, well-build change management process, tailored for the organization, is required if an industrial entity expects to be successful in modernizing their internal processes and procedures.

The concept of adaptation is quite broad and dialectically related to many socio-economic spheres, as a result of which it has a wide meaning and is the subject of independent research.

The view of adaptation as an adaptation is the most common, characterizes its very essence and can be used in any field of science. However, according to the author, such a definition and understanding is most acceptable when it comes to general issues, without going into the essence of the problem.

We can also note a more structured, complex approach to the interpretation of the concept of "adaptation", which is understood as the mechanism of economic and social adjustments that allow the system to maintain (change) the direction and pace of development regardless of the influence of external factors [13–15]. It seems that this approach should be used in the case when the enterprise is able to develop the very mechanism of economic and social adjustments and in the event that the influence of external factors is not significant and does not threaten the existence of the enterprise itself.

Organizational and managerial contour structurally consists of objects, subjects, and tasks. At the same time, we consider it necessary to pay attention to the fact that this circuit should be designed and created on the basis of the theory and practice of project management. All possible organizational structures, according to this theory, are based on the division of labor—vertical (functionally-administrative) and horizontal (project-oriented) [16–19].

The basic principles of adaptation include: complexity, consistency, economic, and environmental feasibility [20–22].

The establishment of resource constraints precedes the implementation of the adaptation system and its corresponding mechanisms in the economic activity of the enterprise. The last elements of the algorithm for the formation of the adaptation system are control over its implementation and assessment of the effectiveness of the measures taken.

The identification of the key elements of the adaptation system and the formalization of the algorithm for its formation make it possible to implement a holistic and systematic approach to its construction and structuring. J. Shaughnessy notes that "in order to understand how the system performs its function, it is necessary to know how all its elements are interconnected with each other and how it is connected with a system that forms its external environment" [23].

2. Literature Foundations

2.1. Management Decision-Making System

The scheme of strategic development of a large industrial enterprise considered as an open stationary socio-economic system, and the action of the mechanisms of its adaptation to the disturbances of the business environment is shown in Figure 1.

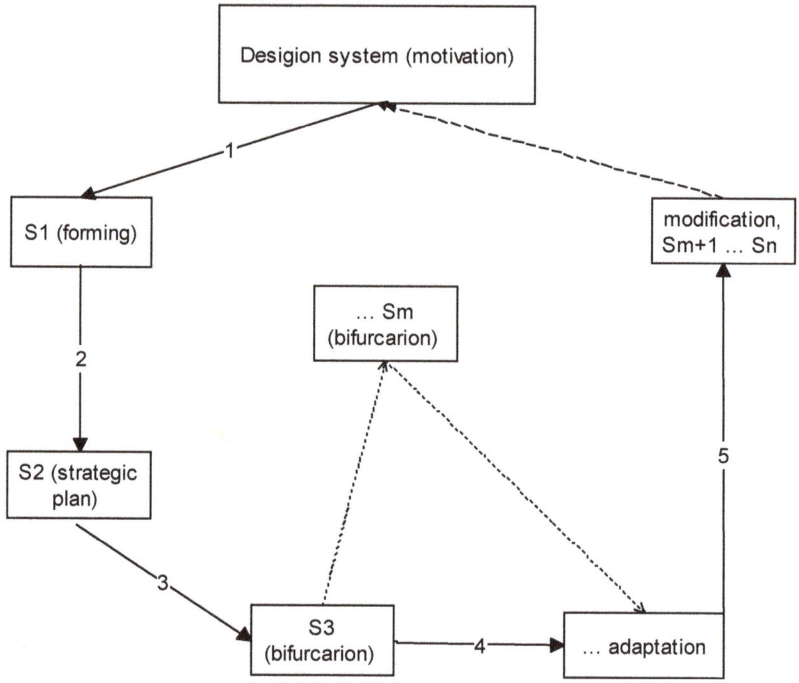

Figure 1. Scheme of strategic development and adaptation of the enterprise to changes in the business environment (Bifurcation-restructuring). Where: S1—the initial state of the industrial holding as an open stationary system; S2, S3, Sn, Sm—the state of the holding after the application of adaptation mechanisms and tools; MDMS—is a management decision making system.

Let us analyze the circuit shown in Figure 1 in more detail. A large enterprise, which is an open stationary system, is in its initial state (S1) under the influence of external and internal environmental factors. Based on the available indicators (characteristics of the S1 system), the management makes a strategic decision on further development and applies the appropriate adaptation mechanisms.

As a result of the implementation of transformational evolutionary adaptation mechanisms, the S1 system adapts to changes in the external environment, while maintaining its fundamental (significant) distinctive signs (integrity), or is transformed into a new state S2, S3, Sn, or Sm, which provides for the nature of changes, in which an open stationary system loses its initial fundamental characteristic features and passes into a new qualitative state, while maintaining a hereditary connection with the previous state.

An important point in the process of using transformational adaptation mechanisms is a clear understanding of the level of socio-economic development of an enterprise for the correct determination of the bifurcation point, which is critical. A. Kusakina and E. Skiperskaya argue that at this point the system becomes unstable with respect to fluctuations and uncertainty of further development arises: the system remains unstable and it collapses or acquires a new, higher level of order. [24,25]

Let us now consider in more detail the specific practical aspects of adapting domestic industrial holdings to the dynamic changes of the modern business environment, which is under the strong influence and pressure of the accelerating digital transformation.

Thus, the hypothesis of the study, subject to mathematical justification, will be confirmed. Namely, MDMS (management decision making system) as a comprehensive toolkit for solving the problem of reducing the competitiveness, profitability, and investment attractiveness of Novatek consists in the competent distribution of leverage from the initial state (S1) in the short and long period:

The main mechanism in a short period is the visual presentation of their services based on high-tech digital technologies (Sm, Sm−1, . . . , S3), in the average (from one month to two years) period, the revision of the capital structure in favor of increasing equity capital, and reducing the borrowed capital to the maximum allowable limits (without reducing the total amount of borrowed and equity capital) has a stronger effect (S3, S2, S1) [26,27].

2.2. Expert Opinion on the Rating of an Industry Enterprise

Tables 1 and 2 show the average results of expert assessments of the influence of the factors of the modern business environment on the activities of PJSC NOVATEK.

Table 1. Analysis of the influence of macroenvironmental factors on the activities of PJSC "NOVATEK".

Factors	Factor Weight	Impact Assessment (Maximum—10 Points)	Weighted Score
Economic	0.2	3	0.7
Scientific and technical	0.2	6	0.8
Legal	0.1	6	0.6
Social	0.1	6	0.6
Political	0.1	5	0.5
Technological	0.1	7	0.7
Cultural	0.05	5	0.3
Demographic	0.05	5	0.3
Natural	0.05	5	0.4
International	0.05	5	0.5
Together	1.00	-	5.4

Table 2. Analysis of the influence of microenvironment factors on the activities of PJSC "NOVATEK".

Factors	Factor Weight	Impact Assessment (Maximum—10 Points)	Weighted Score
Supplies	0.3	7	0.9
Clients	0.3	6	1.6
Competitors	0.2	5	1.0
Mediators	0.1	5	0.4
Contact audiences	0.1	5	0.3
Together	1	-	4.2

That is, the general influence of macroenvironmental factors has a negative impact (5.4 points out of 10 possible) on the functioning and development prospects of an industrial holding. This can be explained by economic turbulence, the complexity of the legal framework, and the unstable international situation. At the same time, special attention should be paid to the fact that the digital divide between the industrial sectors of Russia and the leading countries of the world, the obsolescence of equipment and technology, as well as the obvious delay in the introduction of breakthrough innovations have a significant negative impact.

The cumulative influence of microenvironment factors is also negative, although the situation is somewhat better compared to the influence of macroenvironmental factors (4.2 points out of 10 possible).

At the next stage of the study, we will analyze the compliance of the goals set as priorities by PJSC NOVATEK with the current business conditions. The assessment will be carried out using the SMART Russia toolkit. It seems expedient to analyze the initial system of the holding's goals by the main levels of strategic management: corporate strategy, business strategy, functional strategy, and operational strategy (Table 3).

Table 3. Results of SMART-assessment of the goals of PJSC NOVATEK, maximum assessment—10 points.

Direction of Assessment	S (Specific) Uniqueness	M (Measurable) Possibility of Measurement	A (Appropriate) Relevance	P (Realistic) Realistic	T (Time Bound) Limitation on Time
Corporate strategy	8	4	8	7	10
Business strategy	9	6	7	6	10
Functional strategy	7	8	9	9	8
Operational strategy	9	8	6	8	10

The results of evaluating the system of goals of PJSC NOVATEK by the main levels of strategic management indicate a fairly high level of its balance. High marks according to the criterion "Time constraints" reflect the understanding of the holding's management of the need to determine the planning horizons in the process of formulating goals. Low scores on the criterion "measurable" indicate the qualitative nature of the goals, which cannot always be assessed using a system of quantitative indicators.

However, since assessments of the impact of exogenous changes showed negative results, it is advisable for the holding to adapt its functioning program and development strategy in accordance with the dynamics of the business environment and digital transformations.

The realities of today clearly indicate that the intensification of business activities of economic entities, especially in foreign markets, presupposes an increase in the degree of their openness and interaction with partners. The complication, versatility, and complexity of such interaction illustrates of the expediency and effectiveness of the network approach in the process of PJSC NOVATEK's adaptation to the dynamics of the external environment.

2.3. Marketing Strategy of the Enterprise: Network Technologies

In fact, the network approach is a response to the challenges, including changes in the operating environment of enterprises and their expectations in an uncertain economic space saturated with threats and information while undergoing significant transformations in the digital age. In fact, the network approach provides for the use of one of the types of horizontal integration, which will maximize the attraction of available resources; mastering innovations; and build up competencies, competitive advantages, innovation, production, information, and intellectual potential within the framework of a single multipolar information and communication space [28,29].

The logic of the network approach includes: decentralization, synergy, community, free access, maximization of innovation, multidimensional space, lack of discontinuity, ratio of technologies, and expansion of the space of innovative opportunities [30].

Among the most significant advantages of the network approach are: highlight increased organizational ability, more efficient use of resources and digital opportunities, expanding the horizon of tools and mechanisms that allow solving complex business problems, and improving the quality of services and the complexity of services and services for consumers.

According to the author, the use of the network approach of PJSC NOVATEK, which should replace linear interactions, will contribute to the formation of an additional source of value creation for the holding, since it will:

- Initiate design and production with a priority on optimal functionality;
- Intensify technological and product innovations by obtaining new previously disjointed data about the external environment and a better understanding of production processes, supplier capabilities and consumer needs.

The high potential of the network approach is confirmed by the successful functioning of a number of well-known global corporations, for example, Apple, Intel, Samsung, Exxon Mobil, Procter and Gamble, Tata Motors Ltd., Shougang Steel, etc. At the same time, according to experts, the greatest opportunities when using the network approach open up to reduce the time of the research and production cycle (from the development of new products to its introduction to the market) (by 20–50%), reduce equipment downtime (by 30–50%), reducing the cost of maintenance of machines and mechanisms (by 10–40%), and the cost of maintaining inventories (by 20–50%), thus increasing labor productivity by automating its mental component (by 45–55%) [31].

It is obvious that the introduction of a network approach into the activities of PJSC NOVATEK as a tool for adapting its systems to the business environment should be reflected at the strategic and tactical levels.

The strategic nature provides for the solution of development tasks in dynamic market conditions over a long time period, and the issue of restructuring structures and systems in accordance with the current market conditions in the short and medium term. Of course, the tactical aspect of adaptation should be addressed regarding the strategic directions of the holding's development in the market environment.

According to the authors, when drawing up a strategic adaptation plan for PJSC NOVATEK based on a network approach and considering digital transformation, it is necessary to solve several fundamental technological issues:

- Expedite business and production processes via automation and virtualization;
- Ensure the transparency, maturity and reliability of organizational processes at all levels. This will require a distributed architecture for analytics and industrial internet of things solutions;
- Improve production speeds, product quality and reduce operational expenses.

At the same time, information security is a key element of any digital transformation, especially industrial holdings that have unique production processes, flow charts, and other intellectual property at their disposal. If information protection issues are not given due attention, the business entity may lose all potential benefits.

Organizations should prepare for a significant shift in ideology, business process, and communication strategy when introducing new technology to internal and external stakeholders.

During the first stage of digital transformation of PJSC NOVATEK's organizational mechanisms, the following actions will be required:

- Introduce unified programs for use in various departments;
- Unify operational processes in the corporate center and at production units;
- Carry out a pilot implementation of certain digital control technologies.

The next stage of the holding's adaptation to digital transformations should be the development of organizational processes in all structural divisions. A fast, unified, secure, and comfortable shared information exchange system is essential to drive innovation, especially for agile decision-making, flexibility and operational efficiency. At this stage, these key areas of adaptation of organizational mechanisms are considered:

- Technological automation of business processes;
- Automation of business communications;
- Deep development of IT infrastructure;
- Introduction of a budgeting system and unified electronic document management;
- Providing offices and divisions of the holding with modern IT tools and mechanisms, new communication channels, networks, data centers, and servers;
- Transition to cloud server space.

PJSC NOVATEK is actively engaged with global business partners. Protective measures must be taken to ensure the holding remains competitive in the domestic market in addition to the global market.

Table 4 presents the key aspects of the impact of integration processes on the activities of PJSC NOVATEK.

Table 4. Positive and negative aspects of the impact of integration on the international markets of PJSC NOVATEK.

Positive Consequences	Negative Consequences
Removal of trade restrictions	Increase in production costs due to rising raw material prices and environmental costs
Increased presence in the global market due to the strengthening of integration processes. Expansion of presence in the EU market through further integration of capacities	Increased competition in the domestic market due to increased imports of foreign products
Investment attractiveness, receipt of cheaper financial resources	The problem of product certification in accordance with European and world standards
Modernization of production facilities	Disagreements in quality systems

The opportunities and threats indicated in Table 4 make it possible to highlight the following areas of PJSC NOVATEK's adaptation:

1. Harmonization of the quality system to world standards;
2. Implementation of modern environmental protection systems;
3. Product certification;
4. Regulation of production in accordance with the demand in the world market;
5. Creation of a positive image;
6. Development of an improved model of product entry to world markets.

Taking into account the fact that today the Fourth Industrial Revolution, which was named Industry 4.0, is taking place in the world, it seems appropriate to pay attention to the mechanisms and directions of adaptation of industrial holdings in Russia, in particular PJSC NOVATEK, to radical changes in industry.

Industry 4.0 "indicates the transition to the 4, 5 and 6 technological order, within which both high technologies and computerized approaches to the optimal solution of the problem are developing". Due to the gradual decline in the cost of these technologies, they are becoming available, that is, they are increasingly used by industry and business, which ultimately affects existing business models or even creates new business formats.

In this context, it seems that the process of increasing the adaptability of energy holdings should include four key vectors.

1. Replacement and modernization of equipment. To increase the flexibility of production, industrial holdings need to have universal equipment that will allow switching from the production of one product to another on existing production lines. However, it should be noted that the cost of such equipment is an order of magnitude higher than that of narrow-profile machines and mechanisms, and given the economy mode and limited budgetary funds, as well as investor investments, it is obvious that holding companies will have to cover these costs at their own expense. Therefore, it is necessary to carefully analyze the feasibility of purchasing universal equipment in each specific case, for example, such equipment will be advisable in the automotive industry, where the parameters of the goods need to be changed quite often to meet the requirements of consumers. Domestic machine-building holdings have the nature of mass production; therefore, additional costs for such equipment will be justified, and in such industries as the production of missiles, it is more expedient to have highly specialized equipment, since products are manufactured using technologies that have not changed for many years.

2. Implementation of new solutions for the organization of world-class production. It should be noted that a key characteristic of Industry 4.0 is Horizontal and Vertical System Integration. Modern information and communication tools and technologies make it possible to combine and integrate all its subdivisions into a single information space within

one enterprise, within one supply chain of all its participants, etc. However, in domestic practice, even subdivisions of one enterprise do not always work in a single information system, not to mention the horizontal structures of individual economic units. Industry 4.0 technologies make it possible to combine various structures, enterprises, and participants in the value chain and the division of labor into a single information circuit. In addition, the existing rigid hierarchical system of access to information at enterprises or between enterprises in the conditions of Industry 4.0 will be destroyed: objects connected to the industrial Internet of things will be able to receive any information they need directly, regardless of their information level and position in the production hierarchy.

3. Adaptation of personnel to working conditions in conditions of uncertainty, continuous technological updates, increasing the amount of information, and new breakthrough technologies generated by Industry 4.0. When introducing innovative work systems, it will be necessary not only to train workers on how to use new equipment and use network forms of interaction, but also to pay attention to the psychological aspect of work. The average age of those employed in the industrial sector in Russia is 55 years, which means that changes in work will create a strong psychological stress and cause a resistant reaction, which will significantly reduce the efficiency of activities, because the main driving factor in the implementation of any plan is the employees. Special attention should be paid to the psychological aspects of personnel adaptation; therefore, the management of domestic industrial holdings should consider the costs of personnel adaptation and, above all, workers at production sites.

4. Introduction of research and development. To date, government support in this area is insufficient. Therefore, industrial holdings are forced to carry out research mainly at their own expense. In the current situation, it will be expedient only for science-intensive holdings, for example, manufacturers of spacecraft or heavy engineering products. In many industrial sectors, advanced research and development is extremely necessary, since the vast majority of the technologies used are obsolete and do not meet modern market demands, but conducting fundamental research is too expensive an undertaking, which also has a very indefinite payback period. One of the ways out in this situation may be the unification of scientific departments of several holdings to conduct research in which all participants are interested and can lead to a significant economic effect. In this way, research costs will be shared among several stakeholders.

A study of high-tech companies—global giants showed that 97% of competitive failures are associated with insufficient attention to market changes or an inability to respond to vital information [32,33].

Considering that the amount of information circulating today in the market, in the environment and internal divisions of modern business entities has increased significantly, in order for domestic industrial holdings to quickly respond to changes and adapt to them in time, they must be properly informed. In this case, it becomes necessary to constantly update the information flow by using special, innovative methods of collecting and processing information.

It seems that in the process of adapting to a dynamic information environment, domestic industrial holdings should use the following methods of collecting and analyzing information.

Scanning the environment is one of the areas of analytical and predictive work, which is rapidly developing and is widely used in strategic management systems [34,35]. The purpose of the scan is to collect, evaluate and predict the significance to the subject management of important changes. Scanning is usually carried out in the following areas:

- Economic scanning—research of changes in macro- and microeconomic indicators;
- Industry indicators and competition in it;
- The state of financial markets;
- Technical scanning—the study of scientific and technical progress;
- Fundamental technical and technological innovations;

- Political scanning—assessment of the political situation at the level of the country, region;
- Risk assessment of financial investments, etc.

1. When scanning, a variety of tools are used: expert methods, scenarios, comparison, modeling, morphological, and functional-cost analysis.

2. Monitoring the environment is the constant tracking of current and new information. For industrial holdings that use strategic management technology, it is advisable to create a special tracking system. Within the framework of this system, it is possible to conduct not only regular, but also special observations on critical factors of influence.

3. Forecasting is the formation of an idea of the future state of environmental factors. This tool is an integral component of the strategic planning process.

3. Results

3.1. Digital Transformation: Site Improvement Solution (S_3)

When improving information technologies of high-performance computing in the field of electronic merchandising, it is necessary to adapt the graphic images on the web page of the official website of the enterprise to the avalanche-like growth of the flow of information about the company. In this section, a digital technology is created for transforming the visual perception of images on the NOVATEK website in the optimal mathematical screen zoning mode to attract interested parties (partners, investors, and buyers) to the site.

Mathematical method. Consider a minimax model that allows you to optimally place in different images on the web page of a company website [36–38].

Risk indicators (positive quantitative indicators, monthly click-through rates of images) will be denoted by $V_1, \ldots, V_n$.

Let, for definiteness, $V_1 < \ldots < V_n$. We calculate the proportions of the screen area $\theta = (\theta_1, \ldots, \theta_n)$ using the task:

$$\max_{i=\overline{1,n}} V_i\theta_i \rightarrow \min_{\theta\in D}, \text{ where } D = \{\theta = (\theta_1, \ldots, \theta_n) \in R^n : \sum_{i=1}^{n} \theta_i = 1\} \tag{1}$$

In case (1), it is required to find the portions of the image space on the web page of the NOVATEK website, allocated for placing illustrations in order to attract the attention of contractors.

The solution to case (1) is determined by the formulas:

$$\theta_i = 1/\left(V_i \sum_{k=1}^{n} V_k^{-1}\right), \; i = \overline{1,n}. \tag{2}$$

Computational experiment. The study examined n = 4 positions from the NOVATEK website, of the following functional groups:

1. MARCOM AWARDS;
2. PETROLEUM ECONOMIST 2019 AWARD;
3. GASTECH 2019 AWARD;
4. "CHANGE MANAGEMENT. VISIONERS" AWARD.

Risk indicators associated with the lack of interest of clients in the image of each structural element—ranks, from the best "1" to the most risky "4", therefore V1 = 1, . . . , V4 = 4. The calculation of the zoning parameters was carried out according to the formulas (2).

The screen zoning scheme based on the data from Table 5 is shown in Figure 2.

Table 5. Data for analysis.

5.	Ranking Images by Click-Through Rate, V	1	2	3	4
	Placement shares by area	48%	24%	16%	12%

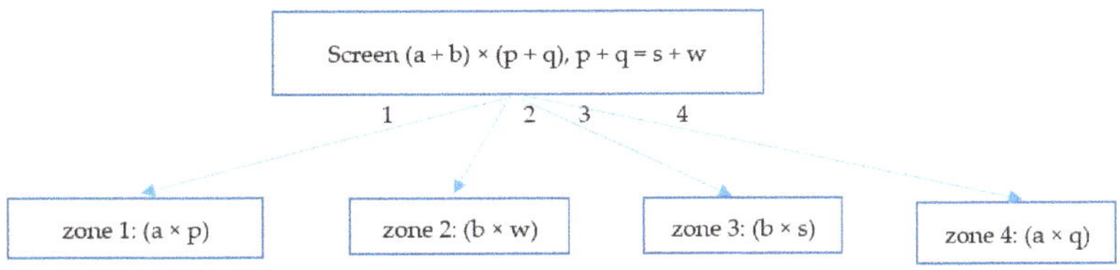

Figure 2. Zoning scheme.

Screen zoning layout based on data from Tables 6 and 7 is shown in Figure 3.

Table 6. Screen zoning.

Proportion	Formula	Meaning
a:b	$\theta_1/(\theta_1+\theta_3):\theta_3/(\theta_1+\theta_3)$	75:25
p:q	$\theta_1/(\theta_1+\theta_4):\theta_4/(\theta_1+\theta_4)$	80:20
s:w	$\theta_3/(\theta_2+\theta_3):\theta_2/(\theta_2+\theta_3)$	40:60

Table 7. Financial analysis of companies (important ratios and ratings).

Company	Year	Net Profit, Thousand Rubles (Index D)	Equity Capital, Thousand Rubles (Index C)	Revenue, Thousand Rubles (Index B)	Assets, Thousand Rubles (Index A)	Debt Capital, Thousand Rubles	Risk (Debt/Equity, Financial Leverage), Shares (%)	Profitability (Net Profit to Equity), Shares (%)	Place, 1-Leader (Including Revenue)
Gazprom	2019	651,124,114	11,334,679,889	4,758,711,459	15,916,355,497	4,581,675,608	40.4%	5.7%	1
Lukoil	2019	405,759,769	957,169,199	444,471,354	2,209,166,567	1,251,997,368	130.8%	42.4%	3
Surgutneftegaz	2019	105,478,643	4,303,834,579	1,555,622,592	4,553,686,428	249,851,849	5.8%	2.5%	2
Novatek	2019	237,224,510	718,557,978	528,544,385	899,787,613	181,229,635	25.2%	33.0%	4

3.2. Company Rating among Oil and Gas Companies (S2)

To build an integral rating, three indicators are used (assets, revenue, equity), after which the coefficients are analyzed.

Financial leverage ratio equal to the ratio of borrowed capital (the amount of short-term and long-term borrowings) to the company's own funds (in the liabilities "capital and reserves", the ratio is considered normal to be less than 1), return on equity ratio equal to the ratio of net profit to equity, and the normal value is more than 0.2.

The results are presented in Table 7.

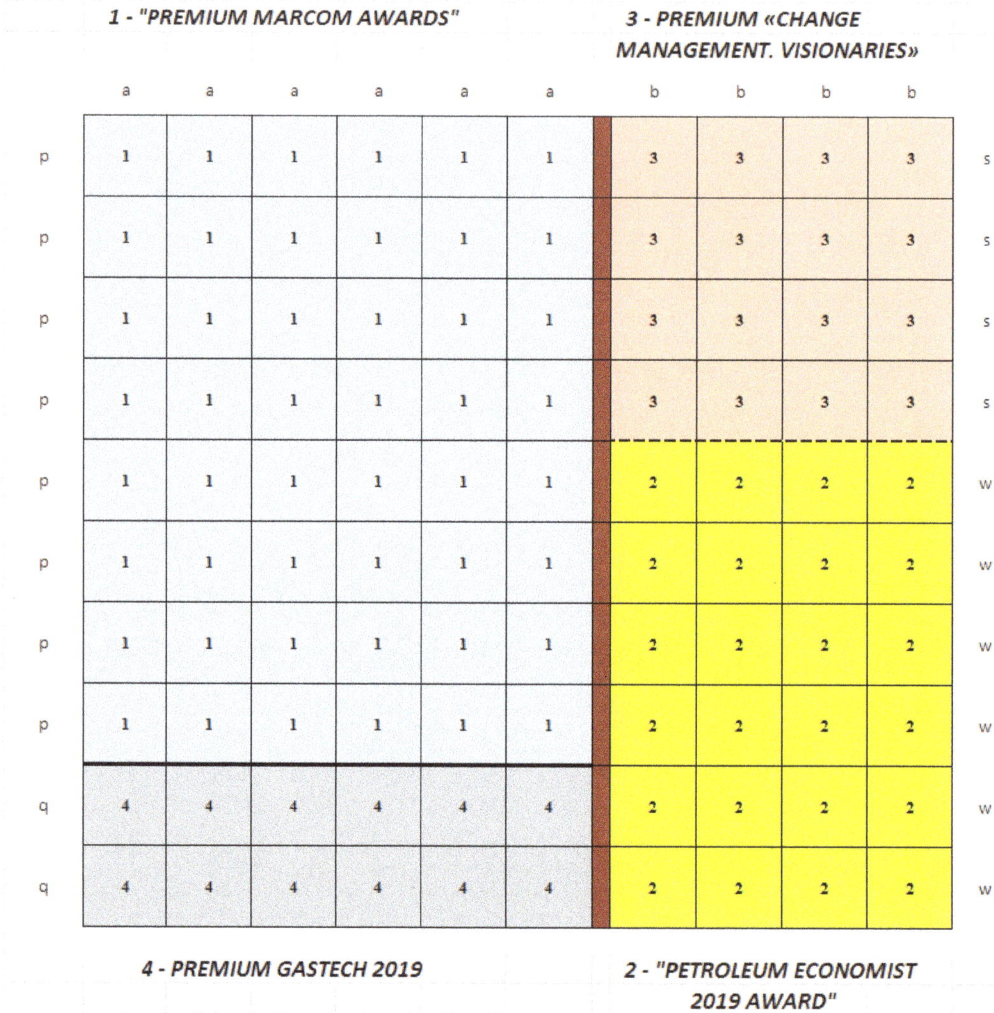

Figure 3. Screen zoning layout for images 1, 2, 3, and 4 (descending rank).

3.3. Financial Portfolio by Minimax Approximation Criterion (S1)

In addition to Novatek, companies are considered the best link in the oil and gas industry according to the rating for the investor.

Step 1. Now risk V is financial leverage, we use the solution to the case (1) to form an investment portfolio. Portfolio shares θ—the distribution of the investor's capital between the companies.

The case is considered (1) and the solution of the case is:

$$\theta_i = 1/\left(V_i \sum_{k=1}^{n} V_k^{-1}\right),\ i = \overline{1,n}.$$

Step 2. The rate of return characterizes the return on equity. It is believed that the results of the activities of the enterprise make a profit, therefore $\eta_1 > 0, \ldots, \eta_n > 0$.

Let the portfolio profitability be the minimum acceptable for the investor $\eta_p = \frac{1}{n}\sum_{i=1}^{n} \eta_i$.

The solution is corrected:

$$\theta_i^* = \eta_i\theta_i / \left(\sum_{k=1}^{n} \eta_k\theta_k \right), \; i = \overline{1,n}. \tag{3}$$

If $\sum_{k=1}^{n} \eta_k\theta_k^* \geq \eta_p = \frac{1}{n}\sum_{i=1}^{n} \eta_i$, then (3) is the desired solution. If not, the asset with the minimum return is excluded from the portfolio, and problem (1)–(3) is solved again. The algorithm is finite, the extreme situation is the presence of the only asset in the portfolio, the return of which is maximum.

The results are presented in Table 8 and Figure 4.

Table 8. Financial analysis of companies (shares in the portfolio).

Company	Year	Share of Shares in the Portfolio (by Risk Level)	Correction (Profitability)
Gazprom	2019	10.1%	6.5%
Lukoil	2019	3.1%	14.7%
Surgutneftegaz	2019	70.5%	19.2%
Novatek	2019	16.2%	59.6%
TOTAL (average)	2019	100.0%	100.0%

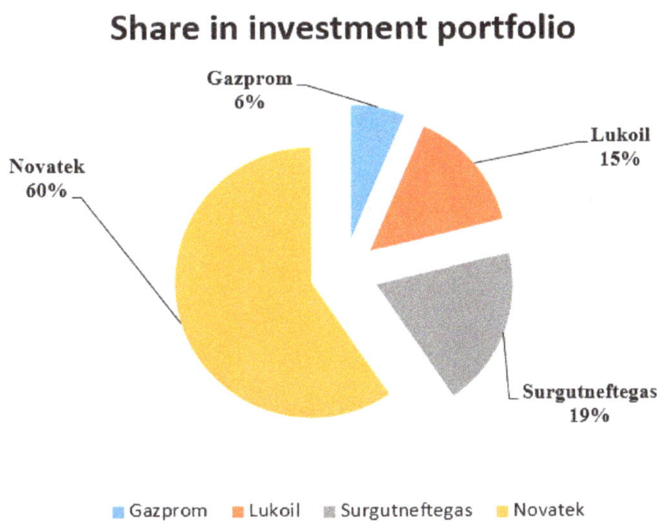

Figure 4. Shares of companies' shares in the investor's portfolio.

4. Discussion

The understanding and analysis of adaptation processes make it possible to establish that the formation of the theory of adaptation is objectively conditioned by changes in environmental factors that act as determinants and driving forces for the development of relations between business entities and their environment. During the study of terminological apparatus revealed the ambiguity of the interpretation of the concept of "adaptation" and on this basis, clarified its content. Under the adaptation of an industrial holding to the modern business environment, the author proposes to understand its ability to determine goals and means of achieving them with the anticipation of future changes using situational (adaptive) analysis to enhance competitive advantages and ensure its sustainable functioning, which implies an analysis of economic, industrial, scientific and technical, financial, and social spheres of activity (taking into account industry characteristics). The structural composition of the adaptation system has also been formalized, which includes: organizational and managerial contour; means, methods, and tools for the implementation of adaptation measures, as well as the functions and key principles of their implementation; management models that allow coordinating the processes of interaction with an unstable environment; and a mechanism for providing adaptation procedures.

Clarification of these elements of the adaptation system made it possible to implement an integrated approach to the selection and substantiation of methods for adapting Russian industrial holdings to changes in the business environment, as well as to highlight the criteria for choosing a management model during the adaptation period.

There is no doubt that the proper organization of adaptation processes at domestic industrial holdings requires the formation of an appropriate system and the use of an adaptive algorithm for its implementation. In the course of the study, the authors have developed a holistic algorithm for the formation of a system for adapting an industrial holding to changes in the business environment, which allows for the implementation of an integrated approach to the implementation and implementation of adaptation measures, which, in turn, will contribute to improving the efficiency of management, ensuring the sustainable functioning and development of the holding in long term.

In considering the practical aspects of adapting domestic industrial holdings to changes in the business environment, special attention was paid to the need to take into account the dynamics and rapid trends in the development of the digital economy, which are occurring at an exponential rate, radically changing the essence of business and transforming all sectors of the national economy.

In this context, the authors proposed to use a network approach, which involves the use of one of the types of horizontal integration, which contributes to the maximum attraction of available resources, the development of innovations, the building of competencies, competitive advantages, innovation, production, information, and intellectual potential within the framework of a single multipolar information and communication space.

Additionally, special emphasis is placed on the need to introduce modern tools for collecting, analyzing, and processing information, which make it possible to form an information flow in real time and track the dynamics of disturbances in the internal and external environment of an industrial holding.

In addition, the expediency and necessity of including in the adaptation system mechanisms and tools for the formation of digital alternatives for manufactured products and services, transformation of the organizational architecture of industrial holdings using modern information, and communication technologies has been substantiated.

5. Conclusions

A marketing system and a quantitative analysis of the strategic change plan within Novatek were developed. The dynamic factors of the external environment were taken into account, a digital merchandising technology for the company was developed, and recommendations on the capital structure were given.

The article deals with a fundamentally new problem: the problem of adapting a large Russian oil and gas company (Novatek) to a rapidly changing external environment, subject to an avalanche of data from competitors, and the need to filter important information for business development as well as the prosperity of the industry as a whole. The authors have developed a system for adapting data flows using mathematical and graphical tools, an optimization model, and criteria for selecting risk values to improve the company's development and develop a technological solution for complex improvement of activities in the following areas: the company's rating in the industry, financial condition, and work with counterparties using merchandising technologies. Computational experiments confirming the theoretical basis of the instrumental solution based on software-oriented technologies are performed.

Recent events have clearly demonstrated the scientific and practical significance of research devoted to accelerating the economic development of any country, which is based on the use of rapidly developing digital platforms [39]. The development of these platforms poses new challenges not only for business processes, but also for their analysis systems, including financial ones. Traditional methods of financial analysis no longer correspond to the changing business environment and require the creation of computer technologies based on mathematical models.

The computer technology proposed by the authors, based on mathematical models, will allow the financial analyst to predict the financial condition of the company based on the analysis of historical information accumulated in the database.

The authors developed a marketing system and conducted a quantitative analysis of the strategic change plan of Novatek. Dynamic factors of the external environment were taken into account, digital merchandising technology was developed for the company, and recommendations on the capital structure were given. The article systematizes the features of the influence of the main formal and informal socio-economic institutions on the effectiveness of the formation and implementation of the financial strategy of modern companies. In addition, the specifics of the key models for analyzing the external environment of the energy sector, such as strategic matrix models, business models, the balanced scorecard model, etc., are highlighted. The limitations of the use of such methods and models of business environment analysis in the practice of financial and economic activities of modern companies are systematized. The implementation of the proposed methodological approaches to improving the efficiency of financial strategy development in the practical activities of the energy holding will allow to identify the state of its external environment, taking into account the level of influence of turbulence factors; to determine the type of stability; and to make an informed choice of financial strategy. For further research, it is proposed to consider algorithms and models that make it possible to form a system for adapting an energy holding to dynamic environmental disturbances, and to improve its internal systems and procedures in accordance with market requirements and consumer requests.

The implementation of the results of the assessment, analysis and forecasting of the factors of the external and internal environment of the holding company using improved tools based on the theory of fuzzy sets will increase the validity of management decisions regarding the choice of financial strategy and will contribute to improving the efficiency of the holding company as a whole.

Author Contributions: Conceptualization, M.T.; methodology, A.B.; software, I.V.; formal analysis, M.S.; writing—original draft preparation, A.K.; writing—review and editing, N.S. All authors have read and agreed to the published version of the manuscript.

Funding: This research received no external funding.

Institutional Review Board Statement: Not applicable.

Informed Consent Statement: Not applicable.

Data Availability Statement: Not applicable.

Conflicts of Interest: The authors declare no conflict of interest.

References

1. Vygodchikova, I. Model of optimal investment portfolio based at minimax approachhikova. *Vestn. Sseu* **2018**, *3*, 170–174.
2. Pham, H. Smooth solutions to optimal investment models with stochastic volatilities and portfolio constraints. *Appl. Math. Optim.* **2002**, *46*, 55–78. [CrossRef]
3. Streimikiene, D. Ranking of Baltic States on progress towards the main energy security goals of European energy union strategy. *J. Int. Stud.* **2020**, *13*, 24–37. [CrossRef]
4. Markovitz, H. Portfolio selection. *J. Financ.* **1952**, *7*, 77–91.
5. Ledoit, O.; Wolf, M. Improved estimation of the covariance matrix of stock returns with an application to portfolio selection. *J. Empir. Financ.* **2003**, *10*, 603–621. [CrossRef]

6. Zhang, W.; Wang, J. Nonlinear stochastic exclusion financial dynamics modeling and complexity behaviors. *Nonlinear Dyn.* **2017**, *88*, 921–935. [CrossRef]
7. Osterwalder, A.; Pinye, I. Building business models. In *Table Book of the Strategist and Innovator. M*; Albina Publisher: Durham, NC, SUA, 2019; p. 112.
8. Khalikov, M.; Maksimov, D. Multi-step optimization of the portfolio of financial assets of a non-institutional investor. *Entrep. Guide* **2017**, *33*, 211–219.
9. Jaakkola, H. Is integration and enterprise migration—Adaptation in global context. *Int. J. Intell. Def. Support Syst.* **2010**, *3*, 78–89. [CrossRef]
10. Shindina, T.; Streimikis, J.; Sukhareva, Y.; Nawrot, Ł. Social and Economic Properties of the Energy Markets. *Econ. Sociol.* **2018**, *11*, 334–344. [CrossRef]
11. Borodin, A.; Mityushina, I.; Streltsova, E.; Kulikov, A.; Yakovenko, I.; Namitulina, A. Mathematical Modeling for Financial Analysis of an Enterprise: Motivating of Not Open Innovation. *J. Open Innov. Technol. Mark. Complex.* **2021**, *7*, 79. [CrossRef]
12. Zollo, M.; Bettinazzi, E.L.; Neumann, K.; Snoeren, P. Toward a Comprehensive Model of Organizational Evolution: Dynamic Capabilities for Innovation and Adaptation of the Enterprise Model. *Glob. Strat. J.* **2016**, *6*, 225–244. [CrossRef]
13. Stavytskyy, A.; Kharlamova, G.; Giedraitis, V.; Šumskis, V. Estimating the interrelation between energy security and macroeconomic factors in European countries. *J. Int. Stud.* **2018**, *11*, 217–238. [CrossRef]
14. Repina, E.; Shiryaeva, L.; Fedorova, E. The Study of Dependence Structure between Small Business Development and Microfinance Security of Russian Regions. *Ekon. I Mat. Metod.* **2019**, *55*, 41–57. [CrossRef]
15. Murray, S. *The Adaptation Industry: The Cultural ECONOMY of Contemporary Literary Adaptation*; Routledge: London, UK, 2012; p. 272.
16. Uchenov, A. Research of Russian and foreign experience of state support of inductrial enterprises for the subseqient adaptation of best practices. *Financ. Econ.* **2019**, *3*, 413–418.
17. Bessarabova, Y.; Evtushenko-Mulukaeva, N. Architectural adaptation of an industrial enterprise to a new function. *Caspain Eng. Construst. Bull.* **2019**, *2*, 28–33.
18. Ruzakova, O.; Kurdyumov, A. Restructuring and adaptation of enterprises in the conditions of information economy development. In Proceedings of the 4th International Scientific Conference on Social Sciences and Arts, Albena, Bulgaria, 24–30 August 2017; p. 795.
19. Keruenbaeva, D. Formation of the concept of enterprise adaptation to environmental conditions. *Econ. Soc.* **2020**, *1*, 482–484.
20. Vo, H.D. Sustainable agriculture & energy in the U.S.: A link between ethanol production and the acreage for corn. *Econ. Sociol.* **2020**, *13*, 259–268. [CrossRef]
21. Shan, S.; Luo, Y.; Zhou, Y.; Wei, Y. Big data analysis adaptation and enterprises' competitive advantages: The perspective of dynamic capability and resource-based theories. *Technol. Anal. Strateg. Manag.* **2019**, *31*, 406–420. [CrossRef]
22. Hnatyshyn, M. Decomposition analysis of the impact of economic growth on ammonia and nitrogen oxides emissions in the European Union. *J. Int. Stud.* **2018**, *11*, 201–209. [CrossRef] [PubMed]
23. O´ Shaughnessy, J. *Principles of Organization of Company Management*, 1st ed.; Routledge: London, UK, 2013.
24. Kusakina, O. Methodological Aooroaches to Assessing the Socio-Economic Efficiency of Integration. Available online: http://www.sworld.com.ua/konfer29/958.pdf (accessed on 15 May 2021).
25. Borodin, A.; Mityushina, I. Evaluating the effectiveness of companies using the DEA method. *Nauk. Visnyk Natsionalnoho Hirnychoho Universytetu* **2020**, *6*, 193–197. [CrossRef]
26. Mokhova, N.; Zinecker, M.; Meluzín, T. Internal factors influencing the cost of equity capital. *Entrep. Sustain. Issues* **2018**, *5*, 827–845. [CrossRef]
27. Ogiugo, H.U.; Adesuyi, I.; Ogbeide, S.O. Empirical test of capital asset pricing model on securities return of listed firms in Nigeria. *Insights Into Reg. Dev.* **2020**, 825–836. [CrossRef]
28. Štreimikienė, D. Externalities of power generation in Visegrad countries and their integration through support of renewables. *Econ. Sociol.* **2021**, *14*, 89–102. [CrossRef] [PubMed]
29. Goltyapin, V. Network approach. *Agro-Bus.* **2019**, *3*, 120–125.
30. Qian, Y. Modelling industry interdependency dynamics in a network context. *Stud. Econ. Financ.* **2020**, *37*, 50–70. [CrossRef]
31. Demircan Keskin, F. A two-stage fuzzy approach for Industry 4.0 project portfolio selection within criteria and project interdependencies context. *J. Multi-Criteria Decis. Anal.* **2020**, *27*, 65–83. [CrossRef]
32. Joseph, A.J. *Health, Safety, and Environmental Data Analysis: A Business Approach*; CRC Press: Boca Raton, FL, USA, 2020; p. 386.
33. Svazas, M.; Navickas, V.; Krajnakova, E.; Nakonieczny, J. Sustainable supply chain of the biomass cluster as a factor for preservation and enhancement of forests. *J. Int. Stud.* **2019**, *12*, 309–321. [CrossRef]
34. Plenkina, V.; Andronova, I.; Deberdieva, E.; Lenkova, O.; Osinovskaya, I. Specifics of strategic managerial decisions-making in Russian oil companies. *Entrep. Sustain. Issues* **2018**, *5*, 858–874. [CrossRef]
35. Vygodchikova, I. Toolkit for making decisions on investing in large Russian companies using a hierarchical ranking procedure and minimax approach. *Appl. Inform.* **2019**, *14*, 123–137.
36. 36. Szczepankiewicz., E.I.; Mućko, P. CSR reporting practices of Polish energy and mining companies. *Sustainability* **2016**, *8*, 126. [CrossRef]

37. Hąbek, P.; Wolniak, R. Assessing the quality of corporate social responsibility reports: The case of reporting practices in selected European Union member states. *Qual. Quant.* **2020**, *50*, 399–420. [CrossRef] [PubMed]
38. Raufflet, E.; Cruz, L.B.; Bres, L. An assessment of corporate social responsibility practices in the mining and oil and gas industries. *J. Clean Prod.* **2014**, *84*, 256–270. [CrossRef]
39. Tvaronavičienė, M.; Prakapienė, D.; Garškaitė-Milvydienė, K.; Prakapas, R.; Nawrot, Ł. Energy Efficiency in the Long-Run in the Selected European Countries. *Econ. Sociol.* **2018**, *11*, 245–254. [CrossRef] [PubMed]

energies

Article

Household Electricity Generation as a Way of Energy Independence of States—Social Context of Energy Management

Shahin Bayramov [1], Iurii Prokazov [2], Sergey Kondrashev [3] and Jan Kowalik [4,*]

[1] Department of Economics, Mingachevir State University (MSU), Mingechevir AZ4500, Azerbaijan; rector@mdu.edu.az
[2] Renaissance Development Russia, 105264 Moscow, Russia; iurii.prokazov@gmail.com
[3] Department of Chemistry, I.M. Sechenov First Moscow State Medical University, 119991 Moscow, Russia; Sk9262718531@gmail.com
[4] Faculty of Management, Technical University of Czestochowa, 42-201 Czestochowa, Poland
* Correspondence: janekowalik@yahoo.pl

Abstract: The purpose of this study was to determine the degree of influence of alternative options for generating electricity by households on the level of energy independence of countries. The research methodology was based on the use of correlation–regression analysis, as well as adapted non-linear optimization by choosing one of three scenarios for electricity generation by households for 20 countries. Regression analysis showed the dependence of a country's energy security on households' energy independence. It is determined that an increase in households' energy production helps to reduce the level of energy dependence in developed countries. However, for developing countries, there is no such interrelation. The solution of the formulated problem of nonlinear optimization for the studied countries has demonstrated that the criterion of energy dependence is superior to the criterion of a country's energy security. In the long term, this study can be deepened in the direction of assessing the effectiveness of household investment in electricity generation projects. The proposed results can be used by responsible persons in the field of economy and energy in order to determine the position of various policies, and use strategic levers and indicators that ensure an effective response to energy security challenges in the regional and global markets.

Keywords: correlation; energy dependence; energy efficiency management; energy security; scenario; social development

Citation: Bayramov, S.; Prokazov, I.; Kondrashev, S.; Kowalik, J. Household Electricity Generation as a Way of Energy Independence of States—Social Context of Energy Management. *Energies* **2021**, *14*, 3407. https://doi.org/10.3390/en14123407

Academic Editors: Miguel-Angel Tarancon and Manuela Tvaronavičienė

Received: 16 March 2021
Accepted: 4 June 2021
Published: 9 June 2021

1. Introduction

Currently, the problem of resource and energy deficit is becoming increasingly urgent in the world. In today's realities with growing global energy problems, the issues of transition to alternative sources of energy supply are becoming increasingly common. Clean energy based on the latest technologies has long been identified as the basis for the future; therefore, the orientation towards oil, gas, and nuclear energy can lead to serious energy dependence on the largest suppliers of raw materials and today threatens the economic security of many countries [1]. Given the terrain complexity, low population concentrations, regulatory and organizational and administrative barriers, or high investment and operating costs, which may be unaffordable for utilities, grid promotion is not as profitable as alternatives, including increased autonomous energy production by households. The IEA estimates that for universal access to electricity by 2030, decentralized solutions are the least-cost option for the 60 percent of people who do not have access to electricity. Currently, decentralized electricity access solutions are scarce, but the pace of their development is accelerating every year. For example, the IEA estimates that 33 million people have access to electricity through off-grid renewables (excluding pico-solar, with 114 million users (according to IRENA)), with the accelerating rate of connection [2].

Energy is a key driver of economic growth, development, and well-being, and power systems need to maintain a constant balance between electricity demand and supply [3]. With the rapid development of distributed power generation technologies such as solar photovoltaic power, household power generation is gaining increasing attention due to decentralized energy autonomy and further economic benefits [4]. In today's world, an increasing number of countries, companies, and households are turning to alternative energy sources. Due to their geographical location, countries have the opportunity to freely use this direction, not to mention its environmental friendliness [5]. A Google Trends search for the key term "renewable energy" indicates that household electricity consumption is decreasing by 16.017 million kWh per unit increase in the search of the keyword "renewable". Google Trends search allows one to assess driving factors that are difficult to identify when analyzing with the use of various economic indicators [6]. However, it should be noted that special patterns of electricity consumption by a small part of electricity consumers, including households, cannot be ignored. It is essential for planning, operation, policy formulation, and decision making of the smart grid [7].

The development of the market for "green" energy production by private households makes a positive contribution to the efficiency of energy supply to states, increasing the environmental friendliness of the energy complex, ensuring the transition to the use of renewable energy sources, and reducing the use of fossil fuels [8]. At the same time, the pace of development of the private sector of renewable energy is not sufficient to make a significant contribution to the achievement of indicators of national plans and programs. The main reasons are doubts about the financial feasibility of such projects requiring state support, and insufficient incomes of the population of some countries, which do not allow accumulating funds for investment in renewable energy, along with the high cost of credit resources [9]. Alternative energy is an important component of the state's energy security, a component of sustainable development, and a means to improve the well-being of the population and the quality of life of people in general [10]. Its penetration into all spheres and industries should be stimulated not only at the national level, but also at the level of individual economic entities, in particular households, since their role in the growth of energy independence and energy efficiency can be very significant. Therefore, the study aimed to determine the role of households not only as potential energy producers, but also as a factor in increasing the level of energy security of countries, taking into account global trends.

2. Literature Review

In the modern scientific literature, there are studies of the effectiveness of power supply systems' technical design using renewable energy in remote settlements, taking into account their social needs. This design is capable of providing electricity not only to households, but also to medical centers, schools, churches, and cultural institutions. This indicates an additional important social effect on society [11]. Despite the global level of progress, there are regions in underdeveloped countries that do not have access to electricity or, if they do, suffer from frequent power outages. At the same time, it has been proven that the lack or low reliability of electricity supply affects the level of GDP [12]. To overcome this problem, government support is necessary, as it is in the interest of society as a whole. Ineffective policy in this direction leads to the fact that the rate of electrification remains low, despite significant investments in the nearby grid infrastructure. This pattern persists over time for both poor and relatively well-off households and businesses [13]. In developed countries, households' adoption of renewable energy is influenced by contextual and behavioral factors. For example, households that use renewable energy have lower energy consumption rates than other households. At the same time, they significantly reduce the load on the power grid. However, the key factor in the development of renewable energy production by households is their environmental motivation. Besides, the difference between households with photovoltaic installation before and after achieving energy system parity is determined, especially with regard to the effects of economic structure containment. A comprehensive support, including efficient technologies and environmentally motivated approaches to energy conservation, is required to reduce energy consumption [14]. Therefore, it is worth considering a different approach of modern researchers to the benefits of autonomous energy production by households as a factor of sustainable development of countries and regions. Regardless of whether electricity has become affordable and massively available for productive use, no country has moved from poverty to prosperity. Household electrification strategies must consider other development goals and opportunities to use electricity access to stimulate inclusive, climate-friendly, and sustainable economic activity [15,16].

There are also studies for assessing the level of sustainable development based on an integrated indicator. At the same time, they demonstrate that the exclusion of the renewables financing indicator as a component of the integrated indicator does not have a significant impact on the final result [17]. To ensure the sustainability of the new energy supply models, it is important, in particular, to ensure that part of the supply chain and its associated benefits are local in nature (building appropriate competencies among households). It is also important that domestic financial institutions reach the necessary level of understanding and capacity to lend for off-grid energy and other types of equipment to households. The role of off-grid mini-grids, which is currently limited, should be expected to increase, especially when access initiatives are aimed at providing electricity to manufacturing and commercial activities, as well as to households. For sustainable development and effective functioning of autonomous mini-grids, an enabling environment is necessary, which includes specific policies and regulations, adapted financing mechanisms, institutional conditions, a focus on capacity building, and adaptation of technology [18].

This study raises the scientific issue of energy security of a country, thus it is necessary to define the qualitative characteristics of modern approaches to its definition and the relationship with the energy security of households.

At the same time, scientists offer their own interpretations of the concept of energy security. Special attention should be paid to those interpretations that are notable for their conciseness and content, such as:

- ensuring uninterrupted access to energy resources at an affordable price [19]; confidence that energy will be available and in the quantity and quality required under given economic conditions [20];
- protection of a state from energy threats [21];
- protection of citizens and a country as a whole from threats of deficiency of all types of energy resources [22];
- reliable and uninterrupted supply of electricity and fuel to consumers [23];
- security of the national economy and population; preventing any threats to the reliable supply of fuel and energy resources [24].

In general, there are two key areas in the modern literature:

- energy security is the timely, complete, and uninterrupted supply of quality fuel and energy to material production, non-production sphere, population, and other consumers [25];
- prevention of harmful effects of transportation, transformation and consumption of fuel and energy resources on the environment in the context of modern market relations, and trends and indicators of the global energy market [26].

Based on the essential characteristics of energy security of a state, the authors propose to define energy security of households as meeting (regardless of circumstances) the growing needs for energy resources (acceptable in price, quality, and assortment), taking into account social values. Thus, the authors emphasize that for countries, depending on their level of development, the value prerogatives of households in the context of energy security may differ. For some, it is a way of ensuring conditions for survival, while for others, it is a way of sustainable development, solving environmental and social problems.

The participation of a state should be manifested in the initiation of the development of investment projects for the use of solar energy based on public–private partnerships [27,28]. At the state and regional levels, the popularization of the use of renewable energy sources among producers and users should be carried out. It is necessary to develop an economic mechanism (soft loans, interest-free loans, and other instruments) that will induce producers and users to replace traditional energy sources with renewable ones. An important financial instrument is the activation of banking programs for financing projects based on renewable energy sources [29].

In social terms, a state can manage the development of energy resources not only at the level of legislative institutions, but also financial ones in the implementation of energy projects with enterprises and receive income from their implementation [30]. Financial partnership in the development and implementation of renewable energy projects will bring income to a state budget, reduce greenhouse gas emissions into the atmosphere, improve the ecological state of the environment, create additional jobs, and contribute to the creation of the energy independence of a state [31].

The renewable energy sector creates different jobs in manufacturing, services, and construction and requires different qualifications and skills. Its development not only increases but also improves the quality of industrial jobs. At the same time, there is a trend towards a reduction in employment in the non-renewable energy sectors [32].

Energy security and energy independence of any country today are an important component of its sustainable development strategy. Political, economic, and infrastructural problems do not make it possible to provide the state with traditional energy sources sufficiently [33]. In such conditions, the role of renewable energy sources and alternative energy is especially growing. At the same time, the use of renewable energy sources helps to preserve traditional resources and the environment, and the introduction of a "green" tariff allows households to have an additional source of income, thus improving their quality of life [34]. Households greatly contribute to the development of alternative energy [35]. That is why it is very important to assess their role and opportunities in further increasing the level of energy independence and security of countries.

Among modern studies, not enough attention is paid to the analysis of independent renewable energy sources, which can become a start for understanding the very process of individual electrification of households. Most researchers mainly consider either the development of alternative energy in the country as a whole, or the peculiarities of its implementation by large enterprises. However, today households should also be taken into account, as they are the most flexible and prone to innovations, especially in the energy sector due to the rise in energy costs in recent years. This study aims to fill this gap by diagnosing the impact of household energy production on the energy efficiency of countries. This contributed to the formation of the research goal, which is to determine the degree of influence of alternative options for generating electricity by households on the level of energy independence of countries. Based on the formed research goal, the following key hypotheses can be identified:

- H1: the level of electricity generation by households affects country's energy security level;
- H2: the level of electricity production by households affects the level of country's energy dependence.

3. Materials and Methods

This study is driven by the need to identify those factors that confirm the impact of household electricity production on the energy dependence of countries. Household Energy Independence Index, Energy Security Index, and Country Energy Independence Index were used as effective factors for the analysis. The study was conducted on the basis of materials from 20 countries.

At the first stage of the study, countries were selected. The study analyzed the impact of households' energy generation on the energy independence of developed and developing countries; therefore, two groups of 10 countries each have been formed. The key criterion for the selection of these countries was belonging to the group of developed or developing countries, according to the United Nations methodology [36], including transition economies. The second criterion for the formation of the choice of countries was the leadership positions in their geographic region in terms of energy security, according to the rating of the Trilemma Energy Index [37,38]. The selection of countries for the study was based on a cross-criteria approach. Each of the selected countries is an important player in the energy market of its region, as well as a representative of developed or developing countries. This made it possible to substantiate the level of development of these countries and present different world regions in the study in the context of energy security. The availability of statistical data on the studied indicators was also taken into account. The selected developed countries were Belgium, Canada, Denmark, France, Hungary, Italy, Poland, Slovakia, the United Kingdom, the USA. The selected developing countries were Albania, Azerbaijan, China, Kenya, Mexico, Moldova, Namibia, Panama, Russia, and Serbia.

At the second stage, a regression analysis was carried out. Based on correlation analysis, to diagnose relationships, the calculation of pair correlation coefficients was carried out. As a result, coefficients were included that revealed the two-way relationships between the indicators under consideration. Besides, in the course of the study, an assessment of the impact of energy resources on the energy dependence of economies was carried out. To substantiate development priorities and formulate energy security strategies for each group of countries, the authors built models of interdependence of the Household Energy Independence Index, as the main monitoring indicator, and indicators of energy resources impact on the level of energy security and independence.

The authors propose the Households Energy Independence Index, which is calculated as the ratio of households' energy consumption volume in a country to the volume of households' renewable energy consumption based on IRENA data [39,40]. Household Energy Independence Index provides an opportunity to determine the level of households' autonomy based on the generation of energy from renewable sources.

The energy security index used in the study corresponds to the rating indicator constituting the Energy Trilemma Index of the World Energy Council [37]. For all the initial indicators, data for 2018 were available. Taking into account the fact that energy security is presented in the Energy Trilemma Index in the form of a rating, the study proposed to transform it into a coefficient using the formula:

$$ESI_i = \frac{1}{ESTI_i} \quad (1)$$

where ESI_i—Energy Security Index of i-th country; and $ESTI_i$—the place of i-th country in the Energy Security Index rating in the context of the Energy Trilemma Index.

The initial data for calculating the Household Energy Independence Index and Energy Security Index are given in Table 1.

Table 1. Initial data for determining the relationship between the level of energy security and household energy independence.

Country	Household Total Final Consumption	Household Final Renewable Energy Consumption	Household Energy Independence Index	Country's Rank in the Trilemma Energy Index (Energy Security)	Energy Security Index
Belgium	336,996	42,863	0.127	41	0.024
Canada	1,504,610	604,100	0.401	3	0.333
Denmark	184,345	115,721	0.628	1	1
France	1,543,673	500,277	0.324	26	0.038
Hungary	243,295	65,091	0.268	22	0.045
Italy	1,337,934	374,367	0.28	20	0.05
Poland	818,268	134,002	0.164	69	0.014
Slovakia	86,164	9183	0.107	29	0.034

Table 1. *Cont.*

Country	Household Total Final Consumption	Household Final Renewable Energy Consumption	Household Energy Independence Index	Country's Rank in the Trilemma Energy Index (Energy Security)	Energy Security Index
The United Kingdom	1,632,475	209,246	0.128	18	0.056
The USA	11,343,716	1,406,490	0.123988	7	0.143
Albania	21,227	16,784	0.790688	98	0.01
Azerbaijan	137,704	2441	0.017726	38	0.026
China	14,516,180	4,358,382	0.300243	43	0.023
Kenya	517,572	122,338	0.236369	34	0.029
Mexico	755,759	287,200	0.380015	65	0.015
Moldova	58,113	31,947	0.549741	115	0.009
Namibia	17,836	12,290	0.689065	93	0.011
Panama	23,907	15,890	0.664669	92	0.011
Russia	6,206,973	16,3371	0.026321	16	0.063
Serbia	118,780	45,591	0.383829	60	0.017

Source: compiled by the authors based on statistical data [37–39]; calculated by the authors.

The authors proposed an Index of a country's energy independence, which integrates the following key indicators: Energy intensity per unit of GDP, Net energy imports, Total energy supply (TES) per capita, Energy share from renewable energy sources, and Energy production to consumption ratio. The integral indicator is the average value of its normalized components:

$$k_i^{norm} = \begin{cases} \frac{k_i - k_i^{min}}{k_i^{max} - k_i^{min}}, & k_i \to max \\ \frac{k_i - k_i^{max}}{k_i^{min} - k_i^{max}}, & k_i \to min \end{cases} \quad (2)$$

where ki is the i-th indicator of integral index, $i = \{1, 2, 3, 4, 5, 6\}$;
k_i^{min} is the minimum value of the i-th indicator;
k_i^{max} is the maximum value of the i-th indicator;
k_i^{norm} is the normalized value of the i-th indicator [41].

The Energy Independence Index was calculated using the formula:

$$SII_i = \sqrt{EI_i^2 + NEI_i^2 + TES_i^2 + RESS_i^2 + REPS_i^2}, \quad (3)$$

where SII_i—Energy Independence Index of i-th country; EI_i—normalized value of Energy intensity per unit of GDP of i-th country; NEI_i—normalized value of Net energy imports of i-th country; TES_i—normalized value of Total energy supply per capita of i-th country; $RESS_i$—normalized value of Energy share from renewable energy sources of i-th country; and $REPS_i$—normalized value of Energy production to consumption ratio of i-th country.

The initial data for calculating the Energy Independence Index are shown in Table 2.

To assess the impact of energy generation by households, the Social Content Index (SCI) is proposed, which is formed on the basis of the social components of The Legatum Prosperity Index (Social Capital–SC, Living Conditions–LC, Health–H, and Natural Environment–NE). These indicators are able to demonstrate the degree of the social context of the development of the studied countries. The social content index is calculated using the formula:

$$SCI_i = \frac{1}{(SC_i + LC_i + H_i + NE_i)/4} \quad (4)$$

Using this indicator, the social context in this study was determined. The initial data are given in Table 3.

Table 2. Initial data for determining the level of energy independence of the studied countries.

Country	Energy Intensity per Unit of GDP, MJ/USD	Net Energy Imports, Mtoe	Total Energy Supply (TES) per Capita, Toe/Capita	Energy Share from Renewable Energy Sources, %	Energy Production to Consumption ratio	Energy Independence Index
Belgium	4.8	52.9	4.7	6.69	0.17	0.70
Canada	7.6	−227.6	8.0	27.57	1.56	1.36
Denmark	2.6	4.5	2.9	26.72	0.82	1.81
France	2.2	119.5	3.7	11.26	0.51	1.10
Hungary	4.2	15.5	2.7	3.40	0.35	0.71
Italy	3.0	121.9	2.5	16.33	0.22	1.00
Poland	4.2	47.2	2.8	6.18	0.55	0.73
Slovakia	4.4	10.9	3.2	7.66	0.33	0.71
The United Kingdom	2.8	66.5	2.6	13.04	0.66	0.88
The USA	5.1	80.7	6.8	8.47	0.95	1.09
Albania	2.9	0.5	0.8	37.20	0.79	1.13
Azerbaijan	3.8	−40.7	1.5	2.87	3.93	1.35
China	6.1	700.5	2.3	12.18	0.80	0.57
Kenya	3.0	6.0	0.5	71.80	0.26	1.40
Mexico	2.0	29.4	1.5	6.47	0.79	1.15
Moldova	7.3	3.3	1.2	26.10	0.00	0.64
Namibia	3.5	1.6	0.8	28.10	0.13	0.98
Panama	2.1	8.3	1.0	22.80	0.19	1.14
Russia	8.3	−701.3	5.3	5.72	1.92	1.50
Serbia	6.1	5.4	2.2	19.90	0.67	0.76

Source: compiled by the authors based on statistical data [40,42]; the authors' own calculations.

Table 3. Initial data for determining the Social Content Index of the studied countries.

Country	Social Capital Rank	Living Conditions Rank	Health Rank	Natural Environment Rank	Social Content Index
Belgium	45	18	24	47	0.032
Canada	10	16	25	15	0.056
Denmark	2	1	8	10	0.158
France	41	17	16	16	0.055
Hungary	90	37	52	30	0.023
Italy	56	24	17	48	0.032
Poland	111	31	40	62	0.022
Slovakia	76	33	43	12	0.034
The United Kingdom	14	8	23	24	0.052
The USA	16	29	59	25	0.029
Albania	120	90	69	73	0.014
Azerbaijan	125	70	66	148	0.010
China	34	66	21	147	0.013
Kenya	64	132	115	129	0.008
Mexico	118	81	37	78	0.016
Moldova	105	74	96	134	0.010
Namibia	60	114	126	77	0.010
Panama	70	78	45	36	0.023
Russia	101	57	103	44	0.014
Serbia	96	47	72	99	0.013

Source: compiled by the authors based on statistical data [43]; the authors' own calculations.

The third stage of the study is a modeling based on a scenario approach to diagnosing the relationship between electricity production by households and the energy dependence of countries with an implemented economic and mathematical method. This method makes it possible to achieve high accuracy in solving nonlinear optimization problems using the hierarchical analysis of Thomas L. Saaty [44]. To conduct the study by solving the problem of nonlinear optimization for the scenario of electricity production by households in the studied countries, the following steps were used:

1. Determination of the regression dependence of energy security (y1), independence (y2), social content (y3) from the Household Energy Independence Index (x) in the studied countries.
2. Paired linear regression models were used, which revealed the relationship between the studied factors: y1 = a1 + b1 × x, y2 = a2 + b2 × x. The variable x was replaced with the obtained values in the equations.
3. Formation of scenarios using t − a time period based on the data being simulated. Three scenarios were proposed. According to each of them, the Household Energy Independence Index is increased by 10%.
4. Determination of the weight of the optimality criteria: energy security and independence were ranked in accordance with their importance for each of the studied groups of countries, according to the Thomas L. Saaty scale.
5. Assessment based on generated scenarios. Moreover, the share of y_{ESI} and y_{EII} for each scenario is a weighted arithmetic mean. Thus, ESI_{sci}^{norm} and EII_{sci}^{norm} were determined.
6. Determination for each scenario of a weighted sum of energy independence indicators: $ESI_{sci}^{ws} = ESI_{sci}^{norm} * \text{weight}$, and the weighted sum of energy security indicators were determined similarly$-ESI_{sci}^{ws} = ESI_{sci}^{norm} * \text{weight}$.
7. Conducting a hierarchical synthesis $(HS_{sci} = ESI_{sci}^{norm} * EII_{sci}^{norm})$. The results obtained were compared, and the scenario with the maximum value was selected according to the hierarchical synthesis [44].

4. Results

Based on a comparison of the level of total energy consumption by households (HTFC) and their consumption of renewable energy (HFREC) (Figure 1), it can be argued that for most of the studied countries, there is a fairly high relationship between these indicators. Analysis of variance results are shown in Table 4.

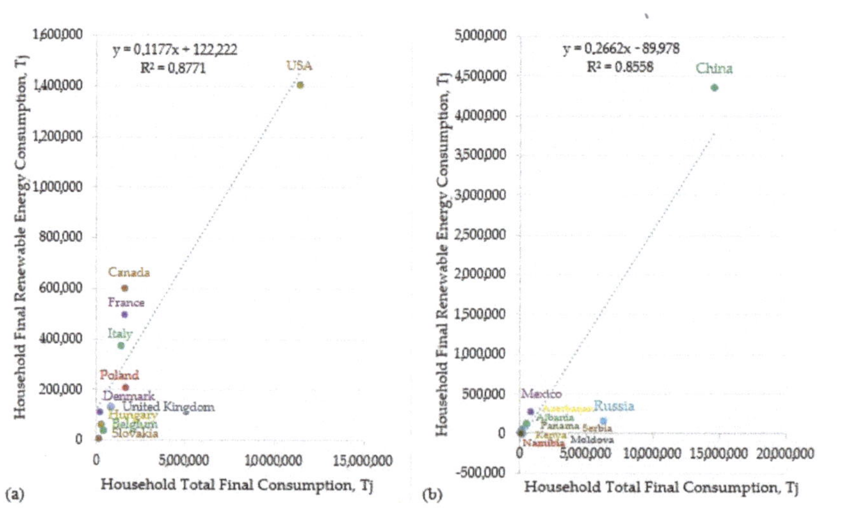

Figure 1. Dependence of total energy consumption and consumption of energy from renewable sources by households in the studied developed (**a**) and developing (**b**) countries. Source: generated by the authors.

Table 4. Indicators of analysis of variance for total energy consumption and consumption of renewable energy by households.

Group	Indicator	df	SS	MS	F	F Sign	
Developed countries	Regression	1	1,418,090,185,502	14,18,090,185,502	57	0.0001	
	Residue	8	1,98,691,189,356	24,836,398,670			
	Total	9	1,616,781,374,858				
Developing countries	Regression	1	14,177,173,812,816	14,177,173,812,816	47	0.0001	
	Residue	8	2,389,700,476,144	298,712,559,518			
	Total	9	16,566,874,288,960				
Group	**Factor**	**Coefficients**	**Standard error**	**t-stat**	***p*-value**	**Lower 95%**	**Higher 95%**
Developed countries	Y-intersection (HFREC)	122,221.9493	57,980.4383	2.1080	0.0681	−11,481.1812	255,925.0799
	X (HTFC)	0.1177	0.0156	7.5563	0.0001	0.0817	0.1536
Developing countries	Y-intersection (HFREC)	−89,978.4280	19,3250.2207	−0.4656	0.6539	−53,5614.2362	355,657.3801
	X (HTFC)	0.2662	0.0386	6.8892	0.0001	0.1771	0.3553

Source: the authors' calculations.

The established relationship between the level of total energy consumption by households and their consumption of renewable energy is confirmed by a number of control points. The *p*-value for variable X (HTFC) is less than 0.05. At the same time, Fcrit < F is also a positive characteristic, namely, for developed countries 5.35 < 22.62, and for developing countries 5.35 < 12.75. The adequacy of the formed equations is confirmed by the Student's criterion (tcrit < tobs): for developed countries 2.31 < 4.76, and for developing countries 2.31 < 3.57.

At the same time, for developed countries, there is a higher degree of dependence ($R^2 = 0.8771$) compared to developing countries ($R^2 = 0.8558$).

The increase in energy consumption by households entails an increase in the consumption of renewable energy, which, presumably, the households themselves generate. Therefore, for both developed and developing countries, the trend is the development of household electricity production from renewable sources. At the same time, the quality of consumption changes, since households in this way substitute their part of the consumed energy. In the two groups of studied countries, the United States and China are clearly distinguished. Indicators of total energy consumption and renewable energy consumption by households of the US and China significantly exceed those of other countries. The US and China are leaders in both household energy consumption and renewable energy production.

To determine the relationship between the level of energy independence of households (Household Energy Independence Index—HEII) and a country's Energy Security Index (ESI), a regression analysis was carried out. Analysis of variance results are shown in Table 5.

Table 5. Indicators of analysis of variance for the Energy Security Index.

Group	Indicator	df	SS	MS	F	F sign	
Developed countries	Regression	1	0.6207	0.6207	22.6179	0.0014	
	Residue	8	0.2195	0.0274			
	Total	9	0.8402				
Developing countries	Regression	1	0.0014	0.0014	12.7470	0.0073	
	Residue	8	0.0009	0.0001			
	Total	9	0.0024				
Group	**Factor**	**Coefficients**	**Standard error**	**t-stat**	***p*-value**	**Lower 95%**	**Higher 95%**
Developed countries	Y-intersection (ESI)	−0.2313	0.1000	−2.3126	0.0495	−0.4619	−0.0007
	X (HEII)	1.5888	0.3341	4.7558	0.0014	0.8184	2.3592
Developing countries	Y-intersection (ESI)	0.0404	0.0063	6.4158	0.0002	0.0259	0.0549
	X (HEII)	−0.0471	0.0132	−3.5703	0.0073	−0.0775	−0.0167

Source: the authors' calculations.

The dependence of the energy security of the studied countries on the level of the Household Energy Independence Index can be graphically interpreted using equations (Figure 2). Their applicability is confirmed by the p-value, the value of which for the variable X (HEII) is less than 0.05. The adequacy of the formed equations is characterized by Fcrit < F, namely, for developed countries 5.32 < 57.1, and for developing countries 5.32 < 47.46. Applicability is also confirmed by the Student's criterion (tcrit < tobs): for developed countries 2.31 < 7.56, and for developing countries 2.31 < 6.89.

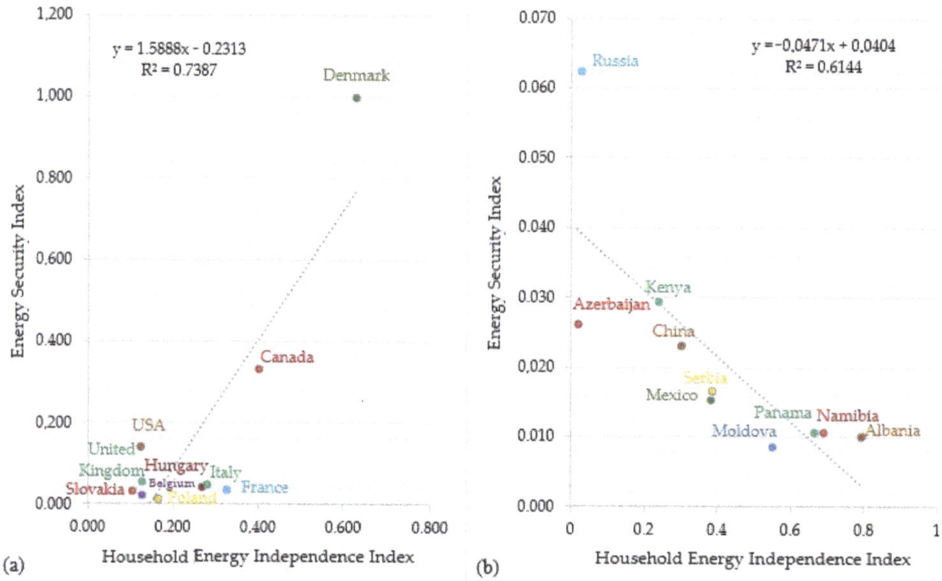

Figure 2. The relationship between a country's energy security and a level of energy independence of households in the studied developed (**a**) and developing (**b**) countries. Source: generated by the authors.

Almost all countries are characterized by a dependence of a country's energy security on households' energy independence. Household energy production influences import independence and energy storage capacity in a country. Moreover, the degree of this dependence for developed and developing countries is approximately at the same level. It is paradoxical that in developed countries an increase in the household energy independence index contributes to an increase in the level of energy security, while in developing countries, an increase in household energy independence leads to a decrease in the energy security index. This is primarily due to the fact that most of the studied developing countries have a fairly high level of energy security, but at the same time, very low indicators of the Household Energy Independence Index (Azerbaijan and Russia). The average household energy independence index in developing countries is almost two times lower (0.255) than in developed countries (0.404).

The results of the analysis of variance for the relationship between the Energy Independence Index (EII) and the Household Energy Independence Index (HEII) are shown in Table 6.

The relationship between the level of energy independence and the Household Energy Independence Index is confirmed by the p-value for the variable X (p-value < 0.05) for developed countries, but for developing countries, the p-value for HEII is greater than 0.05 (0.2455). At the same time, a positive characteristic of the equation for developed countries is the value Fcrit < F, namely for (5.35 < 27.07); for developing countries this criterion has no confirmation (5.35 > 1.57). The adequacy of the equations formed on the basis of the Student's criterion (tcrit < tobs) is confirmed only for developed countries (2.31 < 5.2). For developing countries, the situation is reversed, with 2.31 > 1.25. Based on this, it can be stated that there is no relationship between the factors under study for developing countries. A graphic interpretation of the results obtained is shown in Figure 3.

Table 6. Indicators of analysis of variance for the Energy Independence Index.

Group	Indicator	df	SS	MS	F	F sign	
Developed countries	Regression	1	0.8932	0.8932	27.0745	0.0008	
	Residue	8	0.2639	0.0330			
	Total	9	1.1571				
Developing countries	Regression	1	0.1518	0.1518	1.5705	0.2455	
	Residue	8	0.7734	0.0967			
	Total	9	0.9252				
Group	**Factor**	**Coefficients**	**Standard error**	**t-stat**	***p*-value**	**Lower 95%**	**Higher 95%**
Developed countries	Y-intersection (EII)	0.5232	0.1097	4.7698	0.0014	0.2702	0.7761
	X (HEII)	1.9060	0.3663	5.2033	0.0008	1.0613	2.7507
Developing countries	Y-intersection (EII)	1.2573	0.1840	6.8322	0.0001	0.8329	1.6816
	X (HEII)	−0.4827	0.3852	−1.2532	0.2455	−1.3708	0.4055

Source: the authors' calculations.

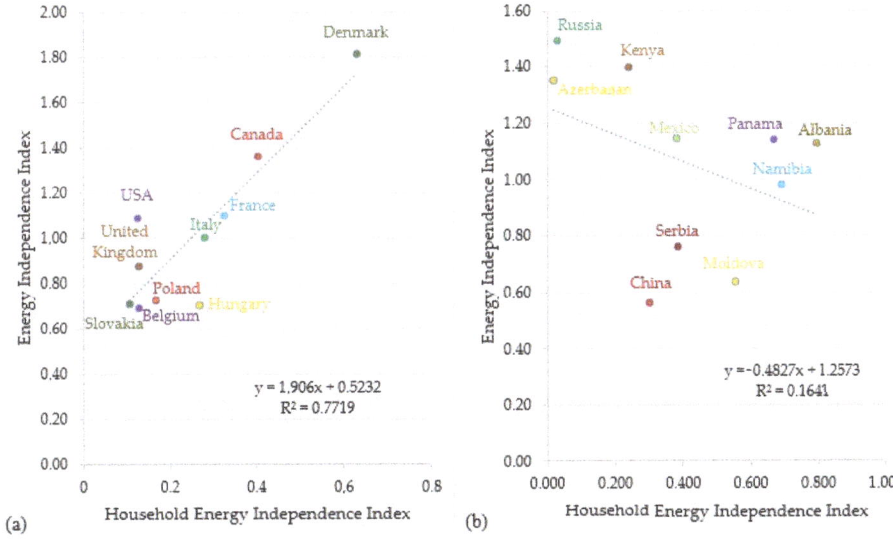

Figure 3. Relationship between the level of energy independence of countries on the index of energy independence of households of the studied developed (**a**) and developing (**b**) countries. Source: generated by the authors.

Increasing energy production by households to meet their needs helps reduce energy dependence in developed countries. This is confirmed by the value $R^2 = 0.7719$. For developing countries, this dependence is absent, since $R^2 = 0.1641$. Therefore, the degree of influence of the studied indicators was determined only for developed countries.

Similarly, the impact of energy generation by households on the level of social development of the studied countries was assessed using the Social Content Index. The results of the analysis of variance are shown in Table 7.

Table 7. Indicators of analysis of variance for the Social Content Index.

Group	Indicator	df	SS	MS	F	F sign	
Developed countries	Regression	1	0.01024	0.01024	19.18578	0.00235	
	Residue	8	0.00427	0.00053			
	Total	9	0.01451				
Developing countries	Regression	1	0.00002	0.00002	1.07689	0.32974	
	Residue	8	0.00014	0.00002			
	Total	9	0.00016				
Group	**Factor**	**Coefficients**	**Standard error**	**t-stat**	***p*-value**	**Lower 95%**	**Higher 95%**
Developed countries	Y-intersection (SCI)	−0.0028	0.0140	−0.2010	0.8457	−0.0350	0.0294
	X (HEII)	0.2041	0.0466	4.3802	0.0023	0.0967	0.3116
Developing countries	Y-intersection (SCI)	0.0110	0.0025	4.4565	0.0021	0.0053	0.0110
	X (HEII)	0.0054	0.0052	1.0377	0.3297	−0.0066	0.0054

Source: the authors' calculations.

The dependence of the social content of the development of the studied countries on the level of the Household Energy Independence Index is graphically interpreted using equations (Figure 4). The applicability of the obtained equation is confirmed only for developed countries, since $R^2 = 0.7057$, and *p*-value for the variable X (HEII) is less than 0.05. The adequacy of the formed equation for developed countries is characterized by Fcrit < F (5.32 < 19.19), as well as Student's tcrit < tobs (2.31 < 4.38). For developing countries, this relationship is practically absent and demonstrates the lack of adequacy of the formed equation.

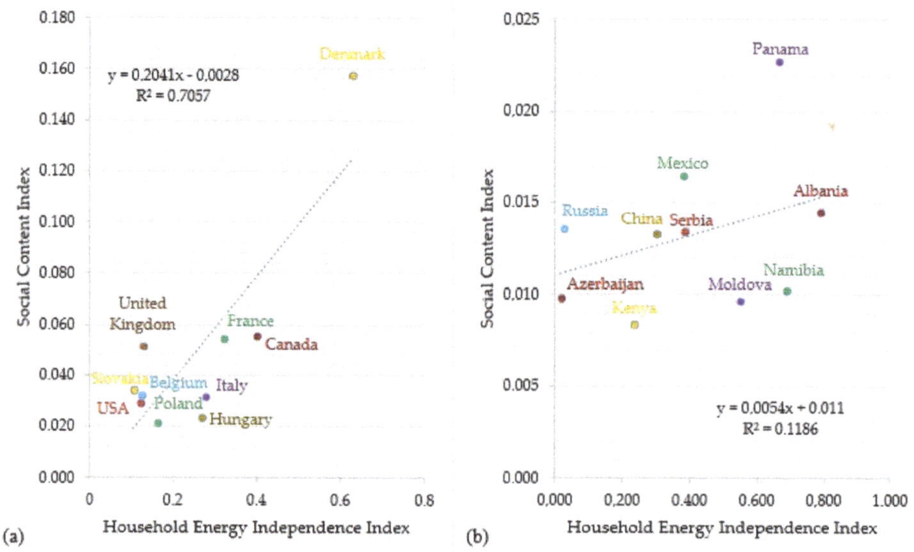

Figure 4. Relationship between Social Content Index of countries and the index of energy independence of households of the studied developed (**a**) and developing (**b**) countries. Source: generated by the authors.

Thus, it can be argued that increasing the volume of renewable energy generation by households is important for developed countries. For developing countries, household energy from renewable sources does not have a significant impact on social capital, health, living conditions, and the external environment.

Key indicators of modeling results for three scenarios of growth in electricity generation by households in the studied countries are shown in Table 8. At the same time, it was assumed that the level of the energy independence index of households corresponds to the level of electricity generation to meet their needs. The first scenario assumes an increase in the average actual index of the energy independence of households (the level of electricity generation) by 10%. The second scenario assumes an increase in electricity generation by households by 20%. The third scenario provides for an increase in household electricity production by 30%.

Table 8. Modeling according to scenarios of electricity generation by households in the studied countries.

Regression Models	a	b	-
y_{ESI}	0.2313	1.5888	-
y_{EII}	0.5232	1.906	-
Scenario modeling	x	y_{ESI}	y_{EII}
Scenario 1 (10%)	0.2805	0.6770	1.0578
Scenario 2 (20%)	0.3060	0.7175	1.1064
Scenario 3 (30%)	0.3315	0.7580	1.1550

Source: generated by the authors.

As shown by the calculations presented in Table 6, the third scenario, assuming an increase in household electricity production by 30%, represents the best option according to the chosen criteria. For example, according to the criterion of energy dependence, one can note a decrease in the level of dependence as a result of electricity generation by households. Moreover, this decline has a rather slow pace. An increase in the volume of electricity generation by households for the studied countries by 1% can lead to a decrease in the level of a country's energy dependence by 11.2%, as well as to an increase in energy security by 0.5%. This is confirmed by the obtained results of the calculations: the hierarchical synthesis of the implementation of the third scenario is the most efficient and amounts to 0.1250. It helps to reduce energy dependence, since the value of the indicator under this scenario demonstrates the lowest level approaching zero (Table 9).

Table 9. Solution of the problem of nonlinear optimization by the Thomas L. Saaty method for the developed countries under study.

Criterion	Assessment	Scenario 1	Scenario 2	Scenario 3
Energy security	Initial (y_{ESI})	0.6770	0.7175	0.7580
	Normalized (ESI_{sci}^{norm})	0.3145	0.3333	0.3522
	Weighted sum ($ESI_{sci}^{ws} = ESI_{sci}^{norm} * \text{weight}$)	1.8871	2.0000	2.1129
Energy dependence	Initial (y_{EII})	1.0578	1.1064	1.1550
	Normalized (EII_{sci}^{norm})	0.3187	0.3333	0.3480
	Weighted sum ($EII_{sci}^{ws} = EII_{sci}^{norm} * \text{weight}$)	0.0542	0.0567	0.0592
Hierarchical synthesis ($HS_{sci} = ESI_{sci}^{norm} * EII_{sci}^{norm}$)		0.1022	0.1133	0.1250

Source: generated by the authors.

Assessment of the proposed methodology on the example of the studied countries demonstrated the excess of the criterion of energy dependence over the criterion of energy security. Household electricity production is based on renewable sources and can replace consumption from other energy sources. In fact, this is a reimbursement of one's own electricity consumption. A higher degree of influence of energy production by households on the level of energy dependence of the studied developed countries should be emphasized. For developing countries, the increase in energy production by households does not have a significant impact on the level of energy dependence due to the very low level of the Household Energy Independence Index and the corresponding volumes of energy production by them, which are not able to cover their needs in general.

5. Discussion

The advantage of the study is the evidence that electricity generation by households is a factor in reducing the energy dependence of households on monopolists and the state as a whole. In addition, it can also have a significant impact on the balance of prices in the country's energy market [45]. At the same time, the state and market monopolists should not have leverage to prohibit or restrain the development of energy in this direction. The results obtained have significant implications for better understanding both the benefits and limitations of household electricity generation, which helps guide practical implementation. For the study countries, the overall benefit that could be derived from household electricity production depends on many factors, and the key is to make these supplies more affordable in the market through proactive energy management solutions for all residents, rather than focusing only on self-consumption of prosumers [4]. There are a lot of levers and mechanisms for solving this problem. Starting from a system of targeted concessional lending or compensation of interest on loans to an increase in electricity tariffs for the population. However, these steps should be woven into the overall strategy of state policy in the energy sector in general and the development of renewable energy sources in particular [46].

From the results of the study, namely the intensity of the impact of electricity generation by households on the level of energy independence, it is clear that alternative energy sources will not solve all energy problems in the coming years. However, focusing on them provides real opportunities to strengthen positions in the future, increase the country's energy security, and reduce the population's expenses for housing services [47]. In addition, despite the relatively high annual rate of increase in the use of renewable energy sources by households, there are a number of problems that impede its intensive development and the formation of household autonomy. Among the main constraints hindering the development of the subjects of this market are the norms for the installed capacity for power generating plants in private households determined by legislation, which slow down the growth rates of this sector. There are also legal restrictions on the types of facilities for renewable energy sources in private households with solar and wind installations [48]. Particularly acute for developing countries, there is the problem of high initial investment in the creation of new capacities on renewable energy sources for households [49]. However, with the development of the scale of electricity generation by households, one more problem is gradually being identified, which will worsen over time. It lies in the fact that an increase in the volume of capacities on renewable energy sources at the existing high rates of the "green" tariff will lead to a gradual increase in electricity prices in the state, since the compensation of the increased "green" tariffs is due to an increase in average prices for electricity obtained from both traditional and renewable sources [50]. It should be noted that these problems arise in many countries of the world, which have introduced economic incentives for the development of electricity generation by households from renewable sources. For example, some European countries have recently curtailed investment programs in renewable energy and reduced rates of "green" tariffs, citing the high cost of energy from renewable energy sources and the inexpediency of a significant increase in electricity tariffs to compensate for the "green" component [51].

This study proves that the development of renewable energy sources for households in developed countries under study is an important factor in increasing the level of energy security, reducing the use of fossil fuel resources (including imported ones), as well as reducing the negative impact of energy on the environment and improving the quality of life of citizens. A similar effect is not typical for developing countries, since renewable energy sources are most often economically more costly than traditional energy sources and fuels. However, along with future technological development, the cost of renewable energy will decrease, and its production will become more and more profitable [52]. Developing countries need to stimulate the development of renewable energy sources at the state level. The focus should be on those sources that have a high probability of economic return in the future and are the most promising from the point of view of production in the territory of a particular country. It is also required to localize the production of the necessary equipment to reduce the cost of renewable energy. It is essential to support the development and implementation of competitive production technologies and installations for renewable energy sources [53]. With reduced costs and favorable social policies, it is possible to develop renewable energy sources and create jobs [54,55].

The scientific contribution of this study is a comprehensive approach to assessing the degree of impact of household energy production on the energy security of countries, taking into account their level of development, as well as the social aspects that are valuable in the process of sustainable development. The study confirms that household energy production can reduce the energy dependence of countries, as well as increase the sustainability of their economic development. At the same time, a significant advantage of this study is the determination of the effect for countries with different levels of development. For most of the studied countries, the dependence of the country's energy security on households' energy independence can be traced. This confirms the impact of household energy production on the level of import-independence and energy storage capacity in a country [56]. For example, the positive impact of household energy production on the level of import-independence and energy storage capacity of developed countries has been identified and is negative for developing countries. The positive side of the research is the study of the relationship between the level of countries' energy independence and the index of households' energy independence. It was proved that an increase in energy production by households contributes to a decrease in energy dependence in developed countries. At the same time, the presence of this connection for developing countries was not observed. The regression analysis in this study was a significant addition to the toolkit for assessing and achieving the set objectives of the study. Thus, it was possible to determine the feasibility for countries with different developmental levels to increase the volume of households' renewable energy production for energy independence [57].

As household income and electricity price are the main determinants of demand, the projected results may vary from country to country. At the same time, the use of replacement fuels also affects the elasticity of electricity demand [58–60]; therefore, there are corresponding limitations of the study. This study confirms that access to free basic electricity and policies to improve access to electricity contribute to the expected outcome. This is due to an increase in the likelihood of households buying electricity and a reduction in overall energy costs [61,62]. Different levels of household income should be considered. Higher demand for electricity is characteristic of households rich in appliances in urban areas, especially if household members are large enough and they live in large dwellings [63]. For example, access to energy technologies in rural areas is significantly limited, which can lead to instability of the economy of this part of households and dependence on the monopoly of energy production by the state. Therefore, a significant advantage of this study is the formation of a basis for drawing more attention of the scientific world to the issue of energy independence of households, which will speed up its solution and the creation of independent sources of electricity in rural areas [64,65]. The solution to power supply of such households makes it possible to develop an antimonopoly system in the field of electrical production and create a basis for future research towards the comprehensive development of such territories. Future research may be based on expanding the number of countries under study in aggregate or for a specific region. Besides, in the future, this study can be deepened in the direction of assessing the effectiveness of household investment in projects for the production of electricity, and the volumes and terms of their payback in various countries and regions of the world. It is important to carry out further scientific research, which must be directed towards a quick and effective solution to the problems of the development of electricity production by households in order to increase their level of autonomy and energy independence of a country. The study can be developed in the field of economics and energy in the context of defining the position of various policies, using strategic levers and indicators that provide an effective response to energy security challenges in the regional and global markets. Expanding the spectrum of this study can help to quickly and efficiently solve the problems of developing electricity production by households in order to increase the level of their autonomy, energy independence of countries, and improve the quality of life of the population. This will make it possible, in the future, to ensure both the development of a country's economy and improve the life of its population.

6. Conclusions

The studied countries are characterized by a high interdependence between the increase in household energy consumption and the increase in household renewable energy consumption. For developed and developing countries, the trend is the development of household electricity production from renewable sources. At the same time, the quality of consumption changes, since households in this way substitute their part of the consumed energy.

For most of the studied countries, there is a dependence of a country's energy security on the energy independence of households. This confirms the impact of household energy production on the level of import independence and the capacity of energy storage in a country. At the same time, the positive impact of this relationship for developed countries and negative impact for developing countries were revealed. The main precondition for this is the very low energy independence of households with a high level of energy security in developing countries. The study of the relationship between the level of energy independence of countries and the index of energy independence of households made it possible to determine that an increase in energy production by households contributes to a decrease in energy dependence in developed countries. However, for developing countries, this relationship is absent. This made it possible to conduct a regression analysis only for a group of developed countries. Increasing the level of energy generation from renewable sources has no significant social impact in developing countries. However, for developed countries, households' renewable energy has an impact on the growth of social capital, living standards, health, and the external environment.

Based on the scenario modeling of the growth of electricity production by households in the studied developed countries, an analysis was carried out, as a result of which the third scenario, assuming an increase in electricity production by households by 30%, is the best option in accordance with the selected criteria. Based on the criterion of Household Energy Independence Index, it becomes evident that countries' energy dependence decreases with the increase of households' electricity generation. At the same time, this decline is occurring at a rather slow pace. An increase in electricity generation by households for the studied countries can lead to a decrease in energy dependence, as well as to an increase in energy security. This is confirmed by the results of calculations based on the hierarchical synthesis of the implementation. The indicators of the third scenario characterize its maximum efficiency. It contributes to the reduction of energy dependence, since the value of the indicator in the scenario under consideration indicates that its level is approaching zero. At the same time, it can be stated that a further increase in electricity production by households will lead to an increase in energy efficiency in the context of energy dependence.

The solution of the formulated problem of nonlinear optimization by Thomas L. Saaty's method for the studied countries has demonstrated that the criterion of energy security is superior to the criterion of energy consumption per capita. Electricity production in households is based on renewable sources and can cover consumption from other energy sources. In fact, this is compensation for one's own energy consumption. At the same time, the level of energy consumption per capita will remain practically unchanged. The higher impact of household energy production on energy dependence should be emphasized. Therefore, it can be argued that the level of electricity production by households can reduce the energy dependence of countries as well as enhance the sustainability of their economic development. Hypothesis H1 is accepted because the influence of the Household Energy Independence Index on the Energy Security Index has been proven. H2 can only be adopted for developed countries. Developing countries have not yet reached the level of sufficient influence of household energy production on their energy independence.

In the future, this study can be deepened in the direction of assessing the effectiveness of household investment in projects for the production of electricity, and the volumes and terms of their payback in various countries and regions of the world. The proposed results can be claimed by responsible persons in the field of economics and energy in order to determine the position of various policies, and use strategic levers and indicators that ensure an effective response to energy security challenges in the regional and global markets. Expanding the range of this research can help to quickly and effectively solve the problems of developing electricity production by households in order to increase their level of autonomy, the energy independence of countries, and improve the quality of life of the populations.

Author Contributions: Conceptualization, I.P. and S.K.; methodology, S.B. and S.K.; software, I.P.; validation, S.B., I.P., and S.K.; formal analysis, S.K. and J.K.; investigation, S.B.; resources, I.P. and S.K.; data curation, S.B.; writing—original draft preparation, S.B.; writing—review and editing, I.P. and J.K.; visualization, S.K.; supervision, S.B.; project administration, I.P.; funding acquisition, S.K. and J.K. All authors have read and agreed to the published version of the manuscript.

Funding: This research received no external funding.

Institutional Review Board Statement: Not applicable.

Informed Consent Statement: Not applicable.

Data Availability Statement: Data will be available on request.

Conflicts of Interest: The authors declare no conflict of interest.

References

1. Mangla, S.K.; Luthra, S.; Jakhar, S.; Gandhi, S.; Muduli, K.; Kumar, A. A step to clean energy-Sustainability in energy system management in an emerging economy context. *J. Clean. Prod.* **2020**, *242*, 118462. [CrossRef]
2. IEA. Energy Access Outlook 2017: World Energy Outlook Special Report. 2017. Available online: https://www.iea.org/reports/energy-access-outlook-2017 (accessed on 2 May 2021).
3. Gołębiowska, B.; Bartczak, A.; Czajkowski, M. Energy demand management and social norms. *Energies* **2020**, *13*, 3779. [CrossRef]
4. Li, Z.; Ma, T. Peer-to-peer electricity trading in grid-connected residential communities with household distributed photovoltaic. *Appl. Energy* **2020**, *278*, 115670. [CrossRef]
5. Katsuba, S.; Shestak, V.; Kvasnikova, T.; Bokov, Y. Liability for Violation of Environmental Legislation in the EU. *Eur. Energ. Environm. Law Rev.* **2021**, *30*, 9–19.
6. Park, S.; Kim, J. The effect of interest in renewable energy on US household electricity consumption: An analysis using Google Trends data. *Renew. Energy* **2018**, *127*, 1004–1010. [CrossRef]
7. Yang, T.; Ren, M.; Zhou, K. Identifying household electricity consumption patterns: A case study of Kunshan, China. *Renew. Sust. Energy Rev.* **2018**, *91*, 861–868. [CrossRef]
8. Taghizadeh-Hesary, F.; Yoshino, N. The way to induce private participation in green finance and investment. *Financ. Res. Lett.* **2019**, *31*, 98–103. [CrossRef]
9. Inderberg, T.H.J.; Tews, K.; Turner, B. Is there a prosumer pathway? Exploring household solar energy development in Germany, Norway, and the United Kingdom. *Energy Res. Soc. Sci.* **2018**, *42*, 258–269. [CrossRef]
10. McKenna, R. The double-edged sword of decentralized energy autonomy. *Energy Policy* **2018**, *113*, 747–750. [CrossRef]
11. Domenech, B.; Ferrer-Martí, L.; Lillo, P.; Pastor, R.; Chiroque, J. A community electrification project: Combination of microgrids and household systems fed by wind, PV or micro-hydro energies according to micro-scale resource evaluation and social constraints. *Energy Sustain. Dev.* **2014**, *23*, 275–285. [CrossRef]
12. Samad, H.; Zhang, F. *Electrification and Household Welfare: Evidence from Pakistan*; The World Bank: Washington, DC, USA, 2018.
13. Lee, K.; Brewer, E.; Christiano, C.; Meyo, F.; Miguel, E.; Podolsky, M.; Rosa, J.; Wolfram, C. Electrification for "under grid" households in rural Kenya. *Dev. Eng.* **2016**, *1*, 26–35. [CrossRef]
14. Wittenberg, I.; Matthies, E. Solar policy and practice in Germany: How do residential households with solar panels use electricity? *Energy Res. Soc. Sci.* **2016**, *21*, 199–211. [CrossRef]
15. Androniceanu, A.; Popescu, C.R. An inclusive model for an effective development of the renewable energies public sector. *Adm. Public Manag. Rev.* **2017**, *28*, 81–96.
16. Tvaronavičienė, M.; Ślusarczyk, B. *Energy Transformation Towards Sustainability*; Elsevier: Amsterdam, The Netherlands, 2019.

17. Siemiątkowski, P.; Tomaszewski, P.; Marszałek-Kawa, J.; Gierszewski, J. The financing of renewable energy sources and the level of sustainable development of Poland's provinces in the area of environmental order. *Energies* **2020**, *13*, 5591. [CrossRef]
18. De Fatima, M.; Arthur, S.R.; Cockerill, A. The roles of government and the public utility in achieving universal access to electricity. *Econ. Energy Environ. Policy* **2019**, *8*, 1–7. [CrossRef]
19. Hasanov, F.J.; Mahmudlu, C.; Deb, K.; Abilov, S.; Hasanov, O. The role of Azeri natural gas in meeting European Union energy security needs. *Energy Strategy Rev.* **2020**, *28*, 100464. [CrossRef]
20. Aslanturk, O.; Kıprızlı, G. The role of renewable energy in ensuring energy security of supply and reducing energy-related import. *Int. J. Energy Econ. Policy* **2020**, *10*, 354. [CrossRef]
21. Rodríguez-Fernández, L.; Carvajal, A.B.F.; Ruiz-Gómez, L.M. Evolution of European Union's energy security in gas supply during Russia–Ukraine gas crises (2006–2009). *Energy Strategy Rev.* **2020**, *30*, 100518. [CrossRef]
22. Azzuni, A.; Breyer, C. Global energy security index and its application on national level. *Energies* **2020**, *13*, 2502. [CrossRef]
23. Azzuni, A.; Breyer, C. Definitions and dimensions of energy security: A literature review. *Wiley Interdiscip. Rev. Energy Environ.* **2018**, *7*, e268. [CrossRef]
24. Šprajc, P.; Bjegović, M.; Vasić, B. Energy security in decision making and governance-Methodological analysis of energy trilemma index. *Renew. Sustain. Energy Rev.* **2019**, *114*, 109341. [CrossRef]
25. Szulecki, K. Securitization and state encroachment on the energy sector: Politics of exception in Poland's energy governance. *Energy Policy* **2020**, *136*, 111066. [CrossRef]
26. Proskuryakova, L. Updating energy security and environmental policy: Energy security theories revisited. *J. Environ. Manage.* **2018**, *223*, 203–214. [CrossRef] [PubMed]
27. Zhao, Z.J.; Su, G.; Li, D. The rise of public-private partnerships in China. *J. Chin. Gov.* **2018**, *3*, 158–176. [CrossRef]
28. Shahbaz, M.; Raghutla, C.; Song, M.; Zameer, H.; Jiao, Z. Public-private partnerships investment in energy as new determinant of CO2 emissions: The role of technological innovations in China. *Energy Econ.* **2020**, *86*, 104664. [CrossRef]
29. Butu, H.M.; Nsafon, B.E.K.; Park, S.W.; Huh, J.S. Leveraging community based organizations and fintech to improve small-scale renewable energy financing in sub-Saharan Africa. *Energy Res. Soc. Sci.* **2021**, *73*, 101949. [CrossRef]
30. Raza, S.A.; Shah, S.H.; Yousufi, S.Q. The impact of public-private partnerships Investment in Energy on carbon emissions: Evidence from nonparametric causality-in-quantiles. *Environ. Sci. Pollut. Res.* **2021**, *28*, 23182–23192. [CrossRef]
31. Wojewnik-Filipkowska, A.; Węgrzyn, J. Understanding of public–private partnership stakeholders as a condition of sustainable development. *Sustainability* **2019**, *11*, 1194. [CrossRef]
32. Zhao, X.; Luo, D. Driving force of rising renewable energy in China: Environment, regulation and employment. *Renew. Sustain. Energy Rev.* **2017**, *68*, 48–56. [CrossRef]
33. Bridge, G.; Özkaynak, B.; Turhan, E. Energy infrastructure and the fate of the nation: Introduction to special issue. *Energy Res. Soc. Sci.* **2018**, *41*, 1–11. [CrossRef]
34. MacDonald, S.; Eyre, N. An international review of markets for voluntary green electricity tariffs. *Renew. Sust. Energy Rev.* **2018**, *91*, 180–192. [CrossRef]
35. Gwerder, Y.V.; Marques, P.; Dias, L.C.; Freire, F. Life beyond the grid: A life-cycle sustainability assessment of household energy needs. *Appl. Energy* **2019**, *255*, 113881.
36. United Nations. World Economic Situation Prospects. 2020. Available online: https://www.un.org/development/desa/dpad/wp-content/uploads/sites/45/WESP2020_Annex.pdf (accessed on 2 May 2021).
37. World Energy Council. Energy Trilemma Index. 2020. Available online: https://trilemma.worldenergy.org/#!/energy-index (accessed on 2 May 2021).
38. Knoema. Energy Trilemma Index. 2020. Available online: https://knoema.com/search?query=trilemma+index&pageIndex=&scope=&term=&correct=&source=Header (accessed on 2 May 2021).
39. International Renewable Energy Agency. Final Renewable Energy Consumption. 2018. Available online: https://www.irena.org/Statistics/View-Data-by-Topic/Renewable-Energy-Balances/Final-Renewable-Energy-Consumption (accessed on 2 May 2021).
40. International Energy Agency. Data and Statistics. 2020. Available online: https://www.iea.org/data-and-statistics?country=WORLD&fuel=Energy%20supply&indicator=TPESbyPop (accessed on 2 May 2021).
41. Reshetnikova, I.; Shvydanenko, H.; Boichenko, K. Determinants to provide the efficiency of integrated development of the light industry enterprises. *Mark. Manag. Innov.* **2020**, *3*, 157–169. [CrossRef]
42. Our World in Data. Share of Primary Energy from Renewable Sources. 2020. Available online: https://ourworldindata.org/grapher/renewable-share-energy?tab=chart&stackMode=absolute&time=earliest..2018&country=AZE~{}MEX&region=World (accessed on 2 May 2021).
43. Legatum Institute Foundation. The Legatum Prosperity Index 2019. A tool for transformation. 2020. Available online: https://li.com/reports/2019-legatum-prosperity-index/#:~{}:text=The%20Legatum%20Prosperity%20Index%E2%84%A2,drive%20tangible%20improvements%20in%20prosperity (accessed on 2 May 2021).
44. Khachaturyan, A.A.; Khachaturyan, K.S.; Shirkin, A.A. Development of Russia's regions in the conditions of the digital economy: Management, effectiveness, and competitiveness. In *State and Corporate Management of Region's Development in the Conditions of the Digital Economy*; Springer: Cham, Switzerland, 2021; pp. 25–29.
45. Dźwigoł, H.; Dźwigoł-Barosz, M.; Zhyvko, Z.; Miśkiewicz, R.; Pushak, H. Evaluation of the energy security as a component of national security of the country. *J. Secur. Sustain. Issues* **2019**, *8*, 307–317. [CrossRef]

46. Galvin, R.; Sunikka-Blank, M. Economic inequality and household energy consumption in high-income countries: A challenge for social science based energy research. *Ecol. Econ.* **2018**, *153*, 78–88. [CrossRef]
47. Wei, T.; Wang, X. Rebound effect from income savings due to an energy efficiency improvement by households: An input–output approach. *Energies* **2020**, *13*, 4044. [CrossRef]
48. Mrówczyńska, M.; Skiba, M.; Bazan-Krzywoszańska, A.; Sztubecka, M. Household standards and socio-economic aspects as a factor determining energy consumption in the city. *Appl. Energy* **2020**, *264*, 114680. [CrossRef]
49. Vasseur, V.; Marique, A.F.; Udalov, V. A conceptual framework to understand households' energy consumption. *Energies* **2019**, *12*, 4250. [CrossRef]
50. Herbes, C.; Rilling, B.; MacDonald, S.; Boutin, N.; Bigerna, S. Are voluntary markets effective in replacing state-led support for the expansion of renewables?—A comparative analysis of voluntary green electricity markets in the UK, Germany, France and Italy. *Energy Policy* **2020**, *141*, 111473. [CrossRef]
51. Hannesson, R. How much do European households pay for green energy? *Energy Policy* **2019**, *131*, 235–239. [CrossRef]
52. Lin, B.; Zhu, J. The role of renewable energy technological innovation on climate change: Empirical evidence from China. *Sci. Total Environ.* **2019**, *659*, 1505–1512. [CrossRef]
53. Ye, B.; Zhang, K.; Jiang, J.; Miao, L.; Li, J. Towards a 90% renewable energy future: A case study of an island in the South China Sea. *Energy Convers. Manag.* **2017**, *142*, 28–41. [CrossRef]
54. Sen, S.; Ganguly, S. Opportunities, barriers and issues with renewable energy development—A discussion. *Renew. Sustain. Energy Rev.* **2017**, *69*, 1170–1181. [CrossRef]
55. Polzin, F.; Egli, F.; Steffen, B.; Schmidt, T.S. How do policies mobilize private finance for renewable energy?—A systematic review with an investor perspective. *Appl. Energy* **2019**, *236*, 1249–1268. [CrossRef]
56. Haseeb, M.; Kot, S.; Iqbal Hussain, H.; Kamarudin, F. The natural resources curse-economic growth hypotheses: Quantile–on–Quantile evidence from top Asian economies. *J. Clean. Prod.* **2021**, *279*, 123596. [CrossRef]
57. Hussain, H.I.; Slusarczyk, B.; Kamarudin, F.; Thaker, H.M.T.; Szczepańska-Woszczyna, K. An investigation of an adaptive neuro-fuzzy inference system to predict the relationship among energy intensity, globalization, and financial development in major ASEAN economies. *Energies* **2020**, *13*, 850. [CrossRef]
58. Pena-Bello, A.; Barbour, E.; Gonzalez, M.C.; Yilmaz, S.; Patel, M.K.; Parra, D. How does the electricity demand profile impact the attractiveness of PV-coupled battery systems combining applications? *Energies* **2020**, *13*, 4038. [CrossRef]
59. Nasr, A.K.; Kashan, M.K.; Maleki, A.; Jafari, N.; Hashemi, H. Assessment of barriers to renewable energy development using stakeholders approach. *Entrep. Sustain. Issues* **2020**, *7*, 2526–2541. [CrossRef]
60. Tvaronavičienė, M.; Plėta, T.; Della Casa, S.; Latvys, J. Cyber security management of critical energy infrastructure in national cybersecurity strategies: Cases of USA, UK, France, Estonia and Lithuania. *Insights Reg. Dev.* **2020**, *2*, 802–813. [CrossRef]
61. Muhumuza, R.; Zacharopoulos, A.; Mondol, J.D.; Smyth, M.; Pugsley, A. Energy consumption levels and technical approaches for supporting development of alternative energy technologies for rural sectors of developing countries. *Renew. Sust. Energ. Rev.* **2018**, *97*, 90–102. [CrossRef]
62. El Iysaouy, L.; El Idrissi, N.E.; Tvaronavičienė, M.; Lahbabi, M.; Oumnad, A. Towards energy efficiency: Case of Morocco. *Insights Reg. Dev.* **2019**, *1*, 259–271. [CrossRef]
63. Ye, Y.; Koch, S.F.; Zhang, J. Determinants of household electricity consumption in South Africa. *Energy Econ.* **2018**, *75*, 120–133. [CrossRef]
64. Grondys, K.; Androniceanu, A.; Dacko-Pikiewicz, Z. Energy management in the operation of enterprises in the light of the applicable provisions of the Energy Efficiency Directive (2012/27/EU). *Energies* **2020**, *13*, 4338. [CrossRef]
65. Tishkov, S.; Shcherbak, A.; Karginova-Gubinova, V.; Volkov, A.; Tleppayev, A.; Pakhomova, A. Assessment the role of renewable energy in socio-economic development of rural and Arctic regions. *Entrep. Sustain. Issues* **2020**, *7*, 3354–3368. [CrossRef]

energies

MDPI

Article

Optimization of the Structure of the Investment Portfolio of High-Tech Companies Based on the Minimax Criterion

Alex Borodin [1,*], Manuela Tvaronavičienė [2,3], Irina Vygodchikova [4], Galina Panaedova [5] and Andrey Kulikov [6]

1 Department of Finance, Plekhanov Russian University of Economics, 117997 Moscow, Russia
2 Vilnius Gediminas Technical University (Vilnius Tech), LT-10223 Vilnius, Lithuania; manuela.tvaronaviciene@vilniustech.lt
3 Institute of Humanities and Social Sciences, Daugavpils University, LV-5401 Daugavpils, Latvia
4 Department of Differential Equations & Mathematic Economics, National Research Saratov State University, Named after N. G. Chernyshevsky, 410012 Saratov, Russia; irinavigod@yandex.ru
5 Department of Tax Policy and Customs Affairs, North-Caucasus Federal University, 355017 Stavropol, Russia; afina-02@rambler.ru
6 Department of Organization of Medical Provision and Pharmacoeconomics, I.M. Sechenov First Moscow State Medical University (Sechenov University), 119991 Moscow, Russia; kulikov_a_yu@staff.sechenov.ru
* Correspondence: aib-2004@yandex.ru

Abstract: A model has been developed for the optimization of the share structure of an investment portfolio in high-tech projects supported by the leaders of the leading industry companies in Russia. Several indicators (financial leverage, integrated rating of companies, industry rating) were applied in the decision support system for the shared distribution of investments. High-tech production is based on innovative technologies for saving resources, the resiliency of systems for transporting and transferring raw materials and finished products within Russia, so the main income will remain within the country. It is possible to export high-tech products, rather than raw materials, which will increase export revenues. Investors will invest in high-tech projects of Russian companies, taking into account the targeting of investment development. The guarantee is the stable financial position of the companies and the competitiveness rating. Methods: The authors propose a new approach that does not contradict modern rating scales, based on a hierarchical rating procedure and fuzzy logical rules that allow you to build an integral rating in the form of portfolio shares from the whole. A higher share shows an indicator of the higher investment attractiveness of companies. The industry rating is obtained based on the principle of the company's first affiliation to the highest rating indicator. The final minimax portfolio is based on the initial ratings in a circular convolution and is then adjusted by industry. A software package has been compiled that allows the testing of the method of capital allocation between investment projects for the largest companies' leaders of high-tech industries in Russia. This software uses the author's method of multi-stage analysis, the evaluation of financial coefficients, the integral ranking and the correction of the solution taking into account the industry attributes. Results: The results are presented with computer-aided design (CAD) in the form of an algorithmized decision support system (DSS). The CAD system is based on a hierarchical algorithm, based on the use of a multi-level redistribution of investment shares of high-tech companies, taking into account the adaptation to the requirements of the return on investment portfolio. When compiling the portfolio, the minimax optimality criterion is applied, which allows the stabilization of the risk by purposefully redistributing funds between the companies involved in the analysis. The authors of the article have compiled an algorithm for the software implementation of the model. Features of the rating approach: the use of the author's mathematical apparatus, which includes a hierarchical analysis of the ranked indicators of the financial and economic activity of companies, taking into account their priority, and the use of a minimax approach to obtain a rating assessment of companies, taking into account the industry attributes. Development: The proposed approach should be used for targeted financing of large industry companies engaged in the implementation of high-tech projects.

Keywords: optimization; investment; financial portfolio; financial leverage; integral rating; industry rating; high-tech company; decision-making; minimax

Citation: Borodin, A.; Tvaronavičienė, M.; Vygodchikova, I.; Panaedova, G.; Kulikov, A. Optimization of the Structure of the Investment Portfolio of High-Tech Companies Based on the Minimax Criterion. *Energies* **2021**, *14*, 4647. https://doi.org/10.3390/en14154647

Academic Editor: Rajender Gupta

Received: 22 June 2021
Accepted: 27 July 2021
Published: 30 July 2021

1. Introduction

The development of high-tech production requires making quick and high-quality decisions on the share structure of investment capital directed to support innovative projects of high-tech industries in Russia [1,2] and other countries [3–7]. The investor is focused on profit; it is the high profit of new projects that can attract him, if he is a competent specialist and understands that you need to expect stable, not ultra-high returns [8,9]. The portfolio approach is the most correct, but uses traditional models of portfolio investment. For example, in the classical problem of Markovitz [10], a number of problems arise, since it is necessary to determine the covariance matrix of asset returns, which is not possible in real time. Obtaining approximate data based on the analysis of financial performance over several years significantly slows down the decision-making process and, in addition, leads to a distortion of the optimization result. Therefore, the current direction of research is to improve the technology of portfolio investment of high-tech projects using rating estimates and parameters of the minimax model.

At the same time, there is the question of obtaining the qualitative parameters of the model task, according to which the formation of the portfolio will be performed optimally and will remain stable for a long time, according to the current investment activity (several months or more).

The authors propose a systematic approach that includes the formation of a portfolio based on important financial indicators (financial leverage, profitability, etc.), on the one hand, and company ratings (linear ranks) on the other hand. To select the leading companies, the author's approach of the circular convolution of financial indicators into an index is used. Two problems are solved on the basis of the minimax criterion. As a result, the correction of the portfolio obtained by the minimum risk problem (financial leverage) is performed on the basis of using the solution of the problem with a circular convolution rating. The investor will receive a balanced system of valuation indicators for his portfolio, which is resistant to noise effects and market news. In principle, it is advisable to revise the model after receiving new data on the financial statements of companies included in the investment portfolio (quarterly report, annual report), depending on the depth of the analytical procedure.

The purpose of the study is to develop a CAD system that contains a methodology and algorithm for optimizing the structure of the investment portfolio of high-tech companies based on the minimax criterion.

The tasks of the work are to obtain a CAD structure for high-tech industries (oil and gas), to obtain a conclusion about the financial condition of companies to be analyzed based on the assessment of capital structure (financial leverage), to obtain an integral rating according to the criteria of volume, dynamics and net profit, to obtain the results of solving two minimax problems and to use the solution of the rating problem as a correction of the solution of the problem of minimizing financial leverage (the ratio of debt capital to equity) in the portfolio, to perform program implementations.

The hypothesis of the work: to obtain the optimal structure of the investment portfolio, it is necessary to create an algorithm for the complex application of two fundamentally different tasks. In the first, the risk criteria will be financial leverage; in the second, the integral rating of the company in the industry, the complex use involves a weighting method based on the results of the risk assessment.

Results: to obtain a CAD-based investment capital structure in the form of a DSS for investment solutions for high-tech projects.

2. Literature Foundations

In the scientific literature, the definitions of a high-tech industry are almost the same, as shown in [11–13], and basically all definitions are reduced to understanding the company's attitude to high-tech industry [14].

Some authors have studied high-tech companies [15], taking into account the industry characteristic [16–22].

The financing of high-tech production in generating cash flow is shown in Reference [23].

The articles [24–28] show the high volatility of shares of high-tech companies that have premiums and the presence of high cash flow.

Portfolio investing is an important and sought-after area of portfolio discussions. The generally accepted terms "do not put all eggs in one basket", "buy cheaper, sell more expensive" cannot be used with new technologies.

So, you need to invest in a stable company of a high-industry complex, and it is not always possible to buy cheaper and sell more expensive; it is advisable to buy shares of companies quite cheap (from the point of view of technical analysis) and own them, not sell them, but receive dividends.

The sale is possible, or the company becomes high-risk and low-profit, or the owner needs money, then you need to look for options for selling at a high price (technical analysis, author's risk indicators) [29,30].

Minimax Approach, Model, Methodology

Let m industries be considered for investment purposes, and a certain number of companies are selected in each industry; in total, n companies are involved in the analysis, distributed across m industries. Estimates of the financial leverage of companies are denoted by $FL_1 > 0, \ldots, FL_n > 0$ (the ratio of debt capital to equity, preferably less).

The integral ranks of companies (hierarchical procedure, author's approach) are denoted by $IR_1 > 0 \geq IR_2 \geq \ldots \geq IR_n > 0$ ("1" is the best, "n" is the worst). It is necessary to determine the investment shares of companies $\theta = (\theta_1, \ldots, \theta_n)$.

A prerequisite for model construction is to take into account the priority of industries to improve the rating of companies using industry-specific adjustments.

Based on the hierarchical analysis of statistical data, an integral rating of companies, V, is constructed, indexed according to the company number in the list (for the i-th company, rating, leverage level, V_i).

In order to obtain the recommended investment shares, a mathematical problem with a non-smooth functional and a linear constraint of the form is used:

$$\max_{i=\overline{1,n}} FL_i\widetilde{\theta}_i \to \min_{\widetilde{\theta}\in\Omega}, \quad \Omega = \{\widetilde{\theta} = (\widetilde{\theta}_1, \ldots, \widetilde{\theta}_n) \in R^n : \sum_{i=1}^{n} \widetilde{\theta}_i = 1\}, \tag{1}$$

the solution of the problem (1) is determined by the formulas (2)

$$\widetilde{\theta}_i = \frac{1}{FL_i \sum\limits_{k=1}^{n} (FL_k)^{-1}}, \; i = \overline{1,n}. \tag{2}$$

When taking into account the profitability as an additional constraint, the model (1) is modified, the set Ω changes to the set

$$\Omega^{\eta} = \{\widetilde{\theta} = (\widetilde{\theta}_1, \ldots, \widetilde{\theta}_n) \in R^n : \sum_{i=1}^{n} \widetilde{\theta}_i = 1, \sum_{i=1}^{n} \eta_i\widetilde{\theta}_i = \eta_p\}. \tag{3}$$

Accordingly, the problem (1) takes the form:

$$\max_{i=\overline{1,n}} FL_i\widetilde{\theta}_i \to \min_{\widetilde{\theta}\in\Omega^{\eta}},. \tag{4}$$

The solution of this problem is described in Reference [31].

If it is impossible to simultaneously maintain the balance of less risk—less profitability, it is necessary to use the model (1)–(2). Profitability will be taken into account when evaluating the ratings of companies at the next stage of data analysis. Profitability is considered as the ratio of profit to equity (return on equity), so the indicators of profit and equity are very important.

According to the results of the analysis of the rank of companies, which is based on a hierarchical analysis of data on their financial and economic activities, the industries are ranked according to the principle, which is a far-reaching generalization [32–34].

The essence of the method consists of a hierarchy of ranking; in essence, the indicators of the balance sheet, which are especially important for the investor.

Similar to problem (1), the problem is set:

$$\max_{i=\overline{1,n}} IR_i\hat{\theta}_i \rightarrow \min_{\hat{\theta}\in\Omega}, \Omega = \{\hat{\theta} = (\hat{\theta}_1, \ldots, \hat{\theta}_n) \in R^n : \sum_{i=1}^{n} \hat{\theta}_i = 1\}, \tag{5}$$

the solution of the problem (5) is determined by the formulas (6):

$$\hat{\theta}_i = \frac{1}{IR_i \sum\limits_{k=1}^{n} (IR_k)^{-1}}, \; i = \overline{1,n}. \tag{6}$$

The final indicator for companies is obtained by adjusting the solution of problem (1) by solving problem (5).

Let $\sum\limits_{i=1}^{n} \hat{\theta}_i \cdot \widetilde{\theta}_i = z$.

The investment shares of the i-th company are obtained according to the formulas (7):

$$\theta_i = \frac{\hat{\theta}_i\widetilde{\theta}_i}{z}, \; i = \overline{1,n}. \tag{7}$$

Accounting for the industry principle, Rank 1 is assigned to the industry whose company has the best rating, then the industries are followed in descending order of the ratings of the leading companies in the rating, and they are assigned an independent rating (rank), from the first (1) to the last (numerically equal to the number of analyzed industries).

Let us denote W_k the rank of the k-th industry according to the number of the industry in the list (by priority). Similar to task (1), for industries, the task is set:

$$\max_{i=\overline{1,n}} W_i \breve{\theta}_i \rightarrow \min_{\hat{\theta}\in\Omega}, \Omega = \{\breve{\theta} = (\breve{\theta}_1, \ldots, \breve{\theta}_n) \in R^n : \sum_{i=1}^{n} \breve{\theta}_i = 1\}, \tag{8}$$

the solution of the problem (8) is determined by the formulas (9):

$$\breve{\theta}_i = \frac{1}{W_i \sum\limits_{k=1}^{n} (W_k)^{-1}}, \; i = \overline{1,n}. \tag{9}$$

The final indicator for the companies is obtained by adjusting the solution of problem (7) by solving problem (8).

Let $\sum\limits_{i=1}^{n} \theta_i \cdot \breve{\theta}_i = zz$.

The investment shares of the i-th company are obtained according to the formulas (11):

$$\theta\theta_i = \frac{\theta_i \breve{\theta}_i}{zz}, \; i = \overline{1,n}. \tag{10}$$

This study uses companies in the oil and gas industry, so the calculations are based on formulas (1–7). The program block contains a three-step procedure.

The common scheme of CAD is shown in Figure 1.

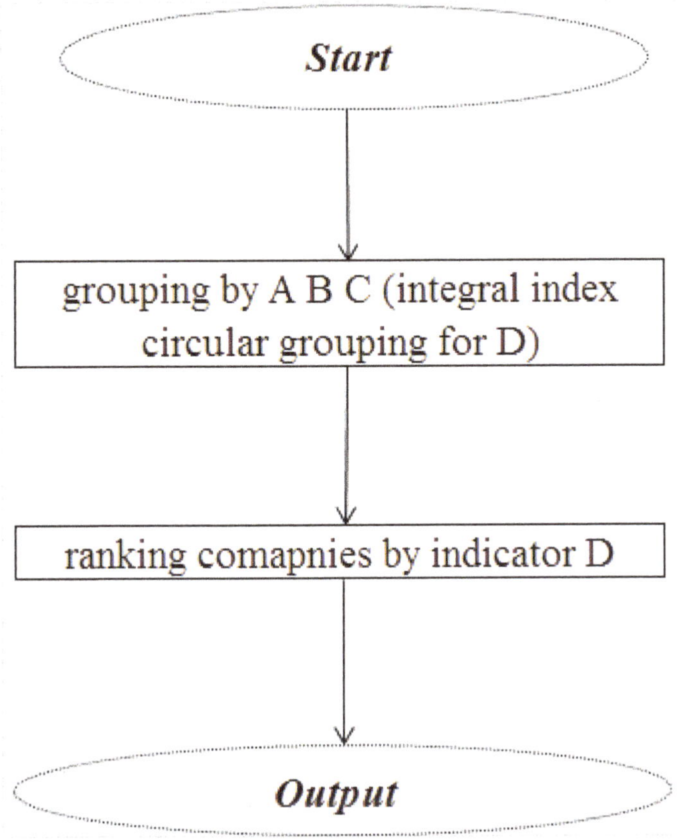

Figure 1. CAD by scheme of information flows, integrated rating.

The method of circular convolution of large companies. At the first stage, two groups are identified: the leaders (ABC indicators are above average) and the last (ABC below average). The remaining companies form a middle group, which is subject to further ranking according to the principle: from the borders to the center of the circle after the signal to stop the process is received: the presence of one or two groups of leaders and the closing ones without the possibility of allocating the middle group.

At the second stage, an integral rating is built. The indicator D is applied, first for all the leading groups, from 1 to the last, then in the central group according to the results of the hierarchical analysis, and then for the closing groups from the last to the first (boundary) (Figure 1).

Programming. Consider the procedure for estimating the parameters of the model (1)–(2) (Figure 2).

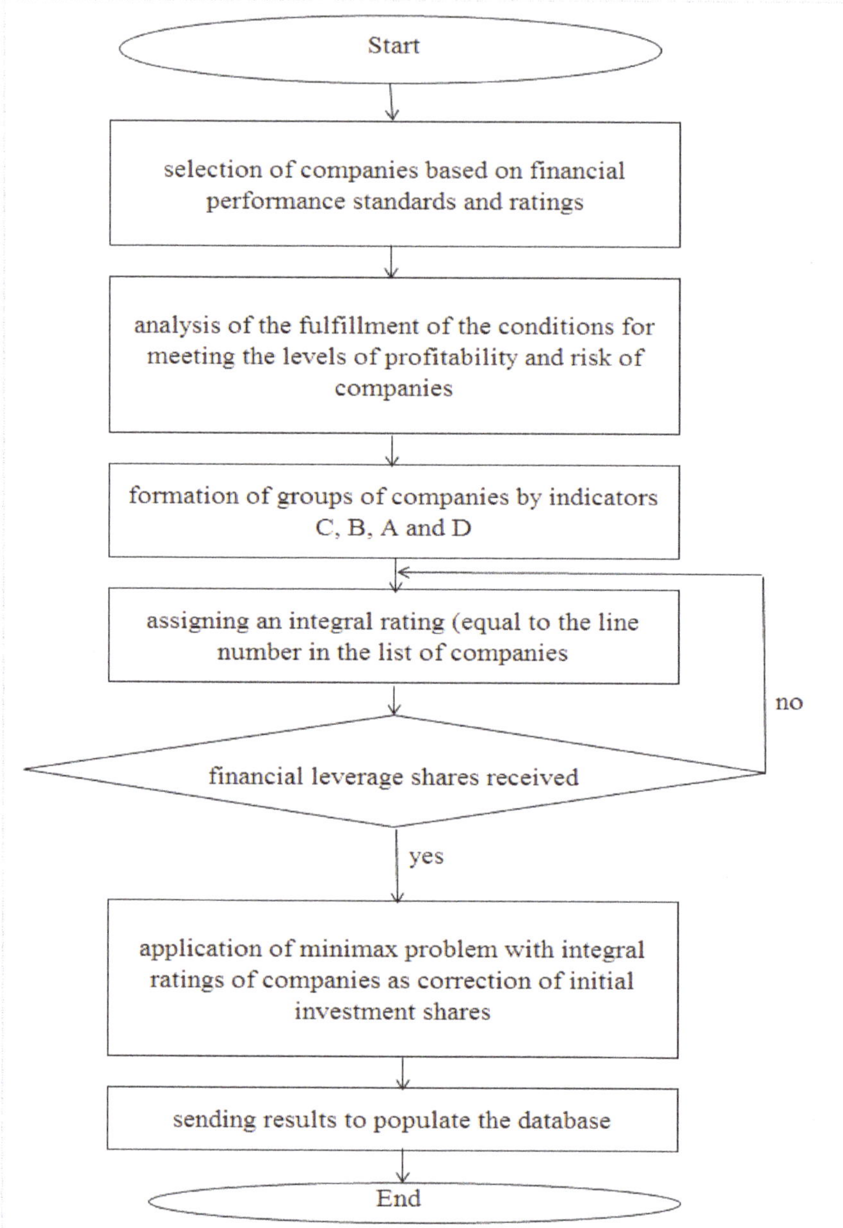

Figure 2. DSS.

3. Results

In Ref. [35], the following data are obtained (2019, Table 1).

Financial data on oil and gas companies that were used in the article can be found in Supplementary Materials.

Profitability and risk are unbalanced; for example, for PJSC Novatek, the profitability is higher than for PJSC Gazprom, and the risk is lower. This does not mean that the portfolio should include only PJSC Novatek, as the volume of production and dynamics of development and market share are important. The undisputed leader is PJSC Gazprom, which has proved itself over the years among the largest companies in Russia from various industries with high-tech development (oil and gas, electricity, banks, high-profile trade sector, metallurgy, telecommunications, etc.). In this analysis, it is necessary to apply the model (1)–(2), the restrictions are used for integral rankings-ratings of companies (hierarchy scheme) (Table 2).

Table 1. Financial analysis parameters of companies for rating.

Company, PJSC	Net Profit, Thousand Rubles	Equity Capital, Thousand Rubles	Revenue, Thousand Rubles	Assets, Thousand Rubles	Debt Capital, Thousand Rubles	Risk (Debt/Equity, Financial Leverage), Shares (%)	Profitability (Net Profit to Equity), Shares (%)
Gazprom	651,124,114	11,334,679,889	4,758,711,459	15,916,355,497	4,581,675,608	40.40%	5.70%
Lukoil	405,759,769	957,169,199	444,471,354	2,209,166,567	1,251,997,368	130.80%	42.40%
Surgutneftegaz	105,478,643	4,303,834,579	1,555,622,592	4,553,686,428	249,851,849	5.80%	2.50%
Novatek	237,224,510	718,557,978	528,544,385	899,787,613	181,229,635	25.20%	33.00%

Table 2. Solving the problem (1)–(2) for ratings.

Company	Net Profit, Thousand Rubles (Index D)	Equity Capital, Thousand Rubles (Index C)	Revenue, Thousand Rubles (Index B)	Assets, Thousand Rubles (Index A)	Place, 1—Leader (Including Revenue)	The Company's Share in The Portfolio by Rating Position
Gazprom	651,124,114	11,334,679,889	4,758,711,459	15,916,355,497	1	48%
Lukoil	405,759,769	957,169,199	444,471,354	2,209,166,567	3	16%
Surgutneftegaz	105,478,643	4,303,834,579	1,555,622,592	4,553,686,428	2	24%
Novatek	237,224,510	718,557,978	528,544,385	899,787,613	4	12%

The industry analysis in the group of the 20 largest Russian companies for the selected companies is presented in Table 3.

Table 3. Industry rating in the group of leaders of the 20 Russian companies.

Company	Rating (C, A)	Rating (C, A, B)
PJSC Gazprom	3	1
PJSC Lukoil	4	3
PJSC Surgutneftegaz	15	15
PJSC Novatek	6	6

The investment shares were estimated for the company indicators, taking into account the formulas (6)–(7). The results are shown in Figure 3.

The estimation of the investment shares, taking into account the formulas (9)–(10) for model (8), is shown in Figure 4.

2019	
Company	Invest, optimal 2019
Gazprom	20%
Lukoil	2%
Surgutneftegaz	70%
Novatek	8%
Total	100,00%

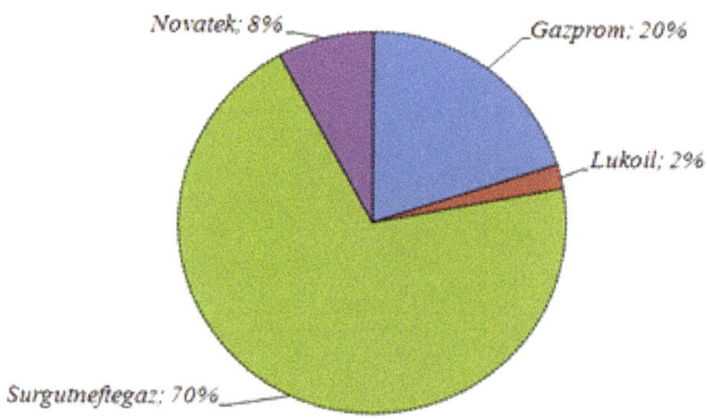

Figure 3. Optimal capital by model (6)–(7).

2019	
Company	Invest, optimal 2019
Gazprom	75%
Lukoil	3%
Surgutneftegaz	17%
Novatek	5%
Total	100%

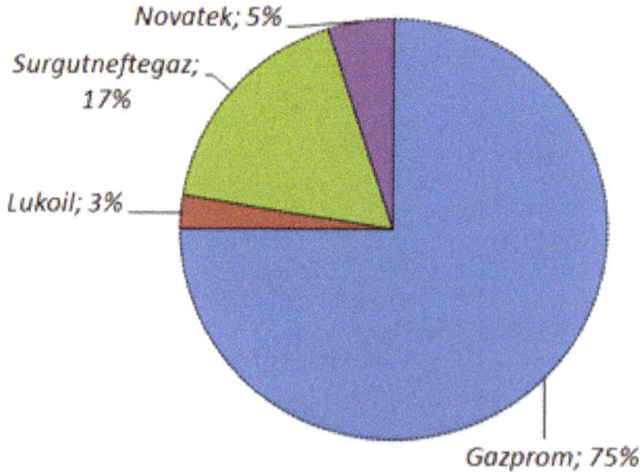

Figure 4. Optimal capital by model (8)–(9).

4. Discussion

Due to the undoubted leadership in the industry principle (among all companies in high-tech industries in Russia), Gazprom is in the first place. According to the author's method of ranking companies, only in 2016 did PJSC Gazprom lose to PJSC Sberbank, which introduced online technologies and convenient services, but in 2017 it again took first place. PJSC Surgutneftegaz and PJSC Novatek should use their industry competitive advantages to develop high-tech production, and not to compete with a leading company.

The goal of development is to balance and properly compete for resources through high-tech processing and environmentally friendly (with minimization and competent disposal of raw material waste, which is an important problem for the oil and gas industry in particular) consumption within the country. The export of raw materials should be replaced by high-tech exports of more expensive products with the competent processing of raw materials into a convenient resource for use. Companies should develop by saving raw materials that are unstructured and dangerous to the health of citizens, and by saving waste through competent high-tech processing, testing and implementation within the country, which is the main competitive advantage of the companies under consideration.

5. Conclusions

This paper considers several companies of the oil and gas complex, selected by the investor on the principle of a comprehensive solution to the problem of high-tech production and the sales of products in Russia. Since the purpose of the work is a comprehensive assessment of the stable development of companies, the profit return on investment goes by the wayside in comparison with the volume of sales of goods (resource availability), with a stable level of other indicators. It is Gazprom that comes out on top, and it is necessary to add high-tech financing for this company in the volume of high-tech products that it will master and therefore be able to produce a productive report. Projects have faded into the background, and the rapid implementation of high-tech projects using computerized CAD systems, in particular DSS in this industry, is coming to the fore. This development is associated with the subsequent growth of the investment activity of interested actors from scientifically oriented and practically financially literate citizens and persons making high-tech decisions on the status and level of the development of structures and industry complexes that are necessary for the life and high-tech development of Russia.

Supplementary Materials: A document with financial statements of the oil companies considered in the article, is available on the Internet at https://e-ecolog.ru/buh (accessed on 30 July 2021).

Author Contributions: Conceptualization, A.B.; methodology, M.T.; software, I.V.; formal analysis, I.V.; writing—original draft preparation, A.K.; writing—review and editing, G.P. All authors have read and agreed to the published version of the manuscript.

Funding: This research received no external funding.

Data Availability Statement: Not applicable.

Conflicts of Interest: The authors declare no conflict of interest.

Nomenclature

The following symbols are used in the article:

DSS	Decision Support System;
SAD	computer-aided design system;
m	number of industries;
n	the number of companies participating in the analysis (from all selected industries for analysis);
FL	Estimates of financial leverage, that is, the ratio of borrowed capital to equity expressed in fractions of a unit or as a percentage;
IR	integral ranks of companies (from the best, equal to one) obtained during the implementation of the author's algorithm;
Ω	there are many restrictions on the structure of the portfolio (the sum of the shares is one);

Ω^{η}	there are many restrictions on the structure of the portfolio (the sum of the shares is one and the required yield is fixed at the level of η_p, taking into account the returns of the assets included in the portfolio η_i, $i = 1, \ldots, n$;
$\widetilde{\theta}$	vector (with components that make up the investment shares, giving a total of one), for a problem with risk assessment in terms of financial leverage;
$\hat{\theta}$	vector (with components that make up the investment shares, giving a total of one), for a problem with an assessment in terms of an integral rating of companies;
θ	vector (with components that make up the investment shares, giving one in total), for the optimal portfolio structure, correction of the solution of the problem with risk assessment in terms of financial leverage by solving the problem for risk assessment in terms of integral rating, the solution of the second problem is the correction coefficients for optimizing the solution of the first problem;
W_i	industry rank, for further optimization of a portfolio solution containing companies from several industries;
$\breve{\theta}$	vector (with components that make up the investment shares, giving a total of one), for a problem with an assessment in terms of an integral rating of industries;
$\theta\theta$	vector (with components that make up the investment shares, giving one in total), for the optimal portfolio structure, correction of the solution of the problem with risk assessment in terms of financial leverage by solving the problem for risk assessment in terms of integral rating, the solution of the second problem is the correction coefficients for optimizing the solution of the first problem, as well as taking into account the correction of the solution for each company taking into account the industry attribute (solving the problem of optimizing the rating of industry positions in the portfolio), if there is one industry, then the result coincides with the calculation;
z, zz	intermediate indicators that do not have significant significance in terms of their interpretation.

References

1. Borodin, A.; Mityushina, I.; Streltsova, E.; Kulikov, A.; Yakovenko, I.; Namitulina, A. Mathematical Modeling for Financial Analysis of an Enterprise: Motivating of Not Open Innovation. *J. Open Innov. Technol. Mark. Complexity* **2021**, *7*, 79. [CrossRef]
2. Kwon, S.; Yin, J. Executive Compensation, Investment Opportunities, and Earnings Management: High-Tech Firms Versus Low-Tech Firms. *J. Account. Audit. Finance* **2006**, *21*, 119–148. [CrossRef]
3. Aw, B.Y.; Roberts, M.J.; Xu, D.Y. Rand. Investments, Exporting, and the Evolution of Firm Productivity. *Am. Econ. Rev. Pap. Proc.* **2008**, *98*, 451–456. [CrossRef]
4. Kohers, N.; Kohers, T. The value creation potential of high-tech mergers. *Financ. Anal. J.* **2000**, *56*, 40–50. [CrossRef]
5. Hąbek, P.; Wolniak, R. Assessing the quality of corporate social responsibility reports: The case of reporting practices in selected. *Eur. Union Memb. States Qual. Quant.* **2020**, *50*, 399–420. [CrossRef]
6. Paryzkyi, I.V.; Ivanchov, P.V.; Antonova, O.M. Privatisation in Ukraine as a deterrent to the innovative development of the national economy. *Estud. Econ. Aplicada* **2021**, *39*. [CrossRef]
7. Zollo, M.; Bettinazzi, E.L.M.; Neumann, K.; Snoeren, P. Toward a Comprehensive Model of Organizational Evolution: Dynamic Capabilities for Innovation and Adaptation of the Enterprise Model. *Global Strategy J.* **2016**, *6*, 225–244. [CrossRef]
8. Coad, A.; Rao, R. Innovation and Firm Growth in High-Tech Sectors: A Quantile Regression Approach. *Res. Policy* **2008**, *37*, 633–648. [CrossRef]
9. Repina, E.; Shiryaeva, L.; Fedorova, E. The Study of Dependence Structure between Small Business Development and Microfinance Security of Russian Regions. *Ekon. Mat. Metody* **2019**, *55*, 41–57. [CrossRef]
10. Markovitz, H.M. Portfolio selection. *J. Financ.* **1952**, *7*, 77–91.
11. Mohd, E. Accounting for Software Development Costs and Information Asymmetry. *Account. Rev.* **2005**, *80*, 1211–1231. [CrossRef]
12. Dehninga, B.; Richardsonb, V.J.; Stratopoulosc, T. Information technology investments and firm value. *Inf. Manag.* **2005**, *42*, 989–1008. [CrossRef]
13. Lendel, V.; Moravčíková, D.; Martin Latka, M. Organizing Innovation Activities in Company. *Procedia Eng.* **2017**, *192*, 615–620. [CrossRef]
14. Ravenscraft, D.J.; Long, W.F. Paths to Creating Value in Pharmaceutical Mergers, Mergers and Productivity. *Mergers Product.* **2000**, 287–326. [CrossRef]
15. Klaus, M.; Bård, M. Investment and uncertainty in the international oil and gas industry. *Energy Econ.* **2009**, *31*, 240–248.
16. Bollerslev, T. A conditionally heteroskedastic time series model for speculative prices and rates of return. *Rev. Econ. Stat.* **1987**, *69*, 542–547. [CrossRef]
17. Donker, H. Purchasing reserves and commodity market timing as takeover motives in the oil and gas industry. *Energy Econ.* **2013**, *37*, 167–181.

18. Engle, R.F.; Bollerslev, T. Modelling the persistence of conditional variances. *Econom. Rev.* **1986**, *5*, 1–50. [CrossRef]
19. Taghizadeh-Hesarya, F.; Rasoulinezhadb, E.; Yoshinoc, N.; Sarkerd, T.; Mirzae, N. Determinants of the Russia and Asia–Pacific energy trade. *Energy Strategy Rev.* **2021**, *38*, 100681. [CrossRef]
20. Kohers, N.; Kohers, T. Takeovers of technology firms: Expectations vs. reality. *Financ. Manag.* **2001**, *30*, 5–30. [CrossRef]
21. Burkynskyi, B.; Iermakova, O.; Laiko, O. Strategic directions for increasing the gross value added of the high-tech sector (on the example of the Ukrainian economy). *Entrep. Sustain. Issues* **2021**, *8*, 508–523. [CrossRef]
22. Bernardo, A.E.; Chowdhry, B.; Goyal, A. Growth Options, Beta, and the Cost of Capital. *Financ. Manag.* **2007**, *36*, 1–13. [CrossRef]
23. Boyer, M.M.; Filion, D. Common and fundamental factors in stock returns of Canadian oil and gas companies. *Energy Econ.* **2007**, *29*, 428–453. [CrossRef]
24. Brown, S.; Warner, J. Using Daily Stock Returns: The Case of Event Studies. *J. Financ. Econ.* **1985**, *14*, 3–31. [CrossRef]
25. Andres-Alonso, P.; Azofra-Palenzuela, V.; Fuente-Herrero, G. The real options component of firm market value: The case of the technological corporation. *J. Bus. Financ. Account.* **2006**, *33*, 203–219. [CrossRef]
26. Berk, J.B.; Green, R.C.; Naik, V. Optimal Investment, Growth Options, and Security Returns. *J. Financ.* **1999**, *54*, 1553–1607. [CrossRef]
27. Cao, C.; Simin, T.; Zhao, J. Can Growth Options Explain the Trend in Idiosyncratic Risk? *Rev. Financ. Stud.* **2008**, *21*, 2599–2633. [CrossRef]
28. Dittmar, A.K.; Dittmar, R.F. The timing of financing decisions: An examination of the correlation of financing waves. *J. Financ. Econ.* **2008**, *90*, 59–83. [CrossRef]
29. Masood, O.; Tvaronavičienė, M.; Javaria, K. Impact of oil prices on stock return: Evidence from G7 countries. *Insights Reg. Dev.* **2019**, *1*, 129–137. [CrossRef]
30. Raufflet, E.; Cruz, L.B.; Bres, L. An assessment of corporate social responsibility practices in the mining and oil and gas industries. *J. Clean Prod.* **2014**, *84*, 256–270. [CrossRef]
31. Vygodchikova, I. Toolkit for making decisions on investing in large Russian companies using a hierarchical ranking procedure and minimax approach. *Appl. Inform.* **2019**, *14*, 123–137.
32. Borodin, A.; Vygodchikova, I.; Dzyuba, E.; Panaedova, G. Food security: State Financial support Measures for sustainable Devel-opment of Agriculture in Russian Regions. *Financ. Theory Pract.* **2021**, *25*, 35–52. [CrossRef]
33. Wu, H.; Xu, Z.S. Fuzzy Logic in Decision Support: Methods, Applications and Future Trends. *Int. J. Comput. Commun. Control* **2021**, *16*, 1–28. [CrossRef]
34. Plenkina, V.; Andronova, I.; Deberdieva, E.; Lenkova, O.; Osinovskaya, I. Specifics of strategic managerial decisions-making in Russian oil companies. *Entrep. Sustain. Issues* **2018**, *5*, 858–874. [CrossRef]
35. Borodin, A.; Tvaronavičienė, M.; Vygodcyikova, I.; Kulikov, A.; Skuratova, M.; Shchegolevatykh, N. Improving the Development Technology of an Oil and Gas Company Using the Minimax Optimality Criterion. *Energies* **2021**, *14*, 3177. [CrossRef]

MDPI
St. Alban-Anlage 66
4052 Basel
Switzerland
Tel. +41 61 683 77 34
Fax +41 61 302 89 18
www.mdpi.com

Energies Editorial Office
E-mail: energies@mdpi.com
www.mdpi.com/journal/energies

www.ingramcontent.com/pod-product-compliance
Lightning Source LLC
Chambersburg PA
CBHW040810250926
51850CB00007BA/1132
* 9 7 8 3 0 3 6 5 4 6 6 1 2 *